T0199320

BIOGRAPHIES OF SCIENTIFIC OBJECTS

BIOGRAPHIES OF
SCIENTIFIC OBJECTS

EDITED BY

LORRAINE DASTON

THE UNIVERSITY OF CHICAGO PRESS
CHICAGO AND LONDON

LORRAINE DASTON is director at the Max Planck Institute for the History of Science, Berlin. She is author of *Classical Probability in the Enlightenment*, *The Empire of Chance* (with Gerd Gigerenzer et al.) and *Wonders and the Order of Nature* (with Katharine Park).

The University of Chicago Press, Chicago 60637
The University of Chicago Press, Ltd., London
© 2000 by The University of Chicago
All rights reserved. Published 2000
Printed in the United States of America
09 08 07 06 05 04 03 02 01 00   1 2 3 4 5
ISBN: 0-226-13670-1 (cloth)
ISBN: 0-226-13672-8 (paper)

Library of Congress Cataloging-in-Publication Data

Biographies of scientific objects / edited by Lorraine Daston.
        p.   cm.
    Includes bibliographical references and index.
    ISBN 0-226-13670-1 (cloth : alk. paper) — ISBN 0-226-13672-8 (pbk. : alk. paper)
    1. Knowledge, Theory of.   2. Science—Philosophy.   3. Science—History.   I. Daston, Lorraine, 1951–
Q175.32.K45 B56 2000
121 21—dc21   99-043745

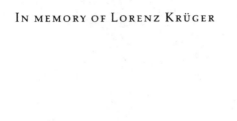

In memory of Lorenz Krüger

CONTENTS

Preface   ix

Introduction
The Coming into Being of Scientific Objects   1
Lorraine Daston

1   **Preternatural Philosophy**   15
     Lorraine Daston

2   **Mathematical Entities in Scientific Discourse**   42
     PAULUS GULDIN AND HIS *DISSERTATIO*
     DE MOTU TERRAE
     Rivka Feldhay

3   **Dreams and Self-consciousness**   67
     MAPPING THE MIND IN THE LATE EIGHTEENTH
     AND EARLY NINETEENTH CENTURIES
     Doris Kaufmann

4   **Mutations of the Self in Old Regime
     and Postrevolutionary France**   86
     FROM *AME* TO *MOI* TO *LE MOI*
     Jan Goldstein

5   **The Coming into Being and Passing Away of
     Value Theories in Economics (1776–1976)**   117
     Gérard Jorland

vii

6   **"An Entirely New Object of Consciousness,**
     of Volition, of Thought"   132
     THE COMING INTO BEING AND (ALMOST) PASSING AWAY
     OF "SOCIETY" AS A SCIENTIFIC OBJECT
     Peter Wagner

7   **"Sentimental Pessimism" and Ethnographic Experience**   158
     OR, WHY CULTURE IS NOT A DISAPPEARING "OBJECT"
     Marshall Sahlins

8   **How the Ether Spawned the Microworld**   203
     Jed Z. Buchwald

9   **Life Insurance, Medical Testing, and the**
     Management of Mortality   226
     Theodore M. Porter

10   **On the Partial Existence of Existing *and* Nonexisting Objects** 247
     Bruno Latour

11   **Cytoplasmic Particles**   270
     THE TRAJECTORY OF A SCIENTIFIC OBJECT
     Hans-Jörg Rheinberger

List of Contributors   295
Index   297

PREFACE

The essays in this volume were originally presented as papers at a conference on "The Coming into Being and Passing Away of Scientific Objects," held in September 1995 at the Max Planck Institute for the History of Science in Berlin. This was one of the very first conferences held at the new institute, and was also the first of a series of three conferences devoted to a historical epistemology of the sciences, followed by "The Varieties of Scientific Experience" (June 1997) and "Demonstration, Test, Proof" (June 1998). Each of the conferences took as its theme a fundamental category in the sciences—object, experience, proof—and investigated its forms and development in specific cases across a broad range of disciplines and periods. The aim was to launch a history of the structures of argument, practice, and classification that make science possible; a history that would pose transcendental questions in a highly particularist mode.

Most of the papers have been substantially revised in light of discussion at the conference and subsequent criticism. Hearty thanks are here due to the commentators at the conference—John Carson, Peter Galison, Johan Heilbron, Krzysztof Pomian, Jürgen Renn, M. Norton Wise—as well as to two anonymous referees for the University of Chicago Press.

The idea for the conference and book arose from a number of stimulating conversations with Lorenz Krüger and Jürgen Renn, when the three of us were first sketching out what the new institute for the history of science might be like. To Jürgen's and my great sorrow, Lorenz did not live to see the realization of our plans: after a courageous struggle with a cruel illness, he died on 29 September 1994. This book is dedicated to his memory.

Lorraine Daston

Lorraine Daston

INTRODUCTION

The Coming into Being
of Scientific Objects

This is a book about applied metaphysics. It is about how whole domains of phenomena—dreams, atoms, monsters, culture, mortality, centers of gravity, value, cytoplasmic particles, the self, tuberculosis—come into being and pass away as objects of scientific inquiry. The echo to the title of Aristotle's treatise *On Generation and Corruption* is deliberate: this is a sublunary metaphysics of change, of the "perpetuity of coming-to-be."[1] If pure metaphysics treats the ethereal world of what is always and everywhere from a God's-eye-viewpoint, then applied metaphysics studies the dynamic world of what emerges and disappears from the horizon of working scientists. But the contrast between pure and applied metaphysics is not necessarily just a reformulation of that between ontology and epistemology, between what is really real and what is dimly known, noumena and phenomena. Applied metaphysics assumes that reality is a matter of degree, and that phenomena that are indisputably real in the colloquial sense that they exist may become more or less intensely real, depending on how densely they are woven into scientific thought and practice. Monsters for example we have always had with us, but only sporadically have they attracted the probing curiosity of anatomists, challenged the natural kinds of taxonomists, furnished crucial experiments for embryologists, filled out the collections of naturalists—in short, come into being as scientific objects.

   This way of understanding scientific objects cuts against the grain of assumptions that inform much if not most literature in the history, philosophy, and sociology of science. It is also at odds with the sense entrenched in

1. Aristotle, *On Generation and Corruption*, I.3, 317b34, in *The Complete Works of Aristotle*, ed. Jonathan Barnes, 2 vols. (Princeton: Princeton University Press, 1984), 1:519.

1

the etymology of the very word "object" in several major European languages. Hence it may be of use first to lay out the prevailing assumptions as clearly as possible, and then to explain how the applied metaphysics of scientific objects diverges from them.

The words "object," *objectus, objet, Gegenstand, oggetto, voorwerp* all share the root meaning of a throwing before, a putting against or opposite, an opposing. In the English verb "to object" the oppositional, even accusatory sense of the word is still vivid. In an extended sense, objects throw themselves in front of us, smite the senses, thrust themselves into our consciousness. They are neither subtle nor evanescent nor hidden. Neither effort nor ingenuity nor instruments are required to detect them. They do not need to be discovered or investigated; they possess the self-evidence of a slap in the face. These are the solid, obvious, sharply outlined, in-the-way things of quotidian experience: the walls that obstruct, the rain that falls, the projectile that hits, the stone that stubs. They are all too stable, all too real in the commonsensical meaning of "hard to make go away." They may be the psychological prototype of all objects, as the etymology suggests, but they are rarely the objects of scientific inquiry. As Gaston Bachelard remarked: "In the formation of a scientific mind *[esprit scientifique]*, the first obstacle is primary experience, experience placed before and above the criticism that is necessarily an integral element of the scientific mind."[2] In contrast to quotidian objects, scientific objects are elusive and hard-won.

Historians, philosophers, and sociologists of science do not confuse quotidian with scientific objects. They have however been locked in a debate between realism and constructionism that implicitly draws upon the obduracy of quotidian objects. As in all protracted scholarly controversies, positions splinter and shade into one another.[3] But the underlying ontological intuitions that sustain the debate are simple enough. Realists picture scientific objects as discoveries, unexplored territory waiting to be mapped. Scientific objects may, like dark continents and invisible planets, take centuries of theoretical and empirical effort to find, or be accessible only by means of the most powerful instruments, but in their essence they are as enduring as quotidian objects. On the realist view, it makes sense to talk about a history of scientific discovery, but not a genuine history of scientific objects. Theories about the furniture of the universe may come and go, but

2. Gaston Bachelard, *La formation de l'esprit scientifique* [1957] (Paris: Vrin, 1989), 23; cf. idem, "Rupture avec la connaissance commune," in *Epistémologie* [1971] (Paris: Presses Universitaires de France, 1992), 12–13.

3. The debate has produced a gigantic literature in the past decade. For a lively and lucid recent account, see Ian Hacking, *The Social Construction of What?* (Cambridge: Harvard University Press, 1999).

the furniture stays (to invoke a popular ontological metaphor rich in the stolid associations of quotidian objects). Constructionists assert scientific objects to be inventions, forged in specific historical contexts and molded by local circumstances. Those circumstances may be intellectual or institutional, cultural or philosophical, but they are firmly attached to a particular time and place. The favored metaphors here are those of craftsmanship (and sometimes craftiness): work, fabrication, plasticity. On the constructionist view, scientific objects are eminently historical, but not real. In much of the debate, the opposition between nature and culture shadows that between the real and the constructed: nature stands for the eternal, the inexorable, the universal; culture for the variable, the malleable, the particular. Like the return of the repressed, the supra- and sublunary spheres of Aristotelian cosmology crop up again in new guise, crystalline nature encircling mutable culture. Both sides of the debate accept the oppositions of the real versus the constructed, the natural versus the cultural. Hence arguments are about in which category notions like "race" or "quark" belong—are they real or constructed? discoveries or inventions?—not about the categories themselves.

Applied metaphysics stands orthogonal to the plane of this debate: it posits that scientific objects can be simultaneously real *and* historical. Just how that is possible is the subject matter of this book, and the eleven essays do not speak with one voice on this issue. But they all offer striking evidence of the ontological fecundity of the sciences, natural and human, pure and applied. The examples from physics, economics, psychology, biology, anthropology, demography, medicine, sociology, mechanics, and sciences that no longer have a name undercut any facile idealistic account of the coming into being and passing away of scientific objects. These are not only stories about how interpretations of the world succeed one another, a *vita contemplativa* of scientific objects. They are also stories of the *vita activa*, of practices and products as concrete as the stacking of individual atoms and the profits of insurance companies. Although scientific objects lack the obviousness and obduracy of quotidian objects, they can be just as heavy with consequences for everyday experience. If economic value or centers of gravity are nonetheless to be described as transient ideas, then they are ideas that work for a living, raising obelisks and toppling governments.

Taken together, these essays blur the distinction between invention and discovery and recall the period when these words were synonyms rather than antonyms. The *Oxford English Dictionary* entry "Invention" offers sixteenth- and seventeenth-century examples of sentences we moderns could barely think, much less utter: "That judicial method which serveth best for the invention of the truth"; " . . . the Invention of Longitudes will come to its perfection." The common element of novelty bound these words together for early modern speakers, although the sense of "invention" as a fabrication or contrivance was also available to them. Apparently only in the course of the eighteenth century did the distinction that matters so centrally to us eventually drive a wedge between "invention" and "discovery": was the novelty revealed, as an explorer fills in a blank spot on the world map, or was it contrived, as an artisan manufactures a device? Nor is this the only example of metaphysically charged words that shifted their meanings nearly 180 degrees during the same period. "Realism" originally referred to the philosophical claim that universals, which existed only as abstract mental entities, were as real or more real than the individual particulars of sensation; by the late eighteenth century, it had come to mean nearly the opposite. The word "objective" performed a similar volte-face, from its fourteenth-century meaning referring to objects of consciousness to its late eighteenth-century meaning referring to objects external to consciousness. During the seventeenth and early eighteenth centuries "fact" shed its associations with "doing" and "making" (fossilized in legal phrases like "after the fact") and migrated toward "datum," that which is given rather than made. Following a trajectory parallel to that of "invention" and "discovery," but one all the more striking because of a shared etymology, "fact" and "manufacture" were nearly antonyms by the late eighteenth century.

More research needs to be done in order to chart these semantic transformations, their causes and their import, but the general pattern is suggestive: sometime in the eighteenth century the distinction between what is and what is made became unavoidable, a metaphysical axiom. This is a theme with many variations, as the word cluster outlined above testifies: discovery versus invention, objective versus subjective, fact versus manufacture. None of these now familiar oppositions is identical to the others, but all rely on the same metaphysical intuitions—and suspicions. In the context of the sciences, that which is made edged closer to that which is made up, to fabrication or invention in the pejorative sense. Hence the strong ambivalence toward the faculty of the imagination evident in so much writing on both the arts and the sciences after circa 1750, which on the one hand acknowledged the necessity of the creative imagination in these endeavors, but on the other hand worried that it could trick the mind into confusing its own inventions with authentic discoveries. The French naturalist Georges Cuvier for example warned against those savants who "cannot prevent themselves from mixing true discoveries, [découverts véritables] with fantastic conceptions . . . they laboriously construct vast edifices on imaginary bases, similar to the enchanted palaces in our old romances that disappear when the talisman upon which their existence depends is broken."[4] The opposition between secrets of nature laid bare, or "discovered" and "vast edifices [constructed] on imaginary bases" still haunts our discussions of scientific objects.

The essays in this volume cannot by themselves undo the metaphysics that forces a choice between invention and discovery, but they can shift the focus of attention to the indisputable fact of novelty in science. Whatever their metaphysical status, new scientific objects pour forth, and old ones fade away. Each of the eleven essays documents in detail how a heretofore unknown, ignored, or dispersed set of phenomena is transformed into a scientific object that can be observed and manipulated, that is capable of theoretical ramifications and empirical surprises, and that coheres, at least for a time, as an ontological entity. Some chart not only the birth but also the death of a scientific object. All confront the engrained opposition between the real and the historical with potential counterexamples: if the sciences furnish us with the best candidates for the real, then scientific realism must take the historicity of scientific objects seriously. Although the authors of these essays differ markedly from one another in their willingness to draw metaphysical conclusions, they concur in making history the departure point for any post-Kantian attempt at a new *Prolegomena to Any Future Metaphysics*. The approach is resolutely empirical, again in the spirit of Aristotle's investigations on the coming into being and passing away of things: "Lack of experience diminishes our power of taking a comprehensive view of the admitted facts. Hence those who dwell in intimate association with nature and its phenomena are more able to lay down principles such as to admit of a wide and coherent development; while those whom devotion to abstract discussions has rendered unobservant of facts are too ready to dogmatize on the basis of a few observations."[5]

The remainder of this introduction is devoted to surveying the "phenomena" set forth in the essays, both to give the reader a sense of their range and to draw out and compare their implications for a reformed understanding of scientific objects in history. The book is catholic in scope, taking full advantage of the amplitude of the German *Wissenschaft*, as opposed to the more restrictive English "science." The natural and the human sciences are represented, as are pure and applied disciplines. Chronologically, the essays span the sixteenth through the twentieth centuries.

4. Georges Cuvier, "Eloge de Jean-Baptiste Lamarck," *Recueil des éloges historiques lus dans les séances publiques de l'Institut de France* [1819–27], 3 vols. (Paris: Firmin Didot Frères, Fils, 1861), 3:180.

5. Aristotle, *On Generation*, I.2, 316a5–9; 1:515.

Disciplinary and chronological breadth allows for comparisons—and provides some counterintuitive surprises. For example, the objects of the human sciences do not appear to be more ephemeral than those of the natural sciences. Some essays are firmly situated in local contexts; others chart developments that occurred on many fronts, over longer time periods. Some are bold in advancing the metaphysical implications of historical studies; others prefer more circumscribed conclusions that stay close to the case at hand. Hence there were many dimensions along which the essays might have been ordered; I have chosen a sequence that is roughly chronological, albeit with a thick concentration in the nineteenth and twentieth centuries. Within this sequence I discern four principal approaches (which can be and are combined with one another in several essays) to the historicity of scientific objects: salience, emergence, productivity, embeddedness.

SALIENCE

There is a great difference between phenomena that exist on the fringes or beneath the surface of the scientific collective consciousness and those that coalesce into domains of inquiry. Dreams, personal identity, monsters, comets, bizarre weather, figured stones, human mortality—these phenomena possess an undeniable reality before and after they become scientific objects. But scientific scrutiny nonetheless alters them in significant ways: phenomena that were heretofore scattered (as in the case of monsters and figured stones) amalgamate into a coherent category; criteria of inclusion and exclusion grow sharper (as in the case of identity); new forms of representation stabilize regularities (as in the case of mortality tables); intense investigation renders evanescent phenomena more visible and rich in implications (as in the case of dreams). In her essay "Mutations of the Self in Old Regime and Postrevolutionary France," Jan Goldstein describes the transition that promoted the "humble vernacular *moi*" to the status of an "intensively theorized" object in philosophical psychology in terms of salience: "I will regard it [the self] as a perennial scientific object whose form and degree of cultural salience are prone to extremely wide variation. What is noteworthy about the early nineteenth-century French moment with respect to the self, then, is not its absolute novelty but rather the heightened, almost obsessive attention paid to that object and the dramatic shift in the relevant vocabulary." "Salience" might serve as shorthand for the multifarious ways in which previously unprepossessing phenomena come to rivet scientific attention—and are thereby transformed into scientific objects.

Doris Kaufmann and Jan Goldstein offer striking examples of how familiar features of private experience, dreams and identity, can abruptly become objects of energetic scrutiny, elaborate theories, and cultural significance. Kaufmann argues that the radical empirical program of *Erfahrungsseelenkunde* arose in the late eighteenth and early nineteenth centuries as a response to *bürgerliche* anxieties about self-control, about the limits of reason and the will. Dreams, particularly dreams that dealt with obscene or violent behavior, seemed to demonstrate the soul's powerlessness to police its own ideas. Not only psychiatrists but intellectuals of all stripes hence considered it a moral as well as a scientific duty to candidly report upon and confront the disturbing contents of their own dreams. In contrast to older traditions of dream interpretation as prophecy or divine communication, the German Enlightenment inquiry focused on the boundary between the voluntary and the involuntary, as well as that between the physical and the psychical. There was no theoretical unanimity among the practitioners of *Erfahrungsseelenkunde;* the very variety of dream interpretations—as rebellions of the imagination, as products of nervous stimuli, as "powerfully productive or poetic activity"—testifies to the richness of dreams as scientific objects, capable of sustaining distinct research (and therapeutic) programs. Goldstein also emphasizes how specific cultural and political circumstances—here, the postrevolutionary aspirations of the French bourgeoisie—singled out a commonplace of mental life, the *moi,* as an object of psychological and philosophical inquiry. For Victor Cousin and his followers, the *moi* was a compound of the private (the will) and the public (reason), and hence "a powerful argument in favor of common standards and values and against the kind of social and political contestation that bred instability and revolution . . . [and] an equally powerful argument in favor of private property." The *moi* itself became semiprivatized: neither women nor hoi polloi seemed to possess one. In both the German and French cases, elements of prosaic mental experience were detached from their traditional associations (indigestion, divine inspiration, autobiography), recombined with urgent current concerns like self-control and political stability, and subjected to meticulous introspection, classification, and explanation. Cultural salience made these objects visible, but the techniques of scientific inquiry made them additionally solid, capacious, ordered, intricate, and deep enough to sustain research and theoretical explication.

Theodore Porter's essay "Life Insurance, Medical Testing, and the Management of Mortality" shows how practical pressures can also render an all-too-commonplace occurrence, death, salient as a scientific object. Actuaries not only developed the mathematical and statistical techniques that revealed the structure of human mortality; they also discovered correlations hardly suspected by physicians, such as the link between high blood pressure and early death. A tug-of-war of opposing interests between insurer and insured and between company management and company agents created new kinds of knowledge: because these conflicts of interest spawned distrust, insurance companies pushed for quantitative and instrumental measures of health: "The dangers of hypertension were discovered by insurance companies twenty years before they came to the attention of clinicians. The measurement of blood pressure thus came into medicine not as a consequence of disinterested medical research or of the concern of physicians for their individual patients. Rather, it arose as part of the effort by life insurance companies to develop better and more objective means of mortality prognosis." This is an explicitly historical rather than an ontological argument: Porter does not claim that hypertension came into being only with the use of sphygmomanometers, nor that its dangers could have been discovered using such instruments only in the context of a medical examination performed under the auspices of suspicious insurance companies. But he does contend that "the right kinds of instruments and the right kinds of people" did in fact solidify and extend human mortality as a scientific object, by submitting it to new forms of investigation, both mathematical and instrumental, and by connecting it to other variables, from age to lifestyle to body type. Porter's emphasis on the labor, techniques, and material culture required to firm up human mortality as a scientific object is echoed in several other essays, particularly those of Rheinberger, Buchwald, and Latour.

Dreams, the self, and death existed as entities, picked out by colloquial nouns, long before they became scientific objects. In the case of Lorraine Daston's essay "Preternatural Philosophy," salience required more than highlighting the already extant; an apparent miscellany of phenomena—two-headed cats, three suns in the sky, rains of frogs, landscapes in marble, magnets, rotten wood that glowed in the dark—had further to be consolidated into a coherent category of investigation in early modern natural history and natural philosophy. Once again, specific cultural circumstances charged these strange facts with significance and also forged links among them: many, though not all, took on ominous meaning as divine portents during the political and religious upheavals of sixteenth- and seventeenth-century Europe. But epistemological and aesthetic factors were also needed to weld so unpromising a collection of oddities into a scientific object: preternatural philosophers believed that exceptions were the royal road to the discovery of nature's rules, and they subscribed to a sensibility of wonder that channeled scientific attention to the new, rare, and unusual. They also were the most resolute naturalizers science has ever known, determined to supply a natural explanation for every marvel and even for some miracles. In contrast to dreams and death, preternatural phenomena did not long endure as scientific objects, at least not as a class: monsters were still studied by anatomists and embryologists, astronomers continued to observe comets and parahelions, physicists experimented with magnets, but by the mid–eighteenth century these deviations from nature's ordinary course no longer cohered as an ontological category: "Why then did the category of the preternatural dissolve in the early eighteenth century? Its solvents were a new metaphysics and a new sensibility, which loosened its coherence without destroying its elements." Hence the history of preternatural philosophy is also one of the passing away of a scientific object, or at least of its fragmentation.

EMERGENCE

Salience, be it cultural or economic or epistemological, silhouettes extant objects; scientific inquiry might be said to intensify their reality but not to create them ex nihilo. Emergence posits a more radical form of novelty. To treat mathematical magnitudes as indeterminate rather than determinate, and to equate geometrical lines, planes, and solids with centers of gravity and motion, as Rivka Feldhay shows the Jesuit Paulus Guldin to have done in her essay "Mathematical Entities in Scientific Discourse: Paulus Guldin and His *Dissertatio de motu terrae* (1635)," is a more ambitious ontological project. For Guldin and his contemporaries it meant redefining the nature of mathematical objects, and positing an enormously fruitful but still controversial metaphysics of the physical world, in which mathematical structures become the true essences of things. Guldin's analysis of the true, mathematical center of gravity of the earth was freighted with implications that stretched well beyond mixed mathematics. His subtle claims about terrestrial motion bore on the debate over Copernicanism, and his manipulations of mathematical objects had consequences for the then-raging theological controversy over the boundaries between human and divine knowledge. Context matters to Feldhay's account of how the new scientific object of symbolic number emerged "in a particular institutional setting," even though the object itself is notoriously abstract and universal.

Peter Wagner and Marshall Sahlins deal with the emergence (and putative disappearance) of scientific objects that are themselves the stuff of context for historians: society and culture. It is instructive to note that the most clear-cut cases for the emergence of scientific objects without a quotidian prehistory among the essays of this volume come from mathematical physics on the one hand and the social sciences on the other, *pace* the

Comtean classification of the sciences that would place these disciplines at opposite poles. Wagner considers the intriguing possibility that not only the science of sociology but also its object of study, society, first emerged sometime around the beginning of the nineteenth century. In contrast to the polity of the eighteenth-century moral sciences, civil society "came to be seen as a phenomenon that was different from the state—but different from individual households as well." Wagner explores the view of nineteenth-century observers like Robert von Mohl that the French Revolution had created not just a new terminology of social relationships, but a new entity, the "society." The glue that bound society together was neither political nor familial but a new kind of human connection. The exact nature of that connection—convivial, commercial, ethnic, linguistic, or other—as well as the causes of the "coherence and boundedness" of society were and remain matters of debate, but the existence of such an entity coextensive with neither the nation-state nor the international economy seemed a brute fact—but also a historical fact, a genuine coming into being.

Sahlins takes on the inverse case, the alleged passing away of culture— or rather, cultures in all their pied variety: "Anthropology may be the only discipline founded on the owl of Minerva principle: it began as a professional discipline just as its subject matter was dying out." He argues vigorously that cultures are indeed "forever disappearing," but only because they perpetually renew themselves, and from the most unlikely sources, from the internet to learned monographs. The new cultures are no longer spatially compact—the paradigmatic village or island of traditional anthropology—but rather temporally contiguous through inherited values and identities. Culture, Sahlins concludes, has never been so robust, despite the elegiac mood of so many anthropologists: "But the sequitur is not the end of 'culture.' It is that 'culture' has taken on a variety of new arrangements and relationships, that it is now all kinds of things we have been too slow to recognize." Here is a challenge to the conception of scientific objects as stable and immutable, and therefore real; on Sahlins's account, cultures are real because protean and flexible. They endure because they change.

PRODUCTIVITY

Whether scientific objects are Parmenidean or Heraclitean, they are never inert. In all of the cases treated in this volume, scientific objects attain their heightened ontological status by producing results, implications, surprises, connections, manipulations, explanations, applications. Three essays in particular underscore the central importance of productivity: Gérard Jorland on the career of value theories in economics since the eighteenth century, Jed Buchwald on the microworld of turn-of-the-century physics, and Hans-Jörg Rheinberger on cytoplasmic particles in the contemporary life sciences. Jorland traces how value undergirded price and indeed the entire economy in the most diverse schools of economic thought, from the physiocrats through marginal utility theory. The theory of value expanded to embrace income distribution and resource allocation, the ethics of labor and the shape of the demand curve. But ultimately this keystone of economic theory vanished like a mirage. Arguing that "ideas are objects that one cannot manipulate at will," Jorland contends that the slow unfolding of the implications of Marx's transformation problem led to the realization that the whole lumbering, creaking apparatus of value was otiose: "one could get a theory of price formation and a theory of income distribution . . . independently of any theory of value whatsoever." After a long and illustrious history as a—perhaps the—object of economics, value passed away, in a sense a victim of its own productivity.

Buchwald offers a piquant example of that metaphysical impossibility, something created from nothing, or at least the real generated by the unreal, in his essay on "How the Ether Spawned the Microworld." Viewing the history of physics in the late nineteenth and early twentieth centuries through the lens of practice, understood to encompass the on-paper manipulations of theorists as well as the in-the-laboratory manipulations of experimenters, Buchwald contends that "[i]n that sense—in the sense of practice—the microworld first became strikingly real among physicists during the 1890s. If we are to understand how this transformation occurred, then we must also understand how it was (frequently) bound on paper and (occasionally) in the laboratory to a world that we no longer believe to exist at all, the world of the ether." This is a criterion of reality for scientific objects that depends crucially on their productivity as "tools," ranging from the scanning tunneling microscope to the mathematical derivations of ether models. It is relationships of reproductive (i.e., of known results) and productive (of novel results) practices that determine ontological lineages in Buchwald's story; hence wave optics and the mechanics of the ether stand in closer relationship to the nascent microworld of circa 1890 than do speculations about atoms or even chemical determinations of elements. Reality as measured by productivity admits of degrees and evolves in time: "Nevertheless, the microworld hardly became real all at once. Tools never do. It takes time to forge them, time to learn how to use them, and time to learn their strengths and limitations."

For Rheinberger, time is literally seminal: scientific reality means being pregnant with the future. The essence of a scientific object is its potential for surprise, its capacity to outstrip expectations and imagination framed by the current way of thinking and doing. To exhaust or freeze such objects is to reduce their scientific reality, though they persist as things: "Scientific objects, not things per se, but objects insofar as they are targets of epistemic activity, are unstable concatenations of representations. At best, they become stabilized for some historically bounded period. It is not that there is no materiality there before such objects come into being, or that they would vanish altogether and shrink to nothing on their way into the future. But they can become, within a particular scientific context, altogether marginal, because nobody expects them to be generators of unprecedented events any more." Scientific objects flout the boundaries between scientific disciplines; cytoplasmic particles engaged the attentions of cytomorphologists, biochemists, and molecular biologists, ultimately becoming a tool for the investigation of a new scientific object, the genetic code. In contrast to a Kuhnian account in which anomalies are swept to the margins of workaday science, Rheinberger suggests that microsomes fascinated researchers by baffling all attempts at final functional classification: was the ribosome a template for protein synthesis or a decoder of messenger RNA? Rheinberger muses on how different the history of science would look if it were narrated not as the history of ideas or institutions or disciplines but instead of "epistemic things."

EMBEDDEDNESS

Rheinberger underscores how "[e]xperimental systems embed scientific objects into a broader field of material scientific culture and practice, including the realm of instrumentation and inscription devices as well as the model organisms to which these objects are generally connected, and the fluctuating concepts to which they are bound." In his essay "On the Partial Existence of Existing *and* Nonexisting Objects," Bruno Latour advances embeddedness in "local, material, and practical networks" as the principal criterion for the reality of all objects, scientific as well as technological, natural as well as human. He puzzles over the asymmetry of our customary ontology, which forbids us to transplant artifacts like machine guns to the time of the Egyptian pharaohs, but allows us effortlessly to project the Koch bacillus as the cause of death of Ramses II. "Effortlessly" is the pivotal word here, for Latour insists that scientific objects like the Koch bacillus must be thickly embedded in a support system of equipment and procedures in order to continue to exist: "There is no point in history where a sort of inertial force can be counted on to take over the hard work of scientists and relay it for eternity. For scientists there is no Seventh Day!" The persistence of scientific objects depends on the institutionalization of practices and an impressive array of apparatus. Reality becomes a relative property, depending on the degree of its embeddedness in such organized systems of techniques and instruments. Latour's account is uncompromisingly symmetrical: if humans have biographies, so should things; if artifacts can come into being, so should scientific objects: "What is relative existence? It is an existence that is no longer framed by the choice between never and nowhere on the one hand, and always and everywhere on the other . . . By asking a non-human entity to exist—or more exactly to have existed—either never-nowhere or always-everywhere, the epistemological question limits historicity to humans and artifacts and bans it for nonhumans." Latour calls for a "homogeneous" ontology, one modeled on our intuitions about the historicity of humans and their handiwork.

If a thread of Ariadne runs through these essays, it is the suggestion, backed up by example after example, that scientific objects have a history. The authors diverge sharply from one another in the ontological conclusions they draw from the historicity of scientific objects, but they converge in assigning scientific objects a different kind of reality than that set forth in the conventional two-valued metaphysics that obliges us to choose unequivocally between "x exists" / "x does not exist" or "x is discovered" / "x is invented." Reality for scientific objects instead expands into a continuum, just as degrees of probability opened up between the poles of true and false in seventeenth-century philosophy. Scientific objects may not be invented, but they grow more richly real as they become entangled in webs of cultural significance, material practices, and theoretical derivations. In contrast to quotidian objects, scientific objects broaden and deepen: they become ever more widely connected to other phenomena, and at the same time yield ever more layers of hidden structure. The sciences are fertile in new objects, and the objects in turn are fertile in new techniques, differentiations and associations, representations, empirical and conceptual revelations. The participle "in the becoming" is more than a quaint rendering of Aristotle's Greek *(genesis)*. It captures the distinctively generative, processual sense of the reality of scientific objects, as opposed to the quotidian objects that simply are. But what can be ontologically enriched can also be impoverished; scientific objects can pass away as well as come into being. Sometimes they are banished totally from the realm of the real, as in the case of unicorns, phlogiston, and the ether. More often, they slip back into the wan reality of quotidian objects, which exist but do not thicken and quicken with inquiry.

For many decades the history of science has been dominated by a neo-Kantian epistemology that carefully distinguishes perceptions (or theories or conceptual frameworks or worldviews) from reality. History of science documents what is known, not what is; intellectual categories rather than things in themselves. Insofar as ontology has been a theme at all, the attitude has been cautiously agnostic: science may advance in terms of scope and accuracy of prediction, breadth and unity of explanation, and variety and reliability of predictions, but whether science thereby asymptotically approaches a reality as God might understand it is a question to be handled gingerly. Historians with philosophical inclinations lean toward instrumentalism, conventionalism, Mach-style positivism, or some variety of Kantianism; even philosophers qualify simple realism as "naive." It is almost impossible for historians of science to speak of the reality of scientific objects without slipping back into the epistemological mode. This volume is an attempt to revive ontology for historians. But history notoriously transforms all that it touches. An ontology that is true to objects that are at once real and historical has yet to come into being, but it is already clear that it will be an ontology in motion.

Lorraine Daston

Preternatural Philosophy

> But what are things? Nothing, as we shall abundantly see, but special
> groups of sensible qualities, which happen practically or aesthetically to
> interest us, to which we therefore give substantive names, and which we
> exalt to this exclusive status of independence and dignity. But in itself,
> apart from my interest, a particular dust wreath on a windy day is just as
> much of an individual thing, and just as much or as little deserves an indi-
> vidual name, as my own body does.
> —William James, *Principles of Psychology* (1890)

INTRODUCTION: A SCIENCE OF ANOMALIES

William James was analyzing the psychological nature of the objects of
consciousness, but he might just as well have been pondering the philosophical nature of the objects of science. Why don't we have a science of dust
wreaths on windy days? Why do we have a science of the interior of animal
bodies, or of the shapes of crystals, or of the genealogy of languages? What
ontological, epistemological, methodological, functional, symbolical, and/
or aesthetic features qualify or disqualify the motion of projectiles, dreams,
the waxing and waning of the Gross National Product, monstrous births, or
electron valences as scientific objects?

　　Aristotle's answer to this question is at once the oldest and, in somewhat
dilute form, the most enduring. Sciences can be made only out of regularities, out of "that which is always or for the most part" (*Metaphysics*
1027a20–27). Aristotle's further and stronger condition that these regularities should be not only universal but also demonstrable by a chain of necessary causes, specifying not only what is the case but also what must be the
case, was mostly honored in the breach, even within medieval Scholasti-

15

cism.[1] Nonetheless, the insistence that science ought to be about regularities—be they qualitative or quantitative, manifest to the senses or hidden beneath appearances, causal or statistical, taken from commonplace experience or created by specialized instruments in laboratories—has persisted long after the demise of Aristotelianism.

Yet regularity alone seldom suffices to pick out scientific objects from the ordinary objects of quotidian experience: whether a class of phenomena is quantifiable, manipulable, beautiful, experimentally replicable, universal, useful, publicly observable, explicable, predictable, culturally significant, or metaphysically fundamental are all criteria that have fortified claims to scientific objecthood beyond mere regularity. These criteria sometimes overlap but seldom entirely coincide. The intensity of psychological attitudes may be quantifiable with the aid of rating scales, but it is not publicly observable; evolutionary theory explains without predicting, and statistical forecasts, both economic and meteorological, predict without explaining; the events of high-energy physics may be metaphysically fundamental but are rarely experimentally replicable. A study of what can and cannot become a scientific object must take into account how these multiple grids are superimposed upon raw experience to highlight some phenomena and to occlude others. If we do not have a science of dust wreaths on windy days, it is not solely or even primarily because the phenomenon is irregular.

If regularity is not a sufficient condition for scientific objecthood, is it at least a necessary one? My purpose in this paper will be to dispute even this minimal claim by means of a historical counterexample: in the late sixteenth and seventeenth centuries natural philosophers and even some mathematicians focused their attention on anomalous phenomena, those that Francis Bacon described as "singular instances . . . of an apparently extravagant and separate nature, agreeing but little with other things of the same species" and as "deviating instances: such as the errors of nature, or strange and monstrous objects, in which nature deviates and turns from her ordinary course."[2] These phenomena were, in the language of the day, *praeter naturam,* "beyond nature," being remarkable divergences from "that which is always or for the most part." The category of the preternatural encompassed the appearance of three suns in the sky, the birth of conjoint twins, the tiny fish that could stop a ship in full sail, the antipathy

1. Eileen Serene, "Demonstrative Science," in *The Cambridge History of Late Medieval Philosophy: From the Rediscovery of Aristotle to the Disintegration of Scholasticism, 1100–1600,* ed. Norman Kretzmann, Anthony Kenny, and Jan Pinborg (Cambridge: Cambridge University Press, 1982), 496–517.

2. Francis Bacon, *Novum organum* [1620], in Basil Montagu, ed., *Lord Bacon's Works,* 16 vols. (London: William Pickering, 1825–34), II.28–29, 14:137–38.

between wolf and sheep, landscapes figured in Florentine marble, the occult properties of certain animals, plants, and minerals, exotic species such as crocodiles and birds of paradise, rains of wheat or blood, the force of the imagination to imprint matter—in short, all that happens "extraordinarily, (as to the ordinary course of nature) though not lesse naturally."[3]

The proviso "though not lesse naturally" was key to what I shall call preternatural philosophy, for however marvelous or even incredible its objects might seem, they were, as we shall see, sharply distinguished from the miraculous and supernatural. Among practitioners of preternatural philosophy, it was an inflexible premise that all such anomalies might be ultimately explained by recourse to natural causes. Hence its claim to the title "philosophy," the repository of causal explanations, as opposed to mere "history," an assemblage of disconnected particulars.[4] Indeed, preternatural philosophy set the most ambitious standards for scientific explanation in the early seventeenth century. Even natural philosophers as deeply skeptical of preternatural philosophy as René Descartes accepted the challenge its objects flung down to any systematic account of natural causes, promising that there were "no qualities so occult, no effects of sympathy or antipathy so marvelous or strange, finally no other thing so rare in nature" that his mechanical philosophy could not explain.[5]

The challenge of preternatural philosophy to early modern natural philosophy, both traditional and reformed, was twofold. First, the oddities that were its objects greatly expanded the domain of phenomena requiring philosophical explanation. Although Aristotelians had never disputed the existence of rare exceptions to nature's ordinary course, nor doubted that these could be traced to natural causes, they had excluded such oddities from the purview of natural philosophy as neither regular nor, a fortiori, demonstrable. As Nicole Oresme argued in his treatise *De causis mirabilium* (comp. ca. 1370), people born with six fingers or who went twenty years

3. Meric Casaubon, *A Treatise Concerning Enthusiasm* (London: Printed by R. D. and are to be sold by Tho. Johnson, 1655), 41.

4. The distinction between philosophy and history, especially natural philosophy and natural history, originates in Aristotle (see *Poetics,* 1451b1–7; *On the Parts of Animals,* 639a13–640a10) and continued to be standard throughout the seventeenth and even eighteenth centuries: see for example Thomas Hobbes, *Leviathan* [1651], ed. C. B. Macpherson (Harmondsworth: Penguin, 1968), 1.9, pp. 147–48; Jean le Rond d'Alembert, "Discours préliminaire des éditeurs," in *Encyclopédie, ou Dictionnaire raisonné des sciences, des arts et des métiers,* ed. Denis Diderot, vol. 1 (Paris: Briasson, David l'aîné, Le Breton, Durand, 1751), especially the chart "Système détaillé des connaissances humaines" and the accompanying explanation.

5. René Descartes, *Principia philosophiae* [Latin 1644, French 1647], in *Oeuvres de Descartes,* ed. Charles Adam and Paul Tannery, 12 vols. (Paris: J. Vrin, 1897–1910), 4.187, pp. 8:314–15.

without eating were rare due to a chance concatenation of causes, each natural enough in itself but in conjunction humanly impossible to explain in
detail: "And for such things, who can give the reason why, other than the
general one, namely that their causes are adequate, and no more and no less,
for producing this? Therefore these things are not known point by point
[punctualiter] except by God."[6] He here followed Aristotle in rejecting the
possibility of a science of chance.[7] In contrast, early modern philosophers of
the preternatural such as Pietro Pomponazzi, Girolamo Cardano, Bernard
Palissy, Francis Bacon, and Gaspar Schott shifted the marvels of nature
from the periphery to the center of their philosophy, and attempted explanations of even the most singular phenomena.

Second, preternatural philosophy expanded the range of explanations as
well as that of objects to be explained. Whereas medieval natural philosophers, following Galen, had acknowledged the existence of hidden or "occult" properties in certain animals, herbs, and stones, they had been content
to ascribe these simply to "substantial forms" rather than to the manifest
properties of hot, cold, dry, and wet, whose combinations accounted for the
ordinary course of nature.[8] Borrowing from the Neoplatonism of Marsilio
Ficino, medical and natural history treatises on the secret virtues of herbs
and gems, Avicenna's writings on the soul, and a miscellany of other
sources, the early modern preternatural philosophers introduced new
kinds of causes—astral influences, plastic virtues, the imagination, sympathies and antipathies—to meet the challenge of their new explananda. The
"nature" of preternatural philosophy was thus doubly transformed, in
both its causes and effects. Despite the unflinching commitment of its practitioners to natural explanation, firmly excluding both the demonic and
the divine, preternatural philosophy looked distinctly unnatural from the
standpoints of the natural philosophies that had both preceded and would
succeed it.

It is my aim in this essay to explore how and why the objects of preternatural philosophy came in the mid–sixteenth century to cohere into a category amenable to scientific study, only to dissolve again into scattered
oddities and anomalies largely ignored by scientists from the early eighteenth century on. Preternatural objects continued to exist, but they were

6. Bert Hansen, *Nicole Oresme and the Marvels of Nature: A Study of His "De causis
mirabilium" with Critical Edition, Translation and Commentary* (Toronto: Pontifical Institute of Mediaeval Studies, 1985), 278–79.

7. See for example Aristotle, *Physics*, 2.8, 199b24–26, concerning the impossibility that
events that occur always or for the most part could be due to chance.

8. Brian Copenhaver, "Natural Magic, Hermeticism, and Occultism in Early Modern Science," in *Reappraisals of the Scientific Revolution*, ed. David C. Lindberg and Robert Westman
(Cambridge: Cambridge University Press, 1990), 261–301, esp. 272–73.

no longer scientific objects. I shall argue that the glue that made the category of the preternatural hold together was compounded of a distinctive ontology, epistemology, and sensibility. Although for analytic clarity I shall discuss each theme separately, they were in fact tightly interwoven. When preternatural philosophy disintegrated, it was not because its characteristic objects or forces were summarily discarded—some, like ethereal fluids and the imagination, remained central to Enlightenment science—but rather because its unifying principles came unraveled.

AN INVENTORY OF THE NONNATURAL

There is an unmistakable resemblance between the objects of preternatural philosophy and the contents of the *Wunderkammer* and cabinets of curiosities stocked during the same period. Carved gems with secret powers, the stuffed carcass of an exotic species brought back from the Far East or the Far West, monsters, a unicorn horn to counteract all poisons, stones figured with landscapes or shapes of fish or plants—almost all of the naturalia displayed in the cabinets also featured prominently in the coeval treatises on preternatural philosophy. The contents of both cabinets and treatises seem the very type of a miscellany, a hodgepodge of strange objects still more strangely juxtaposed. But beneath miscellaneous appearances lay certain tacit principles of selection. Neither the cabinets nor the treatises of preternatural philosophy were encyclopedic in the sense of representing the entire universe of things; on the contrary, they were highly selective samples that systematically ignored all that was mundane, commonplace, or ordinary—in short, all that was "natural" in the usual sense of the word.

Here is a scattering of examples culled from sixteenth- and seventeenth-century treatises devoted in whole or in part to the preternatural: the apparition of an image of Saint Celestine to the entire population of Aquila when torrential rains threatened to flood the town;[9] images found in agates or marble;[10] comets presaging the death of kings;[11] a Medusa's head found in a

9. Pietro Pomponazzi, *De naturalium effectum causis sive de incantationibus* [comp. ca. 1520; 1567] (Hildesheim: Georg Olms, 1970), 159–60, 236–39.

10. Girolamo Cardano, *De svbtilitate libri XXI* (Nuremberg: Joh. Petrus, 1550), 184. A second, enlarged edition appeared later in 1550 (in Paris), and a revised edition in 1554, from which a French translation was made: *Les Livres de Hierome Cardanvs Medecin Milannois, intitvles de la subtilité, & subtiles inventions, ensemble les causes occultes, & les raisons d'icelles*, trans. Richard Le Blanc (Paris: Charles l'Angelier, 1556). In 1663 the revised Latin treatise was reprinted in vol. 3 of Cardano's *Opera omnia*, 10 vols. (Lyons: Jean Huguetan, 1663). Because of textual variations pertaining to my topic, I have used all three editions.

11. Scipion Dupleix, *La Physique, ou Science des choses naturelles* [1640], ed. Roger Ariew (Paris: Fayard, 1990), 7.4, pp. 425–26.

hen's egg in Bordeaux;[12] the power of flax seeds to inspire prophetic dreams.[13] The authors of these treatises were of the most diverse theoretical persuasions, from the dedicated Aristotelian Pietro Pomponazzi to the vehemently anti-Aristotelian Francis Bacon, addressed their audiences in both Latin and the vernacular, practiced professions ranging from professor of natural philosophy (Pomponazzi) to physician (Cardano, Liceti) to mathematician (Cardano) to lawyer and statesman (Bacon, Dupleix), and differed wildly over acceptable causal explanations. Pomponazzi invoked astral influences, Liceti preferred formal principles and virtues, Cardano appealed to chance, Bacon believed in subtle vapors and effluvia. What united the preternatural philosophers was a steely commitment to pushing natural explanations "beyond nature," to phenomena so rare or strange as to have eluded the conventional natural philosophy of Aristotelian regularities.

ONTOLOGY: THINGS RARE AND RARIFIED

Preternatural objects were first selected on ontological grounds. Early modern philosophers followed ancient and medieval Scholastic sources in opposing the natural to at least three other categories (in addition to the artificial):[14] the supernatural (*supra naturam*, literally "above nature"), the preternatural (*praeter naturam*, "beyond nature"), and the unnatural (*contra naturam*, "against nature"). The supernatural referred exclusively to God and the genuinely miraculous, i.e., when God suspended his ordinary providence to warn, punish, or reward.[15] The unnatural was also morally freighted and referred to particularly heinous acts like patricide or bestiality that violated the normative order of both nature and human nature.[16] In contrast, the preternatural was, with one exception, a morally neutral category, referring to things or events outside the quotidian order of nature, but still due to natural causes, however oddly concatenated. The exception was the work of demons, who could not usurp the divine prerogative of suspending the order of nature, but who could work marvels if not miracles by cleverly knitting together natural properties and forces ordi-

---

12. Fortunio Liceti, *De monstrorum causis, natura, et differentiis libri dvo* [1616], 2d ed. (Padua: Paulum Frambottum, 1634), 252–53.

13. Francis Bacon, *Sylva Sylvarum: Or, A Natural History in Ten Centuries* [post. 1627], in *Works*, 10.933, p. 4:502.

14. Aristotle, *Physics*, 2.1, 192b9–193b12.

15. Thomas Aquinas, *Summa theologica*, 1a, 105.6–8.

16. Certain sexual acts, including masturbation and sodomy, were regularly attacked by medieval moralists as "sins against nature": see James A. Brundage, *Law, Sex, and Society in Medieval Europe* (Chicago: University of Chicago Press, 1987), 212–14.

narily found asunder to produce preternatural phenomena. Of sharper intelligence, fleeter foot, and lighter touch than humans, demons could manufacture remarkable effects, but they were nonetheless constrained to work by natural causes.

These categories were the handiwork of professional philosophers, and less specialized works sometimes blurred the boundaries between supernatural and preternatural (especially where demons were suspected to be at work), just as homilies, hagiography, and sermons sometimes conflated marvels with miracles, despite the sharp theological distinction between them. Only in the sixteenth and seventeenth centuries did the lines between the supernatural, preternatural, and natural harden, in part due to the new preternatural philosophy, and in part due to the intense scholarly interest in demonology.[17] These newly rigidified boundaries did not, however, emerge in the first instance because of a new concept of natural law in the Humean sense, universal, eternal, and inviolable. Rather, the territory they divided up remained the territory of nature's habits or customs, from which she was on rare occasions diverted by obstacles, chance, or sheer whimsy. Exceptions to nature's laws in the eighteenth-century sense were miracles;[18] exceptions to nature's habits could be either marvels or miracles. Preternatural philosophy was the science of marvels, a bold attempt to push inquiry "until the properties and qualities of those things, which may be deemed miracles, as it were, of nature, be reduced to, and comprehended in, some form or law; so that all irregularity or singularity may be found to depend on some common form,"[19] as Bacon put it.

What could count as a marvel of nature? The criteria were multiple and intertwined, and none held for all members of the class. Some phenomena were marvels because their mode of operation was hidden from perception. Such were magnetic attraction, or poisons, or the properties of certain animals, plants, and minerals: for example, the power of the urine of a wild boar to cure earaches, or of amethysts to repel hail and locusts.[20] Sympathies and

---

17. Stuart Clark, "The Scientific Status of Demonology," in *Occult and Scientific Mentalities in the Renaissance,* ed. Brian Vickers (Cambridge: Cambridge University Press, 1984), 351–74.

18. David Hume, "Of Miracles," *An Inquiry Concerning Human Understanding* [1748], ed. Charles W. Hendel (Indianapolis: Bobbs-Merrill, 1955), 117–41: "There must, therefore, be a uniform experience against every miraculous event, otherwise the event would not merit that appellation" (122–23).

19. Bacon, *Novum organum,* II.28, 14:137.

20. Pliny, *Historia naturalis,* vol. 8, trans. W. H. S. Jones (Cambridge: Harvard University Press, Loeb Classical Library, 1975), bk. 28.48, pp. 118–19; vol. 10, trans. D. E. Eichholz (Cambridge: Harvard University Press, 1962), bk. 37.40, pp. 264–65. There was a long tradition in medieval and Renaissance natural history of "books of secrets" disclosing the occult properties

antipathies between species of animals and plants also belonged to this category of "occult" properties: why wolf and sheep were eternal enemies, so that a drum made out of sheepskin would not sound in the presence of one made of wolfskin,[21] or why "the Ape of all other things cannot abide a Snail"[22]— these were examples of natural attractions and repulsions that could be neither inferred nor predicted from the manifest properties of hot, cold, wet, and dry. Although occult properties were in principle as regular in their operation as manifest ones, they were opaque to observation and intractable to explanation—except by recourse to equally inscrutable "substantial forms"— and therefore beyond the ken of conventional natural philosophy.

Other objects and phenomena belonged to preternatural philosophy because they were rare: bearded grape vines, earthquakes, three suns in the sky, rains of blood, two-headed cats, people who slept for months on end or washed their hands in molten lead, visions of armies battling in the clouds. Not only rare individuals but also rare species might qualify as objects as preternatural philosophy. Just as stuffed crocodiles and birds of paradise dangled from the ceilings of well-stocked *Wunderkammern*, so they also made their appearance in the pages of treatises on preternatural philosophy. The French surgeon Ambroise Paré regaled his readers with illustrated accounts not only of the colt born with a man's head near Verona in 1224, but also of whales, ostriches, giraffes, and other species exotic to Europeans.[23] Of course there was nothing intrinsically rare about these creatures—giraffes would hardly have astonished an African nor elephants an East Indian. Their rarity was an artifact of an ethnocentric European perspective, acquainted with foreign species by at best a single stuffed exemplar (or perhaps only a claw or hoof) and more often by a woodcut drawn from secondhand reports and endlessly plagiarized, as in the case of Dürer's rhinoceros.[24]

of natural objects: see William Eamon, *Science and the Secrets of Nature: Books of Secrets in Medieval and Early Modern Culture* (Princeton: Princeton University Press, 1994). Pliny was also among the earliest and most often reissued of printed books in the fifteenth and sixteenth centuries: Charles Nauert Jr., "Humanists, Scientists, and Pliny: Changing Approaches to a Classical Author," *American Historical Review* 84 (1979): 72–85.

21. Marshall Clagett, ed., *Nicole Oresme and the Medieval Geometry of Qualities and Motions (Tractatus de configurationibus qualitatum et motum)* (Madison: University of Wisconsin Press, 1968), 1.27, p. 243.

22. Giambattista della Porta, *Natural Magick* [1558] (London, 1658; reprinted New York: Basic Books, 1957), 9.

23. Ambroise Paré, *Les monstres et prodiges* [1573], ed. Jean Céard (Geneva: Librairie Droz, 1971), 7, 124, 126, 132.

24. William B. Ashworth Jr., "The Persistent Beast: Recurring Images in Early Zoological Illustration," in *The Natural Sciences and the Arts: Aspects of Interaction from the Renaissance to the Twentieth Century*, ed. Allan Ellenius (Uppsala: Almqvist & Wiksell, 1985), 46–66.

The challenge of explaining individual oddities was twofold. First, many of them, particularly monstrous births and celestial apparitions, had been traditionally interpreted as portents, as signs sent directly from God to herald religious reformation or impending disaster. During the political and religious upheavals of the sixteenth and early seventeenth centuries, the hermeneutics of prodigies flourished in both vernacular broadsides and Latin treatises throughout Europe.[25] Viewed as divine warnings, strange phenomena teetered on the edge of the supernatural, not amenable to natural or even to preternatural explanation. Second, even when prodigies were classified as natural wonders rather than as divine portents, they were ascribed to "chance"; i.e., to a tangled knot of accidents exceptionally conjoined. To unravel such coincidences on a case-by-case basis was the arduous and often insoluble task of the preternatural philosopher.

To put together the causes that nature ordinarily kept asunder was the work of the natural magician—or the demon. Francis Bacon called natural magic the "operative" counterpart to speculative natural philosophy[26], but it was more narrowly linked to preternatural philosophy. The natural magician delved into the "hidden and secret properties" of things, tapped the invisible but powerful forces of the imagination or the stars, and above all imitated the incessant matchmaking of nature, "so desirous to marry and couple her parts together" in knitting together causes to produce sinister wonders and counterfeit miracles.[27] In principle, demons were simply natural magicians par excellence, endowed with more acute minds and more cunning hands than humans, but bound by the same natural order, manifest and occult, as humans were. As Sir Thomas Browne said of Satan and his powers, "being a naturall Magician he may performe many acts in wayes above our knowledge, though not transcending our naturall power, when our knowledge shall direct it."[28] The only distinction between the works of natural magicians and demons on the one hand and the pretergen-

25. See Jean Céard, *La nature et les prodiges: L'insolite au XVIe siècle, en France* (Geneva: Librairie Droz, 1977); Katharine Park and Lorraine Daston, "Unnatural Conceptions: The Study of Monsters in Sixteenth- and Seventeenth-Century France and England," *Past and Present* 92 (August 1981): 20–54; David Warren Sabean, *Power in the Blood: Popular Culture and Village Discourse in Early Modern Germany* (Cambridge: Cambridge University Press, 1984), 61–93; Ottavia Niccoli, *Prophecy and People in Renaissance Italy* [1987], trans. Lydia G. Cochrane (Princeton: Princeton University Press, 1990), 30–88.

26. Bacon, *The Advancement of Learning* [1605], in *Works*, 2:146.

27. Della Porta, *Natural Magick*, 8, 14.

28. Thomas Browne, *Pseudodoxia Epidemica: Or, Enquiries into Very many received Tenents* [sic] *and commonly presumed Truths* [1646], ed. Robin Robbins, 2 vols. (Oxford: Clarendon Press, 1981), 1.10, p. 1:63. The same arguments held for angels: Thomas Aquinas, *Summa theologica*, 1, qu. 110, art. 4.

erations of unassisted nature on the other was the agency of a free will. And the only distinction between the will of a natural magician and the will of a demon was that between benign and malicious intent. No wonder the line between natural and demonic magic, identical in their means and products, was perpetually and dangerously blurred.[29]

Demons were the craftsmen of the preternatural, not the wielders of the supernatural. This is why sixteenth- and seventeenth-century demonologists intent on fixing the boundaries between the possible and impossible in the all-too-concrete context of witchcraft trials became authorities on the limits of the natural and the preternatural: no witch, even with the aid of a legion of demons, could be charged with felonies transgressing these limits.[30] Preternatural philosophers like Pietro Pomponazzi[31]—or for that matter, Descartes[32]—expelled demons from their treatises because they insinuated the wild card of free will into nature's ordinary and extraordinary processes. Volition, be it human, demonic, angelic, or divine, turned nature into art, by "applying to the natural agent materials that nature never or very rarely assembles and conjoins."[33] Demons were the great derandomizers of nature, manufacturing coincidences at a rate far faster than chance. They were summarily evicted from treatises on natural and preternatural philosophy not because their works were supernatural but because they were artificial.[34]

If the preternatural philosophers were dogged in their adherence to exclusively natural explanations, they nonetheless often invoked causes fully as extraordinary as the effects to be accounted for. Celestial influences, subtle effluvia, the *vis imaginativa*, chance, vegetative and sexual

---

29. D. P. Walker, *Spiritual and Demonic Magic: From Ficino to Campanella* [1958] (Notre Dame: University of Notre Dame Press, 1975), 75–84 et passim; Richard Kieckhefer, *Magic in the Middle Ages* (Cambridge: Cambridge University Press, 1989), 12–17, 149–50.

30. Clark, "Scientific Status."

31. Pomponazzi, *De incantationibus*, 19–20.

32. Descartes, *Principes de la philosophie* [1647] in *Oeuvres*, IV.187, 9:309. The qualification concerning free will is not found in the Latin edition of 1644: *Oeuvres*, IV.187, 8:315.

33. Anselme Boece de Boot, *Le parfait ioallier ov Histoire des pierreries* (Lyon: Chez Iean-Antoine Hvgvetan, 1644), 116.

34. Hence writers like Paracelsus and Cornelius Agrippa who appealed to demons and other spirits cannot, despite some overlap in subject matter, be numbered among the preternatural philosophers, for these "darksome authors of magic" strayed from the "clean and pure natural": Bacon, *Sylva*, 10, p. 4:488. This was at once a charge of immorality (dabbling in necromancy) and philosophical unprofessionalism (abandoning the realm of the purely natural). On Pomponazzi's distinction between natural magic and philosophy see Martin L. Pine, *Pietro Pomponazzi: Radical Philosopher in the Renaissance* (Padua: Editrice Antenore, 1986), 245–46.

principles extended to minerals, plastic virtues, and the sheer whimsy of nature were all causes that might be given a somewhat dubious Aristotelian pedigree,[35] but in fact derived at least as much from the writings of Pliny, Avicenna, and Marsilio Ficino as from the *Meteorology* or *On the Heavens*. What was characteristic of many, though not all, of these preternatural causes was the action of rarefied vapors upon soft, pliable matter. By the fervor of their prayers, the inhabitants of Aquila managed to emit a fine vapor that impressed the form of Saint Celestine upon the turgid, rain-laden air, much as the sweet breath of children can cure some ailments or as joy and sadness can be communicated from soul to soul.[36] If comets presaged the death of princes, it was because the same dry exhalations that fed the comet afflicted the high and mighty, whose delicate and luxurious tastes rendered them susceptible to vivid impressions and acute diseases.[37] Women sometimes bore children with horns and tails not because they had actually slept with demons but because their overwrought imaginations had imprinted a diabolical shape upon the soft matter of the fetus.[38] The famous agate of King Pyrrhus depicting Apollo and the nine muses was originally a painting on marble left by chance "where agates are customarily engendered," so that the nascent, waxy stone absorbed the image.[39]

What is striking about the tone of these explanations, if not their content, is their militant naturalism. First and foremost, they were militant in their explanatory ambitions, reaching from the marvelous almost to the miraculous. In the case of the apparition of Saint Celestine, portentous comets, and certain ominous monsters, philosophers stretched the preternatural perilously close to the boundary with the supernatural. In his massive chronological compendium of all prodigies from the talking serpent in Eden in 3959 B.C. to a stop-the-presses monster with a flattened head born in Basel on 7 August 1557, the humanist Conrad Lycosthenes clearly had such naturalizing forays in mind when he cautioned philosophers against seeking natural causes for divine signs.[40] Furthermore, the explanations were sternly matter of fact and materialistic. If the imagina-

35. Copenhaver, "Natural Magic," 398–400.

36. Pomponazzi, *De incantationibus*, 37.

37. Dupleix, *Physique*, 7.4, p. 426.

38. Liceti, *De monstrorum*, p. 254–57.

39. Cardano, *De la subtilité*, 137r.–v. The story of Pyrrhus's agate is from Pliny, *Historia naturalis*, 37.3.

40. Conrad Lycosthenes, *Prodigiorum ac ostentorum chronicon* (Basel: Henri cum Petri, 1557); Conrad Lycosthenes, "Epistola Nuncupatoria," n.p.

tion could work material changes upon a fetus, another person's body, or even an inanimate object, it did so by an invisible but nonetheless material emission of effluvia. Bacon pointedly treated such alleged cases of the power of the imagination alongside "the transmission or emission of the thinner and more airy parts of bodies; as in odors and infections," progressing by degrees to the "emission of immateriate virtues" in the case of sympathies between individuals. "Airy bodies" become gradually attenuated into attractions at a distance such as electricity and magnetism and then into "influxes of the heavenly bodies" such as heat and light, and finally into "the infection from spirit to spirit," as in fascination or blushing. In all cases, the underlying model was that of contagion by miasma, and the implication was that even the most prodigious powers of the imagination operated by principles as mundane as those by which "Guiney-pepper . . . provoketh a continual sneezing in those that are in the room."[41]

The delicate interactions of airy emanations with soft matter displayed considerably more variability in their outcomes than coarser, quotidian natural processes. Bacon cautioned not to "withdraw credit from the operations by transmission of spirits, and force of imagination, because the effects fail sometimes." Just as not everyone exposed to the plague falls ill, so there were degrees of susceptibility among minds, "women, sick persons, superstitious and fearful persons, children, and young creatures" being the most impressionable.[42] Cardano distinguished between "things that are according to nature" and therefore "more often or frequently true," and things that are "remote and far from nature, which have causes wholly obscure and difficult, such as the direction from which comes the wind."[43] Commenting on the reliability of such esoteric remedies as the root of a male peony plant cut when the moon is full applied to gouty feet, Cardano refused to promise that it would work for all cases, although it worked wonders for some.[44] Hidden causes were variable causes, sensitive to the slightest change of texture and consistency. Although they were natural, preternatural phenomena were not robust, nor did they always follow the maxim "like causes, like effects." In the swirling eddies of subtle vapors, tiny perturbations might vastly alter outcomes.

41. Bacon, *Sylva*, 10.902–45, p. 4:490–507.
42. Ibid., 10.901, p. 4:489.
43. Cardano, *De la subtilité*, 368v.; this passage does not appear in the first Latin edition but was inserted into later Latin versions: Cardano, *Opera*, p. 3:648.
44. Cardano, *De la subtilité*, 358v.; Cardano, *Opera*, 3:641. This passage does not appear in the first Latin edition.

EPISTEMOLOGY: THE HIDDEN, THE RARE, AND THE DIFFICULT

The objects of preternatural philosophy were rare and heteroclite, their causes hidden and irregular. Whereas Aristotelian natural philosophy had required only the most lightweight epistemological apparatus to study manifest properties and commonplace regularities, early modern preternatural philosophy needed heavier machinery to warrant knowledge of such elusive and ornery phenomena. First, there was the problem of how preternatural philosophy could be called knowledge at all, since its treatises were crammed with particular instances, rather than with the universals traditionally thought to be the stuff of philosophy. Although preternatural philosophers strove to provide explanations for their odd particulars (in pointed contrast to natural historians), the work of collecting and accounting for even these rarities might well "never come to an end," as Cardano sighed at the end of his four-hundred-page treatise.[45] Bacon's elaborate tables of presence and absence and lists of prerogative instances in Book II of the *Novum organum* (1620) were systematic attempts to delve beneath the welter of particulars in order to discover the "nature-engendering nature" through "latent conformations" and "latent processes,"[46] but this method was so time-consuming that even Bacon's most loyal disciples seldom applied it.[47] When Descartes plumped for a natural philosophy of "common things of which everyone has heard," it was because he recoiled from the laborious and open-ended investigations of preternatural philosophy: "for it would be necessary first of all to have researched all the herbs and stones that come from the Indies, it would be necessary to have seen the Phoenix, and in short not to overlook anything of all that is most strange in nature."[48]

Descartes's reference to the Indies raised the second epistemological quandary for preternatural philosophy: it trafficked in rarities and marvels, but rare and marvelous for whom? What astonished the homebound lay reader might elicit only a yawn from the seasoned traveler or naturalist. In a dynamic that closely paralleled the economics of *Wunderkammer* collecting in the sixteenth and seventeenth centuries, preternatural objects could

---

45. Cardano, *De la subtilité*, 391r.–v.; *Opera*, pp. 3:671–72. Cardano continued his endless project with another tome of approximately the same length: *De rerum varietate libri XVII* [1557], *Opera*, vol. 3.

46. Bacon, *Novum organum*, II.1–2, 14:91–92.

47. See for example Joshua Childrey, *Britannica Baconica: Or, The Natural Rarities of England, Scotland, & Wales* (London: n.p., 1661).

48. Descartes, *La recherche de la verité par la lumière naturelle* [post. 1701], in *Oeuvres*, 10:503.

lose their cachet through overexposure. Just as flooding the market with narwhal horns brought the price of a "unicorn horn" down from six thousand florins in 1492 to about thirty-two florins in 1643,[49] so yesterday's wonder might be today's commonplace. An epistemology of the rare was exquisitely sensitive to local context.

Natural philosophy had its own traditional criterion of the marvelous, if not of the rare: ignorance of causes provokes wonder, which is in turn the origin of philosophy.[50] Conversely, knowledge of causes destroys wonder, just as peeking behind the curtain at a marionette show deflates the marvel of the apparently self-propelled little figures. This image of lifting a curtain or veil to reveal the hidden causes of things was frequently invoked by the early modern preternatural philosophers (Bacon repeats Aristotle's example of the puppet show almost verbatim[51]) to describe their own inquiries. Although their subject matter could hardly have been less Aristotelian— Aristotle thought the first philosophers commenced their wondering with the most obvious phenomena, not the most esoteric—the preternatural philosophers understood their mission in Aristotelian terms: to explain away wonder. Indeed, they in a sense out-Aristotled Aristotle by taking on the phenomena most difficult to explain, and therefore most wondrous— sometimes to the heretical point of tackling not only the marvelous but even the miraculous, as we have seen. Theirs were to be the Herculean labors of a natural philosophy that quenched wonder with knowledge.

Cardano elevated the very difficulty of this undertaking into a principle: the "subtlety" of his title referred to no particular kind of object but rather to "that reason by which things sensible to the senses and intelligible to the intellect are to be comprehended with difficulty."[52] In his *Exotericarum exercitationum liber XV de svbtilitate ad Hieronymum Cardanum* (1557) the neo-Aristotelian Julius Caesar Scaliger ridiculed the arbitrariness of this criterion, including its relativity to the mind of the knower,[53] but Cardano's epistemology of difficulty was framed within an Aristotelian context. All that was not manifest to the senses, all that did not happen always or for the most part, all that partook of the variable and the fortuitous

49. Antoine Schnapper, *Le géant, la licorne et la tulipe: Collections et collectionneurs dans la France du XVIIe siècle.* vol. 1: *Histoire et histoire naturelle* (Paris: Flammarion, 1988), 89–92.

50. Aristotle, *Metaphysics*, 1.2, 982b10–17.

51. Bacon, *Advancement*, 2:81.

52. Cardano, *De svbtilitate*, 1.

53. On the polemic between Cardano and Scaliger, see Ian Maclean, "The Interpretation of Natural Signs: Cardano's *De subtilitate* versus Scaliger's *Exercitationes*," in *Occult and Scientific Mentalities in the Renaissance*, ed. Brian Vickers (Cambridge: Cambridge University Press, 1984), 231–52.

eluded an epistemology of sensory particulars forged easily and accurately into the universals that could serve as the premises and conclusions of demonstrations. Such outcast phenomena posed special epistemological puzzles; hence "difficulty" was one way of defining them as a category. From this standpoint, the preternatural was all that slipped through the meshes of Aristotelian epistemology—the subsensible, the variable, the rare. Bacon's epistemological reflections echo this theme of difficulty by emphasizing that not only the infirmities of the human mind (the idols of tribe, cave, marketplace, and theater) but also the deviousness of nature, full of "deceitful imitations of things and their signs, winding and intricate folds and knots,"[54] impede natural philosophy.

For Bacon, preternatural philosophy was not, however, simply the most difficult part of natural philosophy. He also intended his projected "history of pretergenerations" to serve as an epistemological corrective to the ingrained philosophical habit of hasty generalization from "a scanty handful" of experience to abstract axioms.[55] The "strange and monstrous objects, in which nature deviates and turns from her ordinary course" would "rectify the understanding in opposition to habit, and reveal common forms."[56] The mission of earlier preternatural philosophers like Pomponazzi and Cardano had been to naturalize marvels and thereby to extend the boundaries of natural philosophy beyond its traditional limits. Bacon went still further, and aimed to use preternatural philosophy to reform natural philosophy, by finding new "common forms" that could encompass both regularities and deviations. As we have seen in the case of emanations and the imagination, Bacon was sometimes willing to countenance peculiarly preternatural explanations, but the ultimate goal of his reformed natural philosophy was synthetic rather than expansive: not only to explain nature out of as well as in course, but also to do so by the same causes.

SENSIBILITY: WONDER AND POWER

The explanatory ambitions of preternatural philosophy were a double-edged affair. On the one hand, preternatural philosophers were the virtuosi of their discipline, boldly stretching natural explanations to cover marvels or even miracles. As naturalizers, they were sworn enemies of wonder, dedicated to pulling back the curtain to expose the manipulations of the puppeteers. In this vein, Cardano loftily pronounced rains of frogs

---

54. Bacon, *Novum organum*, "Preface to the Great Instauration," 14:10.
55. Ibid., I.25, 14:35; I.104, 14:73–74.
56. Ibid., II.29, 14:138.

and fish "no wonder," since they could be explained by strong winds that carried animals and even stones to great heights.[57] On the other hand, preternatural philosophers were aficionados of wonder, their treatises overflowing with stories and examples that could and did find their way into unabashedly popular compilations of marvels.[58] Not only the wonders of nature but also the wonders of art—ingenious codes, chariots drawn by fleas, the feats of jugglers and fire eaters—were grist for their mill, because all belonged to the category of secrets, linked by a shared sensibility of wonder.

Among the preternatural philosophers, this sensibility of wonder displayed a nuanced register of responses. So long as wonder was provoked by ignorance of causes, the ontology of hidden properties and epistemology of difficulty willy-nilly selected objects that were wondrous. But wonder could also become an independent criterion of selection, and it did not always dissolve when causes were laid bare. Rather than stamping out wonder entirely, most preternatural philosophers instead became connoisseurs of that emotion, instructing their readers in the shades of ennui, interest, surprise, admiration, or astonishment appropriate to each object. Cardano briskly dismissed the appearance in 1534 of a red cross in the air in Switzerland as "not marvelous," but admitted to standing open-mouthed before apparitions of the dead "even though one could offer a natural reason for them."[59] Sir Thomas Browne opined that "[t]o behold a Rain-bow in the night, is no prodigie unto a Philosopher."[60] Meric Casaubon thought monsters "the most ordinary subject of their admiration, who are not qualified to admire any thing else, though it deserve it more," acknowledged sympathies, antipathies, and "strength of *imagination*" as "worthy objects of admiration," but reserved the full measure of his wonder for the "strange and incredible" properties of the mathematical asymptotes he had been shown at Oxford as a student.[61]

The uses and abuses of wonder in natural philosophy were a theme that received considerable attention in the middle decades of the seventeenth

57. Cardano, *Opera*, 3:605.

58. Such compilations constituted a large and flourishing early modern literary genre: see for example Pierre Boaistuau et al., *Histoires prodigieuses et memorables . . . divisées en six livres* (Lyon: Jean Pillehotte, 1598); Levinus Lemnius, *De miraculis occultis naturae libri IIII* (Antwerp: Christopher Plantin, 1574); Thomas Lupton, *A Thousand Notable Things* (London: Edward White, 1586); or [Etienne Binet], *Essay des merveilles de natvre et des plvs nobles artifices* (Rouen: Chez Romain de Beauvais et Jean Osmont, 1621).

59. Cardano, *Opera*, 3:605, 660.

60. Browne, *Pseudodoxia*, 1.11, p. 1:67.

61. Casaubon, *Of Credulity and Incredulity in Things Natural, and Civil* (London: Thomas Tomkyns, 1668), 8, 9, 25.

century, in part because of the prominence of preternatural philosophy. Bacon claimed that "by the rare and extraordinary works of nature the understanding is excited and raised to the investigation and discovery of forms capable of including them,"[62] but also scorned the empiricists whose aimless trials "ever breaketh off in wondering and not in knowing."[63] Descartes was perhaps the clearest on the delicate balance to be struck between just enough and too much wonder. He recognized the utility of wonder "in making us learn and hold in memory things we have previously been ignorant of."[64] But this serviceable "wonder [admiration]" is to be distinguished from a stupefying "astonishment [estonnement]," which "makes the whole body remain immobile like a statue, such that one cannot perceive any more of the object beyond the first face presented, and therefore cannot acquire any more particular knowledge." Astonishment differs in degree from wonder—"astonishment is an excess of wonder"—but their cognitive effects are diametrically opposed. Whereas wonder stimulates attentive inquiry, astonishment inhibits it, and is therefore, Descartes asserted, always bad.[65]

The management of wonder had social and political as well as cognitive overtones, for wonder was intertwined with secrecy, and secrecy was the province of princes. Since at least the fourteenth century courtly displays of magnificence had featured all manner of wonders to impress subjects and especially foreign guests with the wealth and power of the ruler. Cardano described how the Emperor Charles V was fêted in Milan at the Sforza court with "marvelous things [that] enchanted the eyes of all present";[66] Paolo Morigi reported that the spectacles designed by Giuseppe Arcimboldo for the imperial court "fill[ed] all the great princes who were present with great wonderment, and his lord [Emperor] Maximilian with great contentment";[67] Galileo sought the favor of the Medicis by offering them "[p]articular secrets, as useful as they are curious and admirable."[68] To dazzle

---

62. Bacon, *Novum organum*, II.31, 14:139.

63. Bacon, *Advancement*, 2:12.

64. Descartes, *Les passions de l'âme* [1649], ed. Geneviève Rodis-Lewis (Paris: Librairie Vrin, 1955), art. 75, p. 119.

65. Ibid., art. 63, p. 118.

66. Cardano, *De svbtilitate*, 342.

67. Paolo Morigi, *Historia dell'antichità di Milano* [1592], quoted in Piero Falchetta, ed., *The Arcimboldo Effect: Transformations of the Face from the 16th to the 20th Century* (New York: Abbeville Press, 1987), 172.

68. Letter to Belisario Vinta, 1610, quoted in *Discoveries and Opinions of Galileo*, trans. and ed. Stillman Drake (Garden City: Doubleday Anchor, 1957), 62. On Galileo as purveyor of marvels, see also Mario Biagioli, "Galileo the Emblem Maker," *Isis* 81 (1990): 230–58, at 241–43.

with wonders was a form of courtly competition, particularly at weddings and coronations when ambassadors and visiting potentates would be in attendance.[69] The wonders of art and nature contained in the Prague *Kunstkammer* of the Emperor Rudolf II were similarly displayed to high-ranking visitors, as a visible sign of "princely prestige."[70]

The power of wonder was multilayered. At the most superficial level, to stun others into wonder without losing one's own sangfroid was a form of one-upmanship. Della Porta advised beginners in natural magic that audiences would admire their feats in proportion to their ignorance: "If you would have your works appear more wonderful, you must not let the cause be known."[71] Philosophers dedicated to revealing causes scorned such tricks made "strange by disguisement,"[72] but accomplished much the same effect through their connoisseurship of wonders. Only those well versed in the preternatural could dictate which marvels deserved to be admired and precisely how much. They thereby exercised the power of wonder at a somewhat deeper level. By a kind of transference, the wonder originally excited by the occult properties of things shifted to the philosopher who penetrated their causes—Cardano going so far as to boast of the "admirable and wondrous side" of his own nature.[73] Princes who beguiled their guests with marvels similarly basked in the reflected wonder, perhaps even to the point of inspiring awe as well as admiration. The marvels of the prince could ape the miracles of God. Finally, secrets of all kinds resonated to a courtly culture of dissimulation, intrigue, necromancy, esoterica, and hunting.[74] The archetypal secret was the secret of state, and the long line of medieval and early modern "Mirrors for Princes" derived from the pseudo-Aristotelian treatise entitled *Secretum secretorum*.[75] Hence natural secrets became by association fitting gifts for the prince, master of all secrets—"most excellent Things fit for the Worthiest Nobles."[76]

69. Mark S. Weil, "Love, Monsters, Movement, and Machines: The Marvelous in Theaters, Festivals, and Gardens," in *The Age of the Marvelous*, ed. Joy Kenseth (Hanover: Dartmouth College, 1991), 158–78.

70. Thomas DaCosta Kaufmann, *The Mastery of Nature: Aspects of Art, Science, and Humanism in the Renaissance* (Princeton: Princeton University Press, 1993), 177–79.

71. Della Porta, *Natural Magick*, 4.

72. Bacon, *Advancement*, 2:146.

73. Cardano, *Eigene Lebensbeschreibung* [1576], trans. Hermann Hefele (Munich: Kösel, 1969), ch. 38, p. 139.

74. William Eamon, ""Court, Academy, and Printing House: Patronage and Scientific Careers in Late-Renaissance Italy," in *Patronage and Institutions*, ed. Bruce Moran (Woodbridge: Boydell Press, 1991), 25–50, at 37–38.

75. Eamon, *Science*, 45–53: more than six hundred manuscripts of this immensely popular work are known.

76. Della Porta, *Natural Magick*, preface [not paginated]. On Della Porta's own gifts of

Certain natural secrets were not only fit for nobles; they were in themselves noble. Preternatural philosophy projected upon the natural order a social hierarchy of superior and inferior tiers of being, God having "enjoyned inferiour things to be ruled of their superiors by a set Law."[77] The English physician and natural philosopher Walter Charleton credited all things with "a kind of native Ambition to ennoble its nature, enlarge its powers," and believed that only constant natures prevented an insurrectionary scramble up the ladder of being: "Can we conceive, that a *Plant* would continue fixed and nayled down by its own roots to the earth, and there live a cold, dull, inactive life; if it could give to its self motion and abilities for nobler actions?"[78] Within this hierarchy of things and forces, preternatural philosophers restricted their attention to the "most excellent" and "noblest" exemplars among animals, vegetables, and minerals. That which was worthy of wonder (Latin *admiratio*) was, etymologically and emotionally for Latinate writers, also worthy of admiration, and hence belonged to nature's nobility. Like the marvels purveyed by the *Wunderkammer* or the princely fête, the objects of preternatural philosophy pleased by their very remoteness from the vulgar and shopworn: "For things rare and unusual . . . call forth the Soul to a very quick and grateful attendance, whilst matters of greater worth and moment, of more familiar appearance (like things often handled and blown upon) lose their value and luster in its ey [*sic*]."[79]

THE DEMISE OF THE PRETERNATURAL

Preternatural philosophy did not, so to speak, die a natural death. Its characteristic ontology, epistemology, and sensibility were instead cannibalized by the natural philosophy of the late seventeenth and eighteenth centuries. The fascination with what Bacon called the "new, rare, and unusual" persisted well into the first decades of the eighteenth century, as the early numbers of the *Philosophical Transactions of the Royal Society of London* and the *Histoire et Mémoires de l'Académie Royale des Sciences* in Paris bear ample witness. Titles like "A Girl in Ireland, who has several Horns growing

secrets to Prince Frederico Cesi, see Paula Findlen, *Possessing Nature: Museums, Collecting, and Scientific Culture in Early Modern Italy* (Berkeley and Los Angeles: University of California Press, 1994), 229–32.

77. Della Porta, *Natural Magick*, 7.

78. Walter Charleton, *The Darknes of Atheism Dispelled by the Light of Nature: A physico-Theological Treatise* (London: William Lee, 1652), 133.

79. John Spencer, *A Discourse concerning Prodigies* (London: J. Field, 1665), a4v.

on her Body"[80] or "Rare and Singular New Phenomenon of Celestial Light"[81] or "Description of an Extraordinary Mushroom"[82] could easily have been taken from the treatises of preternatural philosophy published a hundred years earlier. If anything, philosophical ambitions had sunk in the interim, for very few of these reports to fledgling scientific societies on strange phenomena hazarded a causal explanation. Robert Boyle, describing his experiments on an "aerial noctiluca" that glowed eerily in the dark, was typical in his restraint: "it is not easy to know, what phaenomena may, and what cannot, be useful, to frame or verify an hypothesis of a subject new and singular, about which we have not as yet (that I know of) any good hypothesis settled."[83] One can imagine how Pomponazzi and Cardano, men who had ventured to explain miracles and prodigies, must have sneered in their graves.

When late seventeenth- and eighteenth-century natural philosophers did advance causal hypotheses, they often availed themselves of the same subtle spirits and rarefied effluvia that had been the staple explanatory resources of preternatural philosophy. The "Queries" appended to Isaac Newton's Opticks (1704) are perhaps the most celebrated of these latter-day appeals to what were to become "active principles" and "imponderable fluids" to explain everything from electricity and magnetism to perception, but even the mechanical philosophy that had preceded Newton was rife with "occult qualities."[84] Indeed, Descartes's own "first element," divided into "indefinitely little parts" so fine as to fill every interstice between bodies, resembled the effluvia of the preternatural philosophers in function as well as in form, for Descartes revealingly invoked it to explain the mysterious attractions of the magnet and amber and "innumerable other admirable effects."[85] The orthodox theories of electricity, magnetism, light,

80. "A Letter from Mr St. Georg Ash, Sec. of the Dublin Society, to one of the Secretaries of the Royal Society; concerning a Girl in Ireland, who has several Horns growing on her Body," Philosophical Transactions of the Royal Society of London (1685), 1202–4.

81. "Nouveau Phenomene rare et singulier d'une Lumiere Celeste, qui a paru au commencement de Printemps de cette année 1683," Journal des Scavans (1683), 121–30.

82. Joseph Tournefort, "Description d'un champignon extraordinaire," Histoire et Mémoires de l'Académie Royale des Sciences (1682–93): 101–5 (read 3 April 1692).

83. Robert Boyle, "The Aerial Noctiluca: Or, some new Phaenomena, and a Process of a factitious self-shining Substance," in The Works of the Honourable Robert Boyle, ed. Thomas Birch [1772] facsimile reprint with an introduction by Douglas McKie, 6 vols. (Hildesheim: Georg Olms, 1965–66), 4:379–404, at 4:393–94. On Boyle's "epistemological modesty" see Steven Shapin and Simon Schaffer, Leviathan and the Air-Pump: Hobbes, Boyle, and the Experimental Life (Princeton: Princeton University Press, 1985), 146–54 et passim.

84. John Henry, "Occult Qualities and the Experimental Philosophy: Active Principles in pre-Newtonian Matter Theory," History of Science 24 (1986): 335–81.

85. Descartes, Principia philosophiae, 3.52, 4.187; pp. 8:105, 8:314.

and heat as well as the heterodox theories of animal magnetism of the eighteenth and early nineteenth centuries cheerfully recycled the subtle spirits of sixteenth- and early seventeenth-century preternatural philosophy. Benjamin Franklin's electrical fluids and Antoine Lavoisier's caloric were lineal descendants of Bacon's airy emanations.

Nor did the *vis imaginativa* disappear from Enlightenment natural philosophy. Despite—or perhaps because of—the Cartesian chasm yawning between mind and body, the imagination continued to play its crucial role as mediator between the two. Nicholas Malebranche, who pushed Cartesian dualism to the verge of occasionalism, embraced the theory of the maternal imagination without reservation. If a woman who had witnessed the execution of a criminal on the wheel during her pregnancy bore a child whose bones were broken in the same places, it was because "every blow delivered to the wretch forcibly struck the imagination of the mother, and by a kind of counterblow the tender and delicate brain of her child."[86] Moreover, certain "effeminate" minds were of "such softness" that they were susceptible to a kind of contagion from "strong imaginations."[87] Voltaire insisted vehemently on the reality of both "passive" and "active" imaginations, the former responsible for monsters he had seen himself.[88] Just as Pomponazzi had invoked the power of the imagination to naturalize the putative miracle of Aquila, so the Parisian chief of police Hérault and Archbishop Vintmille invoked the power of the imagination to naturalize the well-attested Jansenist miracles that took place at the parish church of Saint-Médard in the 1730s.[89] And when the joint commission of the Académie Royale des Sciences and the Parisian medical faculty issued its 1784 report concluding that mesmeric fluid did not exist, its members (among them Franklin and Lavoisier) attributed Mesmer's well-authenticated cures, especially of impressionable female patients, to the power of the imagination.[90] For Enlightenment natural philosophers the imagination remained the last resort for natural explanations, a carte blanche to cover the most elusive, mysterious, and intractable phenomena.

The epistemology of the hidden also persisted within natural philosophy.

86. Nicholas Malebranche, *De la recherche de la vérité* [1674–75], 6th ed. (Paris: Michael David, 1712), reprinted as vols. 1–2 of *Oeuvres de Malebranche*, ed. Geneviève Rodis-Lewis (Paris: J. Vrin, 1963), 2.1.7.3, 1:238–39.

87. Ibid., 2.2.8.1–2.3.2.1, pp. 1:311–21.

88. Voltaire, "Imagination, Imaginer (*Logique, Métaphys., Litterat. & Beaux-Arts*)," in *Encyclopédie, ou Dictionnaire raisonné des sciences, des arts et des métiers*, ed. Denis Diderot (Paris: Briasson, David l'aîné, Le Breton, Durand, 1751), 8:560–63.

89. B. Robert Kreiser, *Miracles, Convulsions, and Ecclesiastical Politics in Early Eighteenth-Century Paris* (Princeton: Princeton University Press, 1978), 151–52, 204–23.

90. [Jean-Sylvain Bailly], *Rapport des commissaires chargés par le Roi, de l'examen du magnétisme animal* (Paris: Imprimerie Royale, 1785), 29–58.

If difficulty ceased to be an explicit criterion for selecting objects, nature's secrets were still the quarry for late seventeenth-century natural philosophers. The Royal Society's paid experimenter Robert Hooke recommended "taking more special Notice of such Operations and Effects of Nature as seem to be more secret and reserv'd, working on Bodies remov'd at some distance, such strange Effects as our Senses are wholly unable to shew us any probable Cause thereof," and speculated that "the gravity and Attraction of the Earth towards its Center" might illuminate "the true cause" of planetary motions and the tides.[91] Instruments like the microscope raised hopes that Nature might be pursued "Into the privatest recess / Of her imperceptible Littleness," as Abraham Cowley rhapsodized in his ode "To the Royal Society."[92] Although John Locke and other Fellows of the Royal Society eventually abandoned the idea that the microscope might reveal hidden essences,[93] the conviction that natural philosophy was ultimately grounded on what Hume was to call "the hidden springs and principles of things" never faded. From Descartes's microscopic mechanisms to Newton's corpuscles to Leibniz's *vis viva*, the explanatory resources of the new natural philosophy were "occult" in the literal sense of the word.[94] Only with the advent of militant positivism in the nineteenth century did philosophers like Auguste Comte and Ernst Mach once again flirt with the possibility of an epidermal science restricted to manifest properties.

Yet despite these ontological and epistemological survivals, preternatural philosophy itself had disintegrated by the late seventeenth century. Although popular anthologies of wonders continued to pour from the presses in every European vernacular, and although leading scientific societies crammed their annals with strange reports, few natural philosophers reputable enough to belong to these societies thought any longer to collect these oddities and their explanations into a volume.[95] Preternatural philosophy

91. Robert Hooke, "A General Scheme of the Present State of Natural Philosophy," in *The Posthumous Works of Robert Hooke*, ed. Richard Waller [1705] (reprint, with an introduction by Richard S. Westfall, New York: Johnson Reprint Corporation, 1969), 46.

92. Abraham Cowley, "To the Royal Society," in Thomas Sprat, *History of the Royal Society* [1667], ed. Jackson I. Cope and Harold Whitmore Jones (Saint Louis: Washington University Press, 1958), [not paginated].

93. Catherine Wilson, *The Invisible World: Early Modern Philosophy and the Invention of the Microscope* (Princeton: Princeton University Press, 1995), 236–48.

94. On the shift in meaning of the word "occult" in seventeenth-century natural philosophy, from "hidden" to "unintelligible," see Keith Hutchison, "What Happened to Occult Qualities in the Scientific Revolution?" *Isis* 73 (1982): 233–53.

95. On the model of the pseudo-Aristotelian treatise *Of Marvelous Things Heard*, Boyle had compiled a list of "Strange Reports," but only ten items were ever published, none with any attempt at explanation: Boyle, *Strange Reports in Two Parts* [part 2 was never published], in *Works*, 4:604–9.

had ceased to be a genre. It was not that its examples had been discredited as fabulous, at least not in any wholesale fashion. Although early modern naturalists professed skepticism about this or that item from Plinian natural history—Conrad Gesner doubted that mandrake roots screamed when pulled up; Thomas Browne doubted that elephants lacked knee joints (but not that they could talk); Claude Molinet doubted the existence of unicorns—any empirical debunking was of necessity slow and piecemeal.[96] Nor was preternatural philosophy the casualty of a sweeping elimination of what are now called "the occult sciences" by the new experimental philosophy. Aside from the fact that recent scholarship has shown how indebted leading figures like Boyle and Newton were at least to alchemy,[97] the modern category of "occult sciences" lumps together intellectual traditions—astrology, alchemy, Paracelsianism, natural magic, hermeticism, emblematic natural history—that were conceptually (and sometimes morally) distinct for early modern thinkers.[98] Although preternatural philosophy made occasional use of astral influences, the more general rubric of subtle emanations escaped unscathed from the downfall of astrology. Finally, preternatural philosophy was not the target of the late seventeenth-century polemic against secrecy in science:[99] unlike the alchemists, the Paracelsians, and many natural magicians, preternatural philosophers had not tricked out their works in deliberately obscure language or withheld causal conjectures. They studied secrets, but they were not secretive.

Why then did the category of the preternatural dissolve in the early eighteenth century? Its solvents were a new metaphysics and a new sensibility, which loosened its coherence without destroying its elements. The new metaphysics replaced the varied and variable nature of preternatural philosophy with one that was uniform and simple; the new sensibility replaced wonder with diligence, curiosity with utility. Newton's "Rules of Reasoning" appended to Book III of the *Principia* (1687/1713) neatly epitomizes the new metaphysics: in natural philosophy we must assume that like causes produce like effects in both quality and quantity, and that "nature affects not the vain pomp of superfluous causes."[100] Robert Boyle's

96. On the slow and uneven elimination of magical objects from seventeenth- and eighteenth-century natural history, see Copenhaver, "Natural Magic," 279–80.

97. Betty Jo Teeter Dobbs, *The Foundations of Newton's Alchemy, or "The Hunting of the Green Lion"* (Cambridge: Cambridge University Press, 1975); William R. Newman, *Gehennical Fire: The Lives of George Starkey, an American Alchemist in the Scientific Revolution* (Cambridge: Harvard University Press, 1994), 75–80 et passim.

98. Copenhaver, "Natural Magic," 280–81.

99. William Eamon, "From the Secrets of Nature to Public Knowledge," in Lindberg and Westman, eds., *Reappraisals*, 333–66.

100. Isaac Newton, *The Mathematical Principles of Natural Philosophy* [1687/1713],

reservations about wonder foreshadow the new sensibility of sobriety in natural philosophy. He thought it unseemly to admire "corporeal things, how noble and precious soever they be, as stars and gems, [for] the contentment that accompanies our wonder, is allayed by a kind of secret reproach grounded in that very wonder; since it argues a great imperfection in our understandings, to be posed by things, that are but creatures, as well as we, and, which is worse, of a nature very much inferior to ours."[101] It bordered on idolatry to wonder at the works of nature, for men rather owed "their admiration, their praises, and their thanks, directly to God himself."[102]

Bernard de Fontenelle, longtime perpetual secretary of the Académie Royale des Sciences in Paris, was an indefatigable and eloquent spokesman for both metaphysics and sensibility. The children in Fontenelle's island utopia of Ajoia are made to chant an "Ode to the Marvels of Nature" with the refrain, "the same Nature, always similar to herself";[103] the narrator of his urbane dialogue on the plurality of worlds attacks the devotés of the "false marvelous . . . [who] only admire nature, because they believe it to be a kind of magic of which they understand nothing."[104] It is not so much the variety of nature but the simplicity and economy of its underlying principles that should command our admiration: "[nature] has the honor of this great diversity, without having gone to great expense."[105] In his capacity as perpetual secretary, Fontenelle took a severe line with marvel mongers on the occasion of the dissection of a monstrous lamb fetus, lacking head, chest, vertebrae, and tail: "One commonly regards monsters as sports of nature [jeux de la nature], but philosophers are quite persuaded that nature does not play, she always inviolably follows the same rules, and that all her works are, so to speak, equally serious. There may be extraordinary ones among them, but not irregular ones: and it is even often the most extraordinary, which give the most opening to discover the general rules that comprehend all of them."[106] The baroque nature of the preternatural philosophers, profligate in variety and surprise, had been transformed into a frugal bourgeois matron of plain speech and regular habits.

---

trans. Andrew Motte [1729], rev. Florian Cajori, 2 vols. (Berkeley and Los Angeles: University of California Press, 1971), 2:398.

101. Boyle, *Of the High Veneration Man's Intellect Owes to God* [1685], in *Works*, 5:153.

102. Boyle, *A Free Inquiry into the Received Notion of Nature* [1686], in *Works*, 5:253.

103. Bernard de Fontenelle, *La république des philosophes, ou Histoire des Ajaoiens* (Geneva: n.p., 1768), 63.

104. Fontenelle, *Entretiens sur la pluralité des mondes* [1686], ed. François Bott (Paris: Editions de l'Aube, 1990), 24.

105. Ibid., 67.

106. [Fontenelle], "Sur un agneau foetus monstrueux," *Histoire de l'Académie Royale des Sciences* [1703] (Paris: Imprimérie Royale, 1705), 28–32, at 28.

Wonder did not entirely disappear from natural philosophy, but it was a tamed, theological wonder insufficient to bind scattered phenomena into an object of scientific inquiry. Wonder was to be pried apart from its venerable companions—novelty, rarity, and ignorance of causes—and joined instead to parsimony, order, and simplicity, as innumerable eighteenth-century treatises in the natural theology of everything from fish to stars endlessly argued. In defiance of the ancient dictum that wonder was the beginning, not the outcome, of philosophy, Fontenelle remonstrated with those who rejected "natural science" and instead flung themselves into "admiration of nature, which one supposes absolutely incomprehensible. Nature is, however, never so wondrous [admirable], nor so wondered at [admirée], as when she is known."[107] A kindred form of rechanneled wonder can be found in the natural theology of the Boyle Lectures, in passages glorifying God through his works. Once again, transports of wonder were reserved for the intricacy, symmetry, and regularity of the commonplace—the anatomy of insects being a favorite example—rather than for the "new, rare, and unusual." In all such cases, what wonder remained was post hoc, bestowed upon the final results of scientific investigation rather than selecting the objects at the outset of inquiry.

Natural philosophy was not alone in evicting the sensibility of wonder in the first half of the eighteenth century. Men of letters were if anything even more vehement in their distaste for all that smacked of the marvelous. The author of the article on "Marvelous" in the *Encyclopédie* allowed that marvels might have their place in the epic poetry of Homer or even Milton, but not for contemporary Frenchmen, who could not even digest the true unless it was verisimilar [vraisemblable]: "Whatever one says, the marvelous was not made for us."[108] Samuel Johnson, though no Frenchman, agreed, reproaching the metaphysical poets for their excesses: "in all these examples [from Donne and Cowley] it is apparent that whatever is improper or vicious is produced by a voluntary deviation from nature in pursuit of something new and strange, and that the writers fail to give delight by their desire of exciting admiration."[109] The fustian mantle of decorum that settled over nature at the turn of the eighteenth century also covered literature and religion in its ample folds. It is no accident that Enlightenment natural philosophers likened the preternatural philosophy of their predecessors to religious enthusiasm,

107. Fontenelle, *Histoire du renouvellement de l'Académie Royale des Sciences en M.DC.XCIX. et les éloges historiques* (Amsterdam, 1709), 21.

108. "Merveilleux, adj. (Litterat.)," in Diderot, ed., *Encyclopédie*, 10:393–395, at 10:395.

109. Samuel Johnson, *Lives of the English Poets* [1779–81], ed. George Birkbeck Hill, 3 vols. (Oxford: Clarendon Press, 1905), 1:35.

for both seemed to violate the calm, steady, calculated order of newly pacified Europe.[110]

The sensibility that had glued preternatural philosophy into a coherent category of scientific investigation had dissolved by the mid–eighteenth century. But simply to pronounce nature uniform, regular, and simple could not eliminate the anomalies and variability studied by the preternatural philosophers. If Boyle was perhaps the last well-known natural philosopher to concern himself with the hidden properties of gemstones,[111] there were plenty of other striking, capricious, mysterious phenomena to puzzle Enlightenment savants. The annals of the history of electricity, phosphorescence, and magnetism are full of results that could not be stabilized by the original experimenter, much less replicated by others.[112] And if, in retrospect, it seems only rational that Enlightenment natural philosophers began to reject out of hand many of the phenomena credited without demur by the preternatural philosophers, we should also recall that they refused to believe in the existence of meteor showers because such reports reeked of the prodigious.[113]

Enlightenment natural philosophers did not so much explain preternatural phenomena as ignore them. Although, for example, the French physicist Charles Dufay could exclaim privately over "how different bodies behave which seemed so similar, and how many varieties there are in effects which seemed identical!"[114] he summarized results in published memoirs, "in order to avoid boring detail,"[115] and simply abandoned investigations from which he could not extract firm regularities. As for rare phenomena, "one hardly deigns to observe them," remarked Fontenelle, "because they lead to nothing."[116] A new ethos of utility replaced the old

110. Michael Heyd, "The Reaction to Enthusiasm in the Seventeenth Century: Towards an Integrative Approach," *Journal of Modern History* 53 (1981): 258–80.

111. Boyle, "A Short Account of Some Observations Made by Mr. Boyle, about a Diamond, that shines in the Dark," in *Works*, 1:789–99, at 1:794.

112. On electricity, see John L. Heilbron, *Electricity in the 17th and 18th Centuries: A Study in Early Modern Physics* (Berkeley and Los Angeles: University of California Press, 1979); on phosphorescence, E. Newton Harvey, *A History of Luminescence from the Earliest Times until 1900* (Philadelphia: American Philosophical Society, 1957).

113. Ron Westrum, "Science and Social Intelligence about Anomalies: The Case of Meteorites," *Social Studies of Science* 8 (1978): 461–93.

114. Quoted in John L. Heilbron, "Dufay, Charles-François de Cisternai," in *Dictionary of Scientific Biography*, ed. Charles Gillespie, 15 vols. (New York: Charles Scribner's Sons, 1970–78), 4:214–17, at 4:215.

115. Charles Dufay, "Mémoire sur un grand nombre de phosphores nouveaux," *Mémoires de l'Académie Royale des Sciences* (1730) (Paris: Imprimerie Royale, 1732), 524–35, at 527.

116. [Fontenelle], "Sur l'electricité," *Histoire de l'Académie Royale des Sciences* (1730) (Paris: Chez Durand, 1732), 4–13, at 4.

one of curiosity, requiring that phenomena be replicable without respect to the contingencies of local conditions. The case of the phosphors investigated by Dufay is particularly instructive, since these had been preternatural objects par excellence for seventeenth-century investigators. Dufay simply omitted or replaced examples of phosphors that could not be produced at the experimenter's will.[117] Although there was nothing particularly useful about his reliably glowing barometers or clamshells that shined in the dark, Dufay and many of his colleagues in the 1720s nonetheless understood the stabilization of physical phenomena as the necessary, if not sufficient condition for practical applications. The objects of preternatural philosophy did not cease to exist, but they no longer commanded scientific attention.

It was only in cases in which anomalies refused to be swept under the carpet, or smoothed into summarized results, that the preternatural caught the attention of natural philosophers, as in the celebrated case of mesmerism. In such cases, the explanations as well as the objects of preternatural philosophy were briefly revived. Neither the objects nor the explanations had disappeared, but they no longer constituted a coherent category of inquiry, as the highly diverse phenomena of electricity or even color had become by the 1780s. Tarred with the brush of enthusiasm, preternatural philosophers were suspected of imagining the marvels they sought to explain. It is therefore a grating irony that enemies of enthusiasm themselves reached automatically for the naturalizing explanations of preternatural philosophy, as when Shaftesbury suggested that in a crowd stirred by religious enthusiasm, "the Imagination [is] so inflam'd . . . the very Breath and Exhalations of Men are infectious, and the inspiring Disease imparts it self by insensible Transpiration."[118]

117. Christian Licoppe, *La formation de la pratique scientifique: Le discours de l'expérience en France et en Angleterre (1630–1820)* (Paris: Editions La Découverte, 1996), 116–26; Lorraine Daston, "The Cold Light of Facts and the Facts of Cold Light: Luminescence and the Transformation of the Scientific Fact, 1600–1750," *Early Modern France* 3 (1997): 1–28.

118. Anthony, third earl of Shaftesbury, *A Letter Concerning Enthusiasm, to My Lord* ***** (London: J. Morphew, 1708), 69.

Mathematical Entities in Scientific Discourse

PAULUS GULDIN AND HIS

DISSERTATIO DE MOTU TERRAE

INTRODUCTION

In his *Greek Mathematical Thought and the Origin of Algebra*,[1] first published in Germany in 1934, Jacob Klein suggested a new angle from which to interpret the transition from ancient and medieval science to the new mathematical physics of the seventeenth century. His was the seemingly narrow—but only deceptively so—perspective of the ancient concept of *arithmos*, compared to the concept of number in its modern, symbolic sense. In Klein's own words, the underlying thematics of the book never loses sight of the "general transformation, closely connected with the symbolic understanding of number, of the 'scientific' consciousness of later centuries."[2] Without pretending to do justice to many of the subtleties of Klein's thesis, I would like to open this paper by referring to some of his most prominent contentions.

Although the Greek conceptualization of mathematical objects was indeed based upon the notion of *arithmos*, this notion should not be thought of as a concept of "general magnitude." It never means anything other than "a definite number of definite objects,"[3] or an "assemblage of" things

---

1. Jacob Klein, *Greek Mathematical Thought and the Origin of Algebra*, trans. Eva Brann (1968; reprint, New York: Dover Publications, 1992); originally published as *Die griechische Logistik und die Entstehung der Algebra*. All citations are to the 1992 edition.
2. Ibid., 9.
3. Ibid., 7.

counted. Likewise, geometric figures and curves, commensurable and incommensurable magnitudes, ratios, have their own special ontology which
directs mathematical inquiry and its methods

In contradistinction to Greek parlance, "general magnitude," according
to Klein, is clearly a modern concept. Klein's succinct formulation of the
transformation that occurred within modern usage and thinking is worth
quoting at some length:

> Now what is characteristic of this "general magnitude" is its indeterminate-
> ness, of which, as such, a concept can be formed only within the realm of sym-
> bolic procedure. But the Euclidean presentation is *not* symbolic. It always
> intends *determinate* numbers of units of measurement, and it does this *with-
> out any detour through a "general notion" or a concept of a "general magni-
> tude."* In *illustrating* each determinate number of units of measurement by
> measures of distance it does *not* do two things which constitute the heart of
> symbolic procedure: It does *not* identify the object represented with the
> means of its representation, and it does *not* replace the real determinateness
> of an object with a *possibility* of making it determinate, such as would be ex-
> pressed by a sign which, instead of *illustrating* a determinate object, would
> *signify* possible determinacy (emphases in the original).[4]

Klein pointed out Descartes as the first thinker who fully articulated the
major implications of the modern symbolic conceptualization of number:

> From now on the fundamental *ontological* science of the ancients is replaced
> by a *symbolic* discipline whose ontological presuppositions are left unclari-
> fied. This science, which aims from the first at a comprehension of the totality
> of the world, slowly broadens into the system of modern mathematical
> physics (emphases in the original).[5]

Two different trains of thought were combined in Descartes's achievement:

> (1) the conception of algebra as a "general" theory of proportions, whose ob-
> ject, only symbolically comprehensible, acquires its specific characteristics
> from the *numerical* realm . . . , and (2) the identification of this "symbolic"
> mathematical object with the object of the "*true physics*" (emphasis in origi-
> nal).[6]

Put somewhat differently, Klein shows that the concept of "general
magnitude" and its symbolical interpretation by Vieta, Stevin, Descartes,

---

4. Ibid., 123.
5. Ibid., 184.
6. Ibid., 198.

Wallis, and others allow for the collapse of the distinction between discrete number and continuous magnitude. At the same time it allows for a non-problematic symbolization of physical phenomena by mathematical entities. Hence it enables a fundamental restructuring of the boundaries within the mathematical sciences between arithmetics and geometry, and also of the boundaries between mathematics and physics, or natural philosophy.

Klein's book offers more than a particularly sensitive case study in the history of mathematics. In addition, his interpretation implies an insight into some fundamental aspects of scientific discourses. By analyzing the transition from the Greek to the modern concept of number he drew attention to a deep historical transformation that occurred on the level of the object of the most universal of all fields of knowledge. His analysis shows that even mathematical objects may undergo transformations in the course of historical time. Furthermore, Klein shows how one transformation—at the core of the body of knowledge—affected the boundaries among fields of inquiry. In Klein's view, conceptual developments in mathematics cannot be analyzed and understood without paying attention to specific discursive practices and means of representations that are historically and culturally constituted.

My paper is an attempt to exemplify the historical complexity involved in the coming-to-be of a new object of mathematical discourse—symbolic number—in a particular institutional setting. By suggesting a reading of a physico-mathematical treatise on the motion of the earth written by a Jesuit mathematician in the seventeenth century, I shall first point out the conceptual and technical manifestations of the new object in the text. I shall then look more broadly at the conceptual resources available in the Jesuit environment, which supported the transition to the new object and were used in legitimizing the project of Jesuit mathematicians. However, the dissolution of the old boundaries and the constitution of new ones did not result unreflectively from the modification of scientific objects. Rather, they will be treated as strategies in the politics of knowledge, in need of historical reconstruction. By pointing out the persistence of old boundaries as means of controlling Jesuit mathematical discourse—in spite of the assimilation of the new object—I hope to show that Klein's broad and sometimes nondifferentiated claims can be refined and further historicized.

From a methodological point of view, it seems to me that an analysis on the level of the objects of scientific discourses, the practices involved in their transformation and legitimization, and the reconstitution of boundaries on the "globus intellectualis" such practices entail enables a more dynamic reconstruction of the relationship between the inner core of scientific arguments and the authoritative structures that promote and inhibit them. The elaboration of analytical frameworks that do not take the objects and boundaries of scientific discourses as naturally given is necessary in order to show the ways by which science is historically connected to a particular culture and the manner in which its history is part of the history of culture.

The text that offers the opportunity to analyze some crucial aspects of the transition to symbolic number is the *Dissertatio physico-mathematica de motu terrae*, published in Vienna in 1635[7] by the Jesuit mathematician Paulus Guldin (1577–1643). Guldin was first trained in the Jesuit College in Munich. He then spent about nine years in Christopher Clavius's academy of mathematics in Rome, followed by a period in Graz, where he taught mathematics. Between 1622 and 1624 he appears to have lectured on mathematics at the university in Vienna,[8] where he also published his magnum opus, *De centro gravitatis*, including four volumes on the science of statics, with a long introduction on the status, uses, and classification of the mathematical sciences.[9] The *Dissertatio* became part of the first volume of this work.

As we shall soon see, Guldin's text opens a window onto some of the practices that signal the paradigmatic change from ontological to symbolic number, and from qualitative to quantitative physics. At the same time the text exhibits significant constraints expressed in the way it interprets the meaning and scope of the transition it signals. It is this double-layered message that seems to provide an insight into conceptual development, cultural transition, and their interaction in the Jesuit environment.

MATHEMATICAL ENTITIES IN GULDIN'S SCIENTIFIC DISCOURSE

Guldin clearly divides his dissertation into two parts: the first is designated by him *physico-mathematical*, and the second *purely geometrical*.

The first part closely follows a passage from the second book of Aristotle's *On the Heavens* (bk. 2, chap. 14). The context is the place toward which

7. In *Pauli Guldini Sancto-Gallensis et Societate Jesu De centro gravitatis, Liber primus* (Vienna, 1635). All my quotations refer to this edition, and will be marked henceforth in the text by page numbers within parentheses.

8. According to the official Jesuit catalogue of members in the province of Austria. The unpublished manuscript of Joanus Josephus Locher, "Speculum academicum Viennense," however, which I found at the National Library in Vienna, mentions him lecturing in 1626/27. No trace of his stay at the university in 1622/23 exists in Locher's manuscript, which contains lists of all university professors between the sixteenth and the eighteenth century.

9. See note 7.

heavy bodies on Earth naturally move. Aristotle maintains that heavy bodies naturally move toward the center of the universe, and only incidentally towards the center of the earth, which is located at the center of the universe. This, however, raises the question of the exact location of the earth. Aristotle problematizes the earth's location through an imaginary argument: suppose, he says, we add a large weight to one of the hemispheres. In this case, the center of the earth will no longer coincide with the center of the universe. Aristotle resolves the problem by resorting to the case of falling bodies. Falling bodies do not stop their fall when their external surface touches the center, but go on moving until their center coincides with the center of the universe, and so does the earth "until it surrounds the center in a uniform way and the tendencies to movement in the various parts will counterbalance each other."

Aristotle's attempt to cope with the exact location of the earth, and his solution to the problem in terms of a vague concept of "balanced tendencies to movement" is declared by Guldin as the source of his theory. Guldin, however, is careful not to present his argument as simply depending upon Aristotle's authority:

> It is generally accepted, according to *sense evidence* and *experience*, and according to the testimony of the *most educated people*, and is *proved by reasons*, that unimpeded heavy bodies move downwards by their nature towards the center of the universe, and aspire to have their center coincide with the center of the universe. (138)

The evidence of the senses as testified to by the most educated people is a primary source of knowledge. Only then comes the quotation from Aristotle, as a kind of endorsement of general consensus. To elucidate Aristotle's abstract consideration Guldin uses another imaginary trick: he imagines the vast globe of the earth displaced in the concavity of the orbit of the moon, and a heavy body dropped from elsewhere. The body will then directly descend toward the center of the universe, not toward the earth (ibid.). This trick is obviously taken from Albert of Saxony's *Quaestiones* on Aristotle's *Physics*.[10] Guldin, however, never recognizes his debt to the Parisian nominalist. Instead, he moves directly to the implications of the clear distinction suggested in his work between the center of the universe and the center of the earth. These implications result in a theory of the motion of the earth, which Guldin hurries to relate back to Aristotle's teachings:

---

10. *Quaestiones subtilissimae in libros Physicorum,* in P. Duhem, *Les Origines de la Statique,* 2 vols. (Paris, 1905–6), 2:21–33.

He—namely Aristotle—teaches us at last, by *reason* and *experience*, that which I shall demonstrate *geometrically* a little later: that the *center of gravity* of the body can be moved from its place in the figure if something heavy is either added or detracted, or if the parts are somewhat differently constituted. (139)

This is obviously an anachronistic reading of Aristotle that heavily relies on Buridan's and Albert's theories. Aristotle did not use the term "center of gravity," and imagined the displacement of the earth only to deal with the problem of its exact location. Guldin's medieval predecessors did develop a theory of the motion of the earth. However, their main interest did not lie in any of the quantitative aspects of that motion. Instead, they discussed and debated the causal mechanism of that motion in terms of geological changes, the material heterogeneity of the earth, the relations between the motions of its different parts, etc.[11] Guldin's interests, as we shall see in a moment, lay elsewhere. Without mentioning his medieval predecessors, he proceeded to an Archimedean argument, in an attempt to prove geometrically their theories of the motion of the earth. However, unlike the proof, which he deemed original, the theory, he insisted, belonged to Aristotle. Still, he reminded his readers, Aristotle used different methods, namely reason and experience, whereas his demonstration was going to be geometrical.

Guldin's interpretation of Aristotle is very different from the caricaturist portrayal of Aristotelians popular in large parts of the modern literature. Rather, it is constructed as a concrete and direct reliance on experience, supported by rational arguments, and by the testimony of the most reliable witnesses. Furthermore, it is certainly not perceived as incommensurable with his geometrical approach.

The geometrical demonstration starts from a basic premise, well anchored in ordinary experience: Archimedes' law of equilibrium, illustrated by the drawing of a concrete balance (figure 2.1).

Two heavy bodies A and B, applied on the straight line CD, which passes through their centers of gravity, are in equilibrium. If they are equal, they would balance each other on point E, which is the middle of CD. If they are not equal, they would balance each other when their respective distances from E are inversely proportional to their weight (140–41).

---

11. Duhem, *Les Origines*; E. Grant, *Planets, Stars, and Orbs: The Medieval Cosmos, 1200–1687* (Cambridge: Cambridge University Press, 1994; E. Grant, "In Defense of the Earth's Centrality and Immobility: Scholastic Reaction to Copernicanism in the 17th Century," *Transactions of the American Philosophical Society* 74, no. 4 (1984): 1–69.

Figure 2.1.   Archimedes' law of equilibrium, illustrated by the drawing of a concrete balance. (From *Pauli Guldini Sancto-Gallensis et Societate Jesu De centro gravitatis, Liber primus* [Vienna, 1635])

From this law Guldin deduces the centers of gravity of solids, which are constructed separately, and then composed. Guldin constructs two cubes, so that: CD = the other CD; DF = the other EC; FC = the other DE. Then he produces GC equal to FC; and he produces DH equal to DF; so that he lets KN be similar to A and ML be similar to B. He quotes from Commandino's treatise of centers of gravity, stating that D is the center of gravity of ML, and C is the center of gravity of KN; but joined together to become a composed KL the center of gravity E of KN moves to C. From this he concludes, following Luca Valerio, that:

> In each heavy body the center of gravity is *removed* from its place in the figure, if the same weight is added or subtracted or its parts are differently constituted. The center of gravity C in KN moves, after the addition of ML from C to E. And the same E, which is the center of the whole KL, after the subtraction of part ML, is changed from E to C (141). (emphasis added)

Guldin then applies this proof to the terrestrial globe, and the displacement of its center of gravity, which he identifies as the motion of the earth. This is done by imagining part of the globe (DFCH) transferred from point F to point G. The two parts together amount to the figure of the cone AMB. Now Guldin aims to discover the "species of magnitude" of this cone, representing, in his demonstration, some mountain on the earth's surface (142)

On the basis of the most accurate estimations of the diameter of the earth (1,500 German miles), and by transferring a segment of the sphere whose height is 1, Guldin shows how the distance by which the center of the earth was displaced can be calculated. He computes the respective volumes of the cone and the sphere, the ratios among which is like the ratio NE:LE. LE—the required distance—is then calculated through manipulation of that proportion. Thus, he confirms not only the possibility of proving the motion of the globe, but also of measuring the distance by which the center of gravity is displaced, always in proportion to the ratio between the sphere and its truncated part. For his particular example he argues that "the center of the earth moves by 40 feet." Finally he concludes: "I have demonstrated, and it is my opinion that the center changes; as a result of which the earth can move" (143).

The attempt to combine Aristotelian and Archimedean theories in order to gain Aristotelian legitimization for a thoroughly non-Aristotelian idea is one of the most outstanding features of the text. That, however, does not exhaust Guldin's strategy. From the very beginning he is also keen to narrow down the significance of the motion of the earth, reducing it to mere local trepidation, and differentiating it from the bold claims made by his Copernican contemporaries, whose name, however, he avoids mentioning:

> I would not like the motion of the earth to be destructive to us, or shock you totally by what I said I intended to do with demonstrations and reasons . . . I do not want to say that the globe of the earth moves with that most speedy motion, which many indicate, that it moves around its center according to the different parts of the day; nor that it moves around the sun and makes one whole circle around it in one year. I spread around no such motion of the earth. (138)

And yet, a strong, rather repetitive voice insists on the claim that the earth moves physically. Such motion exists, and it is physically and mathematically demonstrable.

The motion to which Guldin refers is of a very peculiar kind: a displacement whose motive force, velocity, and cause is consciously and determinately banished from discourse. At the very beginning of his treatise, Guldin declares not only "I shall teach nothing about the cause by which the immense mass of the earth is moved," but also "I will not dwell either on the facts or the use of difficult machines"—namely on mechanical questions concerned with the relationship of motion, force, and weight. This approach is being further emphasized in the dedicatory letter to the abbot of Melk, which opens the first volume of *De centro gravitatis:* "Let others concentrate on Archimedean masses," he insists, cautiously drawing his patron's attention to his reductive approach: "Do not wonder that I adhere to narrowed down things, reducing them into narrow passages: there is no greater art than that which is totally minimal" (3–4).

The *Dissertatio de motu terrae* neutralizes the concept of motion from any physical aspects, while still attempting to prove the existence of this motion in the real physical world. This approach is not peculiar to the dissertation. It characterizes other parts of Guldin's work and is epitomized in the first chapter of *De centro gravitatis*. After deploring the general confusion concerning centers of gravity, Guldin suggests three definitions for the three types of continuous magnitudes: lines, planes, and solids, in respect to their figures or magnitudes, and gravity.

In his attempt to stabilize the concept of the center of gravity of solids he quotes three definitions: those of Aristotle, Pappus, and Commandino,[12] which he presents as continuous and complementary, without showing any awareness of the deep contradictory nature distinguishing Aristotelian concepts of gravity from Archimedean ones. Then he raises the crucial issue concerning the entities with which he deals. Guldin is aware that these definitions of centers of gravity should in fact be appropriate to bodies only, for he says: "In as much as these three definitions are appropriate to bodies alone, indeed only to those to which physically speaking gravity is fitting. . . . " However, maintaining that mathematicians can enjoy the liberty of abstracting from physical matter those dimentions in which they are interested and treating them separately, he claims the right to do the same with gravity: "And just as mathematicians, with such freedom and privilege, pull asunder from those very bodies, surfaces, and lines, though they cannot separate them, avoiding their three dimensions and considering only two or even one, it should be allowed to us to deal with gravity similarly." And this procedure is universally valid, if one remembers that bodies, or solid figures, in fact are *finite (terminate) quantities* and should always be represented by this term (23).

For Guldin, then, lines, planes, and solids are magnitudes, each one conceived by mathematicians as *quantitas terminata* = terminate quantity, the subject matter of mathematicians. Centers of gravity are similar kinds of entities. The text betrays a certain discomfort, a sense of violation performed by mathematicians, who choose to treat heavy bodies as if they

12. Pappus: "The center of gravity of every body is a certain point located within the body. If one imagines the body suspended from that point, while suspended it will remain immobile and retain the initial orientation and will not rotate." Commandino: "The center of gravity of every solid figure is that point located within, around which the parts have moment equilibrium; if indeed a plane is drawn through such a center, no matter how it cuts the figure it will always divide it into parts of equal weight."

were geometrical figures, and to speak about the center of gravity of geometrical lines. Still, he hastens to conclude his discussion of that point by simply declaring that: "The center of gravity is [this point from which] a body *is only imagined* to be suspended [*sola cogitatione*, suspensum corpus]" (ibid.).

A few preliminary remarks concerning the objects of Guldin's physicomathematical dissertation can now be made. Examining the dissertation from this specific point of view may throw light on the radical potential of the arguments presented. Guldin's radical move consists in an actual attempt to modify the rules of the game that used to govern the field of arguments about the motion of the earth in a Scholastic environment. In an Aristotelian framework of thought *gravity*, or more specifically the *gravity* of the earth, is the most substantial argument against its motion. The *center of gravity* is the center of the *universe*, the place toward which all heavy bodies are attracted in their striving for rest. In a very subtle rhetorical gesture Guldin suggests *equilibrium* on the earth's *center of gravity* as a condition of possibility of the earth's *immobility*. However, any commonsensical knowledge about the *actual* physical conditions on the surface of the earth is enough to alert one's attention that such equilibrium is very unlikely indeed, a rare—if ever actually fulfilled—condition of possibility. Thus, from something close to logical and physical *impossibility* the motion of the earth becomes a most commonsensical *probability*. Under such a radical transposition of the conditions of possibility of the earth's motion the causes of that motion lose their primary significance, and their discussion can be left for a subsequent, much less prominent discourse. Guldin's contention that the causes of motion are not of interest to him should be understood against this background. At the same time the measurement of motion assumes a much more prominent role as the center of a new physico-mathematical discourse.

For Guldin geometrical lines, planes, and solids are the model for thinking of centers of gravity, moments, and, strangest of all, even motion. Just as mathematicians can think of lines and planes in abstraction from their concrete-physical instantiations, such is also the case with centers of gravity, moments, and motions. All of them are entities abstracted by mathematicians from their concrete, physical manifestations, transplanted to the space of mathematical discourse, where they are endowed with the status of a "quantitas terminata"—both abstract and real. On the one hand they are subject to rigorous mathematical treatment; on the other hand they pertain to physical reality. A further glance at Guldin's drawing may elucidate his strategy. The point of departure is a balance, upon which two pairs of even and uneven weights are suspended. This represents a very concrete, ordinary experience. However, the law of equilibrium is already stated in much more abstract terms as the inverse proportion between weight and distance from the common center. The abstraction then increases as Guldin speaks about the addition of weight to a body that changes its center of gravity, and is intensified in the discussion of the terrestrial globe, whose heterogeneity is represented in terms of the cone AMB (see figure 2.1).

The possibility of thinking of centers of gravity, or even motion, in terms of "quantitas terminata" is perceived by Guldin as an act of a new boundary making: differentiating himself from mechanicians who deal with forces and weights, from Copernicans who deal with the rotational and orbital motion of the earth, and from Aristotelians who deal with moments in dynamic terms. Guldin stresses his difference from all those, and attempts to justify himself not in front of his colleagues and readers, but in front of the abbot of Melk, claiming that his solutions have aesthetic superiority, and are useful for the community. The authority to do so, however, comes from the professional privileges of mathematicians.

But these remarks are still very preliminary. For it is not yet clear what kind of abstraction is performed by mathematicians, and what such abstraction actually involves. A deeper conceptual analysis is necessary before any generalization about mathematical entities in Guldin's discourse can be made.

A clue to the conceptual skeleton underlying Guldin's typical techniques can be found in one term he chooses to invoke while applying the theory of centers of gravity to the specific case of the motion of the earth. What he is looking for, he claims, is the *species* (142) of the mountain representing the heterogeneous nature of the earth. This terminology already alludes to the framework of mathematical symbolism within which Guldin's treatise should be read and interpreted, for the notion of the *species* is invented by Vieta in his *In Artem Analyticam Isagoge*[13] to denote "general magnitudes" common to geometry and arithmetics, namely symbolic numbers. A further look at the various steps through which the proof develops only strengthens this first impression. True, on one level it may be argued that the proof develops along perfectly traditional lines, heavily relying on orthodox Euclidean and Archimedean methods. Thus, a segment of the sphere truncated and transferred from one side to another is claimed to be equal to a cone according to one Euclidean proposition. Then an Archimedean theorem about the proportion between the ratios of the vol-

---

13. F. Vieta, *In Artem Analyticem [sic] Isagoge*, Seorsim excussa ab *opere restitutae Mathematicae Analyseos, seu*, Algebra Nova, (Tours, 1591), quoted by Klein, *Greek Mathematical Thought*, 315.

umes of a sphere and a cone to their respective distances from the center of an equilibrium system is used to express the displacement of the center of gravity of the earth due to its heterogeneity. Within this orthodox framework, however, certain untraditional steps also take place.

First, in order to calculate the volume of the sphere Guldin uses a theorem from Villalpando[14] that states that the ratio of three diameters to half the circumference of a circle equals that of the diameter of a sphere to the third power to the volume of the sphere. This is clearly a proportion between nonhomogeneous magnitudes (invoking a ratio between diameter and volume), unaccepted within the strict rule of homogeneity guiding Euclidean discourse.

Second, such deviation is possible, however, since the proportion seems actually to be treated as an equation, the volume of the sphere being explicitly defined as an unknown magnitude to be discovered through the manipulation of three known ones.

Finally, the demonstration is not exhausted by stating proportions, as is usually the case in Euclidean and Archimedean discourse. Rather, the volumes of the sphere and the cone, as well as the distances from the center of the system, are all calculated in numerical terms leading to the measurement of the displacement of the center of gravity of the sphere in a specific case, while this displacement is being interpreted as the measure of the motion of the earth.

In the light of this analysis my claim is that the techniques used by Guldin signal a conceptual framework that implies new options for interpreting mathematical entities and their relation to physical reality. As stated above, Guldin declares that the main target of his geometrical demonstration is to discover the species of a mountain, which accounts for the motion of the earth, argued for in the first physico-mathematical part of the treatise. The use of the concept "species," however, alludes to every possible physical phenomenon capable of changing the earth's equilibrium. A mountain is just one instance among a variety of other possibilities that might bring about the same effect. At the same time "species" also stands for every possible number that might enter into a relation with the volume of the earth and enable the calculation of the displacement of its center of gravity. Such a concept is meaningful only in a symbolic framework that allows "general magnitudes" to be interpreted in numerical and physical terms.

14. The Jesuit Juan Baptista Villalpando in collaboration with Jeronimo del Prado wrote a three-volume commentary on the prophecy of Ezekiel, which was published in Rome between 1596–1604. The second volume included a reconstruction of the Temple of Solomon and included remarks on centers of gravity relevant for problems of construction.

In order to further explain the logic of Guldin's text let me return, for a moment, to Klein's conceptual framework.

The concept of the "species" *undergoes a universalizing extension while preserving its tie to the realm of numbers. In the light of this general procedure, the species,* or as Vieta also says, the "forms of things" . . . *represent general magnitudes simply.* (emphasis in original)[15]

But what is the inherent meaning of the universalizing, or symbolizing, extension of the notion of species in Vieta's new technique?

Two aspects of the symbolic treatment of mathematical entities emerge from Klein's discussion of Vieta's text. The first relates to the technical side of their operation, characterized by three main stages: the construction of an equation; its transformations until it has acquired a canonical form that immediately supplies the "indeterminate" solution; and the computation of numbers.[16]

As I have shown, Guldin's text contains enough traces of the algebraic operations involved in his demonstration. The actual construction of an equation and its manipulation, however, remain hidden from the eyes of the reader. On the one hand, the text refers to an "unknown" that Guldin strives to discover. The equation itself, however, is never actually being written down. The drawing contains the precise numerical values that are being manipulated, but in fact, what we are left with is material for a historical reconstruction, not a complete mathematical argument. This may be interpreted in one of two ways. Guldin, while addressing professionals, may not have deemed it necessary to write down all the stages of his proof. Another possibility is that he still felt uncomfortable about mixing the language of proportions with algebraic equations. The knowledge and skills required for using the techniques of symbolic mathematics, however, were undoubtedly part of his intellectual baggage. Even a superficial look at the table representing the division of the mathematical sciences appended to Guldin's *Prolegomena* (20) is enough to discover the full integration of Vieta's text into his scheme. Algebra—divided, after Vieta, into the zetetic, the poristic, and the exegetical—is granted an honorable place between arithmetic and geometry, and bears witness to the reception of Vieta in Jesuit circles.

The second aspect of Klein's discussion concerns his conceptual analysis of the symbolic framework of mind, an analysis that remains rather compact, in need of unpacking. But let me quote him first:

15. Klein, *Greek Mathematical Thought,* 166.
16. Ibid., 156.

[T]he "being" of the objects of "general analytic" is to be understood neither as independent in the Pythagorean and Platonic sense nor as attained "by abstraction," . . . i.e. as "reduced" in the Aristotelian sense, but as *symbolic. The species are in themselves symbolic formations, namely formations whose merely potential objectivity is understood as an actual objectivity.* (emphasis in original)[17]

Potential objectivity, according to Klein, has to do with the kind of indeterminateness associated with the modern concept of number. Klein compares it with the determinateness of ancient "arithmos" and "magnitude," always referring to concrete "assemblages of" entities, or to concrete magnitudes. Now, as Klein never tires of pointing out, the differences between concrete and abstract number do not capture what is at stake in the transition from an ontological to a symbolical interpretation of mathematical beings. In fact, what such transition really entails is the creation of new units of calculation, while the real computation takes place in terms of number. Thus, the calculation with species is shifted into the domain of the indeterminate. In Klein's words:

The letter sign intends *directly the general character of being a number* which belongs to every possible number, that is to say, it intends "number in general." (emphasis in original)[18]

This means, however, that becoming a sign in a mathematical symbolic system already presupposed a systematic context, a system of rules that "defines," so to speak, the "object," or, as Klein puts it:

The letter sign designates the intentional object of "a second intention" [intentio secunda], namely of a concept which itself directly intends another concept and not a being.[19]

The object thus defined through a network of other concepts, however, has more than one and only one "assemblage of things counted" as its term of reference. Therefore, Klein speaks of its *potential* or *possible* determinateness:

This "general number" in all its indeterminateness, that is, in its merely *possible determinateness*, is accorded a certain independence which permits it to be the subject of "calculational" operations. (emphasis added)[20]

17. Ibid., 175.
18. Ibid., 174.
19. Ibid.
20. Ibid.

Klein's discussion of the transition from an ontological to a symbolical interpretation of being is confined to the field of mathematics. One cryptic remark, however, indicates his awareness of a symbolic framework of mind, which probably accompanied the revival of Greek mathematics and the integration of algebraic techniques into the sphere of mathematics as a theoretical science: "[T]he revival and assimilation of Greek logistic in the sixteenth century," he claims, "are themselves prompted by an already current *symbolic* understanding of number" (emphasis in original).[21] At this stage, it may be useful to elaborate a bit about the conceptual resources that seem to have facilitated a symbolic understanding of being in Jesuit culture and the adoption of the traditional canons of knowledge to such understanding.

CONCEPTUAL RESOURCES AND LEGITIMIZATION FOR THE SYMBOLIC INTERPRETATION OF NUMBER AND BEING

According to Klein symbolic number is conceived as an entity of "possible determinateness" that allows it to be subject to calculational operations, and to represent physical phenomena unproblematically. We have seen how Guldin's use of the term "species" implies an entity of "possible determinateness" and how his calculation of the displacement of the center of gravity of the earth is interpreted by him in physical terms as a measurement of the motion of the earth. It will now be argued that sixteenth-century Jesuit Thomism—in contrast to traditional Thomism—could accommodate the notion of true knowledge of "possibles"—including symbolically conceived numbers—which facilitated the reception of Guldin's type of physico-mathematics in the Jesuit environment.

Within the framework of traditional Thomism there existed a clear relation between the ontological status of the objects of knowledge and the status of the knowledge acquired. *True* knowledge consisted in knowledge of *real* beings, of which there were three kinds: the objects of philosophy, abstracted from individuals and consisting in "universals"; mathematical entities, abstracted from matter and time, and consisting in "intelligibles"; and the objects of metaphysics, separated both from sensible matter and from intelligible matter.[22] In this framework of thought it made no sense to talk about real knowledge of "possibles." Speaking in such terms meant committing a categorical mistake.

---

21. Ibid., 9.
22. Thomas Aquinas, *The Division and Methods of the Sciences, Questions V and VI of His Commentary on the De Trinitate of Boethius,* trans. with introduction and notes by A. Maurer (Toronto: Pontifical Institute of Mediaeval Studies, 1953; reprint 1986), q. VI, art. 1.

Sixteenth-century Thomism, however, developed in new directions, partly in response to the intellectual and existential challenges of the period. Franciscus Suarez's attempt to conceptualize "possibles" (possible beings) as real beings, and hence as objects of true knowledge, in his *Disputationes metaphysicae*[23] throws some light on the new orientations that could provide conceptual resources for developing new interpretations of being. Such interpretations could be used—and, I assume, were used—to legitimize the mathematicians' claims to real knowledge within a symbolic framework of mind. At the same time Suarez's metaphysics was probably inspired by the new Molinist theology, a special brand of Jesuit Thomistic theology that I will discuss below—and also provided for it firm philosophical foundations. Thus, the transition to symbolic number may prove to be but one aspect of a much broader transition from ontological to symbolic modes of thought, which affected different segments of Jesuit and non-Jesuit culture.

Suarez's positions on "possibles" have become subjects of many misunderstandings and disagreements among scholars, particularly in the last thirty years. This is deep water into which I cannot delve now.[24] For the sake of my argument, however, suffice it to draw attention to a few points that appear to be accepted by all. In his attempts to carve a space for being that is neither pure *essence prior* to any existence, nor just *actual existence*, Suarez conceived of the category of *possible beings* that have aptitude for existence, or nonrepugnance to it, but do not *actually* exist, since they have not *actually* been created by God. In many places in the *Disputationes metaphysicae* Suarez insisted that things had no true reality before their creation in actual existence. This contention buttressed the Thomists' positions, for whom God is the source of all truth, even necessary truths. Thus essences—the objects of true and real knowledge—do not have any reality prior to God's willful creation. Suarez wrote:

> First and foremost, it must be stated that the essence of the created thing, or the created thing by its own nature, has no genuine reality in itself prior to its creation by God, and that in this sense, when existence is excluded, essence is not a kind of object but absolutely nothing.[25]

23. My treatment of Suarez in this paper is still very preliminary. I have relied heavily on J. P. Doyle, "Suarez on the Reality of the Possibles," *Modern Schoolman* 45 (1967–68): 29–47. On Suarez and the Jesuits' position on the question of mathematical entities, see also P. Dear, *Mersenne and the Learning of the Schools* (Ithaca: Cornell University Press, 1988), chap. 4.

24. I do not pretend to take any position on Doyle's controversy with T. J. Cronin, *Objective Being in Descartes and Suarez* (Rome: Gregorian University Press, 1966), and N. J. Wells, "Old Bottles and New Wine," *New Scholasticism* 53 (1979):515–23.

25. Suarez, *Disputationes metaphysicae*, disp. 31, sec. 2, no. 1, in *Opera Omnia*, ed. Juan Luis Vives (Paris, 1856–77), 25:754, cited by Doyle, "Suarez," 31–32.

In spite of his negation of the reality of essences prior to creation by God, Suarez made a distinction between actually existing things (and their essences) on the one hand and *possible beings* that do not actually exist, but have *no repugnance* toward existence, and that, he maintained, do have *reality* in themselves, in contradistinction with *beings of reason*, on the other:

> the objective *potential* essence of the created thing of divine science does not exist in conflict with the mind, but is actually a *possible being* capable of existence in the real world *[realis existentiae capax]*; therefore [essence] must not be understood as a *being of reason* but as some kind of *real being*. I already stated earlier that the essence of the created thing, even when not actually existing in the real world is in some way a real essence. (emphasis added)[26]

According to Suarez, then, *beings of reason* have no aptitude to exist; they are repugnant to existence. *Possible beings*, though not actually existing, have the aptitude to do so. Therefore they are also subject to true knowledge.

Many scholars[27] emphasize that the conceptual framework in which the distinction between *real beings*, *possible beings*, and *beings of reason* came into existence after the acceptance, within the Jesuit environment, of another distinction between the *formal* concept and the *objective* concept, both playing an essential role in Suarez's metaphysical account of the conditions of knowledge. A formal concept is the inner word by which the intellect signals to itself the thing that is to be known:

> By formal concept we must understand the act or the word (which rather amounts to the same thing) by which the mind conceives any thing or [any] general principle.[28]

The objective concept is that to which individually and without mediation the formal concept refers:

> By objective concept we must understand the thing itself, or the principle, that is properly and immediately known and represented by the formal concept.[29]

26. Suarez, *Disputationes metaphysicae*, disp. 31, sec. 2, no. 10.
27. Doyle, "Suarez"; N. J. Wells, "Objective Being: Descartes and His Sources," *Modern Schoolman* 45 (1967–68): 49–61; Dear, *Mersenne and the Learning of the Schools*, 50.
28. Suarez, *Disputationes metaphysicae*, disp. 2, sec. 1, no. 1.
29. Ibid.

This distinction opened the door for the understanding of real knowledge in terms of concepts *referring* to, or *signifying* other concepts. Suarez's language bears witness to such a development:

> We are not concerned with signs but with the signified thing; not with formal, but with objective concept.[30]

But a concept, which is the object of real knowledge, does not necessarily point to things in actual existence. It might as well be a concept of *possible beings*. That does not, however, detract from the *reality* of the science it gives rise to. Thus Suarez concluded that sciences that abstract from existence (in order to consider things in the mind) do not concern *beings of reason* but *real* beings:

> And it follows in the same manner that sciences which consider things abstracted from existence, are not concerned with *rational* but with *real* beings, because they consider essences to be real, not by their objective status in the intellect but by their own nature *[secundum se]*, or as far as they are apt to exist with certain characteristics or properties.[31]

In contradistinction to the traditional-Thomistic position, which granted the status of real knowledge only to knowledge of existing beings—whether actual creatures or created essences—Suarez's metaphysical reflections allowed for knowledge of "possibles" to be endorsed as knowledge of true beings. This in itself should not be hastily read as a symbolical interpretation of beings. But it created an intellectual space for thinking of *indeterminate* or only *potentially determinate* entities (such as "general magnitudes") as objects of necessary and true knowledge.

The interpretation of objects of knowledge in relation to the status of the knowledge acquired was not a problem confined to metaphysics. At the end of the sixteenth century similar discourses emerged in two other areas of Jesuit culture: the mathematical disciplines on the one hand and moral theology on the other.

A few influential Jesuit philosophers who argued against the broadening claims of mathematicians to understand physical reality by means of mathematical concepts used a distinction between "real" and "rational" beings to contend that the objects of mathematical discourse had no actual existence in the world and could not therefore produce true and real knowledge. Pedro da Fonseca, for example treated number as a prototype of a nonreal be-

30. Ibid., disp. 29, sec. 3. no. 34.
31. Ibid., disp. 31, sec. 2. no. 10.

ing. He argued that a number is not a real being (ens reale) but only a being of reason (ens rationalis). This view of number formed the background to the verdict of the commentators of Coimbra, according to which mathematicians "consider the nature and essence of no real being,"[32] an idea shared by the prominent philosopher Benedictus Perera, who wrote that "the mathematical sciences are not real sciences."[33] Since mathematical entities referred only to concepts in the intellect, mathematics was limited in its claims for knowledge of the real world.[34]

The response of Jesuit mathematicians was to develop a discourse on mathematical entities that aimed to show that they could provide true and real knowledge of the world. This discourse served to legitimize their aspirations to a higher professional status and to enlarging the scope of mathematical teaching.

In the *Prolegomena* to his commentary on Euclid's *Elements* Christopher Clavius, who held the chair of mathematics at the Collegio Romano for almost thirty years, argued that the peculiar ontological status of mathematical entities enabled them to mediate between material things (the subject of physics) and spiritual reality (the subject of metaphysics): "Because the mathematical disciplines discuss things which are considered apart from any sensible matter, although they are immersed in material things, it is evident that they hold a place intermediate between metaphysics and natural science."[35] It is this intermediary position which secured their place among the sciences. Clavius, however, did not have a well-developed metaphysical view in which he could anchor his claims about mathematical entities. But his student Josephus Blancanus could have relied on Suarez's metaphysics in elaborating his own arguments on mathematical entities. Blancanus's main strategy was to insist on the materiality, essentiality, and reality of mathematical entities, from which the truthfulness, causality, and certainty of mathematical demonstrations was inferred. True, the materiality he claimed for his subject was the materiality of "intelligible matter," abstracted from time and place. Following Suarez, however, Blancanus contended that abstraction did not detract from the reality of an object, since the objects of all sciences were abstracted from existence.[36] Suarez's argu-

32. See Dear, *Mersenne and the Learning of the Schools*, 65.

33. See Rivka Feldhay, *Galileo and the Church: Political Inquisition or Critical Dialogue?* (New York: Cambridge University Press, 1995), 217–18.

34. Ibid., 214 n. 2; 217–18; see also Dear, *Mersenne and the Learning of the Schools*, chap. 4.

35. Feldhay, *Galileo and the Church*, 215.

36. See Dear, *Mersenne and the Learning of the Schools*, 67–68; Feldhay, *Galileo and the Church*, 165–69.

ment in the *Disputationes metaphysicae* that "three and four are seven is perpetually true, even if there be nothing which is numbered" is thus echoed in Blancanus's assertion that mathematical entities are not figments of the intellect associable with physical objects but archetypes in the mind of God that find realization in sensible matter.

But moral theology was an even more sensitive area of debate over the interpretation of objects of knowledge and their relationship to the status of the knowledge produced.[37] The new interpretation of the Thomist doctrine of salvation was mostly associated with Louis Molina (1536–1600), a Jesuit theologian from the University of Evora, who published his *Concordia liberi arbitrii* in 1588. Molina's originality lay in his conception of God's "scientia media" ("middle science"), which allowed him (Molina) to compromise the principle of human free will and the principles of divine grace, foreknowledge, and predestination. Before every act of grace, God can discern, by means of his scientia media, those individuals who are able to cooperate with him, through the exercise of their free will. It is this divine "science" of man's future actions that finally guides the choice of grace imparted to the elect, and necessarily and inevitably brings them to salvation. The crucial question upon which Molina's concept of God's scientia media hinged concerned the status of the entities presumably known by God of man's future acts, not yet determined by his will. The traditional Thomists, especially among the Dominicans, contended that prior to God's determination of these acts through his will, these acts were but hypothetical. Therefore, the divine knowledge of them had to be understood as hypothetical, and hence conditioned by human will alone. This was obviously heretical. Jesuit theologians who defended Molina, however, insisted that God's knowledge of the future acts of man, prior to their determination by his will, was necessary, certain, and infallible. The status of the entities known to God through his scientia media was not hypothetical but *possible*. Their canons of knowledge—as against the traditional ones—accepted the notion of *"real knowledge of possibles."* Thus, in addition to the subtle metaphysics of Suarez, the theological discourse of Molina could also provide conceptual resources to support the construction of the symbolic number as the new object of mathematicians.

Up to now my discussion focused on the traces left by a new type of object that seems to have emerged in Jesuit mathematical discourse in the seventeenth century. I have then looked at possible conceptual resources that could have supported the emergence of the new object, especially in the work of Franciscus Suarez. My remarks on Suarez, however, as well as

37. See Feldhay, *Galileo and the Church,* chap. 9.

those concerning discourses on real beings, possible beings, and rational beings in the mathematical and theological contexts, are meant only to delineate a possible direction for further research. If this direction is found to be valid, then a much deeper research of a paradigmatic shift from an ontological to a symbolical framework of mind should be conducted. My aim in this paper, however, is to further delineate the conditions of possibility of such shift, which depended not only on the nature of the object of mathematical and other discourses, but also on the politics of knowledge that was associated with it. As I have already stated, Klein's view of the connection between the object and the boundaries of scientific discourse seems in need of modification, for the construction of boundaries is not likely to stem automatically from the emergence of a new object. Rather, new boundaries are always the product of complex negotiations among different groups carrying professional, cognitive, and institutional interests. Klein's type of history of ideas is not likely to take such processes as an object of research. A complementary approach is here needed even for the mere sketch of the problem of the boundaries (in their connection to the new object) within the Jesuit educational system.

The circumstances in which discourses on real beings, rational beings, and possible beings emerged among mathematicians, philosophers, and theologians were those of struggle for professional status and cultural hegemony among groups within the Society of Jesus or between the society and other parts of the church establishment.

One struggle was fought between mathematicians and philosophers over the epistemological status of the mathematical sciences, their boundaries, their relevance for philosophy, and their authority in the cultural field. The militant mood of Jesuit mathematicians was reflected in Clavius's treatises from the 1580, among the first documents in the history of the society to treat the problem of the instruction of mathematics as a problem of cultural policy.[38] The treatises contained an expanded program of mathematical studies accompanied by a propaganda campaign for the status of those disciplines and their professors. The claim of mathematicians that their discipline should be given the same high status as was enjoyed by natural philosophy was justified by a conviction of their equal relevance for an understanding of reality, the traditional goal of philosophy: "It is necessary that the pupils should understand that these sciences are necessary and useful for a correct understanding of the rest of philosophy." The social status of the professors of mathematics within the framework of the colleges was

---

38. The history of the *Ratio studiorum* told in the rest of the paper is based on chap. 11 of Feldhay, *Galileo and the Church*.

to be reaffirmed by their participation in all official occasions such as graduation and public disputations. In addition, passing an examination in mathematics was to become a condition for acquiring a degree not only in philosophy but also in theology.

Much more acute, however, was the struggle between the Dominicans and the Jesuits over the interpretation of the Catholic doctrine of salvation. The *Disputationes* was published in 1597, the year in which the debate over predestination and free will was intensified to the point of a major cultural crisis, splitting the Catholic establishment into two rival intellectual elites.[39] Initially the debate was confined to Spain, where the pope had sent for the opinions of theologians of the two orders (the Dominicans and the Jesuits), professors in Spanish universities, and bishops in an attempt to reach some kind of consensus on the question of grace and free will. This consensus, however, was not forthcoming. After the publications of Banez's *Apologia Fratrum Praedicatorum,* he was invited to Rome, where a committee of theologians was set up to examine the claims of both sides. The committee of 1597 marks the beginning of a second stage—remembered as the controversy *de auxiliis,* in which the two strongest and most influential orders in the Catholic world engaged in a public struggle for hegemony that lasted actively for ten years until silenced by the pope in 1607, but not resolved. Sometime in 1597 Suarez was called to write in defense of Molinism, and composed his *Opuscula theologica,* three treatises of which were sent to Rome, representing the official Jesuit position on the questions debated. Suarez's support of Molina's concept of God's scientia media was expressed in his *De scientia qua Deus habet de futuris contingentibus.*

It seems, then, that one may speak of the combined efforts of some Jesuit metaphysicians, mathematicians, and theologians to break through the boundaries of traditional Thomism in the 1580s. Traces of the effects of this process in the sphere of mathematics may be found in the first version of the *Ratio studiorum,* composed as a creed and a common curriculum for all Jesuit educational institutions.

The chapter on mathematics in the *Ratio* of 1586 was written in Clavius's spirit and contained many of his suggestions. It opened with an apology intended to prove the relevance of mathematics for all other spheres of activity in which the Jesuits were engaged: salvation through the study of theology, considered as the ultimate goal of the Society of Jesus; the teaching of all other sciences to which mathematics is necessary; and the dissemination of practical knowledge useful for civil and religious life. Even without a detailed analysis of the program delineated by this document of

---

39. See Feldhay, *Galileo and the Church,* chap. 9.

1586, its deviation from the Thomistic attitude toward mathematics is obvious: the relevance of mathematics for both the ascent toward theology and for the descent toward more practical spheres of knowledge such as mechanics provided a justification for placing it at the center of the curriculum, more relevant, in fact, than traditional philosophy. On the one hand mathematics was seen as the key to the understanding of reality. On the other hand it was considered a model for correct rational procedures.

EPILOGUE: CULTURAL AND INSTITUTIONAL CONSTRAINTS ON A PARADIGMATIC SHIFT AMONG JESUITS

My reading of Guldin's treatise in the first part of this paper has shown the emergence of a new object of mathematical discourse that signaled the possibilities for the development of physico-mathematical science by Jesuit mathematicians. A glance into Suarez's metaphysical writings, into the mathematicians' discourse on mathematical entities, and into some of the directions taken by the architects of the Jesuit educational system further points out the conditions that seemed to favor the institutionalization of the project of Jesuit mathematicians along nontraditional lines, in spite of strong opposition within Jesuit intellectual circles, and perhaps other parts of the church establishment as well.

Guldin's computation of the distance by which the earth's center of gravity must be displaced (forty geometrical steps, according to him) may be seen as a vindication of Blancanus's claims that mathematicians are able to demonstrate unambiguously matters about which the philosophers can only argue dialectically. It was much better, Blancanus argued, to arrive at numerous and marvelous truths about such entities than it was to be concerned with a thousand differences of opinions about material substance, a true knowledge of which will never be attained. Furthermore, Guldin's reference to mathematical methods as the source of special privileges of the mathematicians is another indication of a sense of professional identity and authority cherished in his environment.

Guldin's narrow interpretation of motion in terms of quantitas terminata—abstracted from forces, moments, and time—is also a prominent feature of the text, however, and thus in need of explanation. This narrow interpretation on the conceptual level is paralleled by the delineation of very narrow boundaries to his discourse, differentiating it from that of Copernicans dealing with rotational or orbital motions, from Aristotelians interested in a dynamic approach to motion, and even from those Archimedeans interested in any way in the study of machines. Seeking for a legitimization in Aristotle's On the Heavens, ignoring the debt to nominalist writers, insisting on the compatibility between the Aristotelian and Archimedean approaches may all signal a difficulty in breaking through the status of mechanics as a "mixed science," interpreted in traditional Thomistic terms as more mathematical than philosophical or physical, and depending on traditional natural philosophy for the principles concerning natural substances. In this sense the dissertation may signal not only the options opened for Jesuit mathematicians in the 1580s, but also the limits of their ability to develop arguments and approaches when dealing with a problem traditionally pertaining to philosophy. Again, the context provides hints for the need to pursue further this hypothesis.

The censure of the *Ratio* by the Inquisition in the 1580s, and the rejection of Molinist theology by the traditionalists in the 1590s, signaled the great vulnerability of the Jesuits within the church establishment. This vulnerability, I believe, is the key to understanding the Jesuits' attempt to gain legitimization through the last version of the *Ratio* from 1599. The last *Ratio* was a conservative document, exhibiting a sophisticated use of mechanisms of exclusion and control. The construction of boundaries between the disciplines—especially between mathematics and philosophy—and the socialization of students into thinking within the limits allowed by those boundaries characterize the cultural policy implemented by the society at the beginning of the seventeenth century. A glance at the chapters on philosophy and mathematics confirms this impression.

Not much of Clavius's grand project to improve the status of the mathematical sciences and raise it to equal philosophy remained in the last version of the *Ratio*. Instead of involving a year and a half of study, the mathematics course was shortened to one year only. The relevance of mathematics to physical problems was not reinforced. On the contrary, the policy of the *Ratio* was to isolate carefully philosophical problems from mathematical problems and vice versa. The "mixed sciences," namely those specific areas in which physical problems were treated with mathematical methods, were still subalternated to mathematics, and hence their status as true *scientiae* remained ambiguous. Above all, contrary to Clavius's recommendations, no examination in mathematics was required of the students of philosophy and theology. In the absence of any clear external indication of merit, mathematics remained relatively marginal to the curriculum.

While the *Ratio studiorum* should not be read as a simple reflection of the practices of Jesuit philosophers and mathematicians, which often deviated widely from the official educational policies of the society, the *Ratio* did represent compromises reached among different kinds of pressures and interests that shaped intellectual choices to a certain degree. Guldin's strategies in dealing with the motion of the earth, while expressing some

tendencies toward a new kind of physical-mathematics also manifests reluctance to reinterpret the boundary between mathematics and physics, reluctance that echoes the Jesuit policies recommended by the *Ratio
studiorum*.

My reading of Guldin's treatise from the perspective of the discourse of
mathematical entities suggests an alternative for traditional history of
ideas, which tends to classify scientific writings in terms of progressive or
reactionary texts. Guldin's treatise is obviously anti-Copernican or non-
Copernican in its contents, and as such has fallen into total oblivion as reactionary and nonrelevant for the new science. However, it clearly represents
a genre of practicing physical-mathematics popular among Jesuit mathematicians in the seventeenth century. The text manifests typical tensions
between the tendency to incorporate innovations and the necessity to adhere to tradition that pervaded Jesuit science of the period.

A close reading of texts, which points out the emergence of new objects
and their effects on the politics of knowledge without reducing them to each
other, may produce complex historical arguments about the vicissitudes of
the development of the new science.

3   Doris Kaufmann

Dreams and Self-consciousness

MAPPING THE MIND IN THE LATE
EIGHTEENTH AND EARLY NINETEENTH
CENTURIES

This chapter tackles the coming into being of dreams as an object of *Erfahrungsseelenkunde* (empirical psychology or science of the soul, covering the still unseparated fields of psychology and psychiatry) in German thought of the late Enlightenment at the end of the eighteenth and the beginning of the nineteenth centuries. This dream research then stopped for almost one century. It was Sigmund Freud who once again attended to dreams, and made them the starting point and key object of his scientific approach. Though he took up questions similar to those of the Enlightenment psychological and psychiatric discourse on dreams, he was not aware of his predecessors.[1] They were not only forgotten by Freud and his contemporaries, but also by the later historiography on dream theories—such as the psychoanalyst Ludwig Binswanger's 1928 *Wandlungen in der Auffassung und Deutung des Traums* or the literary scholar Albert Béguin's 1937 *L'âme romantique et le rêve: Essai sur le romantisme allemand et la poésie française.* The early twentieth-century dream historiography discovered in the eighteenth century only the dominance of a "mechanistic psychology,"[2] transforming the "individual's living self" into a "mechanical-dynamic play of forces." Such a view of the human being, Binswanger wrote, was "not favorable" to the investigation of

1. Sigmund Freud, *The Interpretation of Dreams*, trans. James Strachey, the Pelican Freud Library, vol. 4 (Harmondsworth: Penguin), 1976.

2. Albert Béguin, *Traumwelt und Romantik: Versuch über die romantische Seele in Deutschland und in der Dichtung Frankreichs* (1937; reprint, Bern: Francke, 1972), 71.

dreams.[3] More recent studies also assign scientific dream theories a later beginning, namely as part of the Romantic period and its central interest in dream images and in a universal language of symbols, which was different from that of the Enlightenment dream discourse.[4]

I propose to rewrite this historiography of dreams. I shall investigate the Enlightenment discourse on dreams, and shall focus on the following questions. Why did the last three decades of the eighteenth century witness a broad Enlightenment discussion of dreams? Who recounted and discussed dreams, and for what reasons? Where was this need articulated? Did interrelations and interactions exist between everyday knowledge and scientific knowledge in the field of empirical psychology *(Erfahrungsseelenkunde)*? What importance did the different emergent dream theories of German *Erfahrungsseelenkunde* have for the differentiation and the future development of this field? Why were these dream theories thereafter dismissed for such a long time? Did the Enlightenment dream discourse already contain a possible anticipation of this demise and of basic controversies that later dominated the fin de siècle discussion on dreams?

1

From the last third of the eighteenth century until the first decades of the nineteenth century a discourse on self-knowledge and knowledge of human nature preoccupied the writers of the German Enlightenment. In the emerging bourgeois public sphere discussions on the external, i.e., social and political, constraints on reason were matched by an anguished concern with the internal forces and passions that disabled individual reason. The examination of the "other" of reason or the "dark sides" in oneself and one's fellow human beings was considered to be the key to deciphering the inner forces and workings of human nature, ultimately the key to a rational way of life. Collective anxieties like losing control or feeling endangered by a threat to one's own ego expressed the painful experiences and uncertainties experienced by the members of the new middle-class strata in their attempts to establish civil society and culture.[5] The efforts to create a

---

3. Ludwig Binswanger, *Wandlungen in der Auffassung und Deutung des Traums von den Griechen bis zur Gegenwart* (Berlin: Springer, 1928), 27.

4. For example, Henry F. Ellenberger, *The Discovery of the Unconscious: The History and Evolution of Dynamic Psychiatry* (New York: Basic Books, 1971), draws a direct line from the Romantic period to Freud and C. G. Jung.

5. See Doris Kaufmann, *Aufklärung, bürgerliche Selbsterfahrung und die "Erfindung" der Psychiatrie in Deutschland, 1770–1850* (Göttingen: Vandenhoeck & Ruprecht, 1995), 25–109.

bürgerliche identity and constitution of the self, clearly drawing the line between socially acceptable and deviant behavior, were articulated and discussed mainly in the new genre of the psychological periodical,[6] such as the *Magazin zur Erfahrungsseelenkunde* (Journal for the experience and knowledge of the soul), which emerged in the last third of the eighteenth century.

This journal, edited from 1783 to 1793 by the author, educator, and former Pietist Karl Philipp Moritz, was probably the best-known organ of the discourse on the unveiling of inner nature at the time.[7] Moritz organized a broadly supported project that the German philosopher Johann Gottfried Herder, among others, had already suggested in his treatise *Vom Erkennen und Empfinden der menschlichen Seele* of 1778 (On the thoughts and sensations of the human mind). Herder had proposed collecting empirical sources both on everyday expressions of the mind, such as dreaming or remembering, and on signs of mental deviance in order to discover how thinking and feeling functioned. The methodological model was taken from the sciences of anatomy and physiology, which had already made the internal workings of the human body visible and comprehensible.[8] At the very beginning of his "Vorschlag zu einem Magazin einer Erfahrungsseelenkunde" (Proposal for a journal of the experience and knowledge of the soul), which appeared in 1782 in the Enlightenment journal *Deutsches Museum*, Moritz emphasized the similarity between the study of the body and the study of inner nature.[9] Knowledge of the body, Moritz noted, had

6. For full references, see Johann Baptist Friedreich, *Systematische Literatur der ärztlichen und gerichtlichen Psychologie* (Berlin: Th. Enslin, 1833), 1–5.

7. *Gnothi sauton oder Magazin zur Erfahrungsseelenkunde*, 10 vols. (1783–1793; reprint, Nördlingen: Franz Greno, 1976 [referred to henceforth as *MzE*]). The literature on the journal includes Hans Joachim Schrimpf, "Das Magazin zur Erfahrungsseelenkunde und sein Herausgeber," *Zeitschrift für deutsche Philologie* 99 (1980): 161–87; Schrimpf, *Karl Philipp Moritz* (Stuttgart: Metzler, 1980); Raimund Bezold, *Popularphilosophie und Erfahrungsseelenkunde im Werk von Karl Philipp Moritz* (Würzburg: Königshausen & Neumann, 1984); Werner Leibbrand, "Karl Philipp Moritz und die Erfahrungsseelenkunde," *Allgemeine Zeitschrift für Psychiatrie und ihre Grenzgebiete* 118 (1941): 392–414; Ulrich Herrmann, "Karl Philipp Moritz: Die innere Geschichte des Menschen," in *Wegbereiter der Historischen Psychologie*, ed. Gerd Jüttemann (Munich: Beltz, 1988), 48–55.

8. Michel Foucault, *Die Geburt der Klinik: Eine Archäologie des ärztlichen Blicks*, trans. Walter Seitter (Frankfurt am Main: Ullstein, 1976), 38–68; Georges Canguilhem, *Das Normale und das Pathologische*, trans. Monika Noll and Rolf Schubert (Frankfurt am Main: Ullstein, 1977), 75–156.

9. Karl Philipp Moritz, "Vorschlag zu einem Magazin einer Erfahrungsseelenkunde," *Deutsches Museum*, 1782. The program of the Société des observateurs de l'homme, founded in 1799, demonstrated the simultaneity of such initiatives in the European Enlightenment. See Sergio Moravia, *Beobachtende Vernunft: Philosophie und Anthropologie in der Aufklärung* (Frankfurt am Main: Ullstein, 1977).

been advanced by its diseases. Under the present circumstances he was convinced of an urgent need for knowledge in the field of the experience of the soul as well. "The maladies of the soul" were "far more various, pernicious, and widespread than any physical ailment" and the yet unestablished science of mental disorders "more indispensable than any medicine for the body."[10]

Sacrifices had to be made, however. The general accessibility of case histories, i.e., their publication as a necessary precondition for their use, might, after all, in some cases expose their subjects to "public shame." Moritz nevertheless demanded this sacrifice. He compared it to leaving one's corpse to be dissected by anatomists, a highly controversial act at the time.[11] Becoming a "calm, cold self-observer" was, therefore, on the one hand, a sacrifice to be made for science. On the other hand, Moritz—with autobiographical overtones—assumed that those interested in self-observation would be driven by a certain degree of inner suffering. So he promised a positive therapeutic effect: "Comfort and a refuge from our own particular grief."[12]

The discourse on threats to the equilibrium of the faculties of the soul in the Enlightenment press arose against the background of a new consciousness of a coherent self that "belonged to oneself," was separate from that of one's fellow human beings,[13] and could, in principle, be studied with the methods of the natural sciences. This process of "naturalizing" human inner life was not restricted to the level of philosophical, medical, and literary reflection. The relationship between physical and psychological states became a central theme of public discussion in the Enlightenment.[14] In his *Anthropologie in pragmatischer Hinsicht* (Anthropology from a pragmatic point of view), the philosopher Immanuel Kant posed the essential question in the discussion on self-knowledge: what rules and purposes had been given to mankind by nature, and how great was the part played by per-

10. Moritz, "Vorschlag," *Deutsches Museum*, 1782, p. 486.

11. For a detailed discussion, see Ruth Richardson, *Death, Dissection, and the Destitute* (London: Penguin, 1989).

12. Moritz, "Vorschlag," quoting 492–95. See Moritz's *Anton Reiser*, which he introduced as "a 'biography' in the truest sense of the word, a truthful and faithful presentation of a human life down to its tiniest nuances." Karl Philipp Moritz, *Anton Reiser: Ein psychologischer Roman* (1785; reprint, Munich: C. H. Beck, 1987), 93.

13. Norbert Elias, *Über den Prozeß der Zivilisation: Soziogenetische und psychogenetische Untersuchungen*, vol. 1 (Frankfurt am Main: Suhrkamp, 1976).

14. See Roger Smith, "The Language of Human Nature," in *Inventing Human Science: Eighteenth-Century Domains*, ed. Christopher Fox, Roy Porter, and Robert Wokler (Berkeley and Los Angeles: University of California Press, 1995), 88–111, for the importance of the category of human nature for the organization of knowledge about the human subject.

sonal freedom, that is, "that which he [man] can or should make of himself as a being capable of acting freely."[15] Each individual must endeavor to reduce as much as possible the scope of his or her "involuntary" nature in relation to the scope of his or her own voluntary and calculated goals. Enlightenment thinkers thus combined two objectives. First, they sought to establish "healthy, purified, unclouded reason," for the "universal good of humanity." Second, they sought to find the way to the "greatest possible satisfaction of one's personal inclinations" by means of individual knowledge of one's own faculties of the soul. On the level of middle-class everyday experience, the last programmatic point in particular, however, tended to be inverted into a fear of not being able to establish the desired balance of psychic and physical powers.

The experience that "the soul's own power over its ideas"[16] did not function during certain periods was shared above all by the many self-observers who reported their dreams in the *Magazin zur Erfahrungsseelenkunde*. This potentially everyday experience, often associated with fear, made the reporting of dreams in the form of case histories the largest rubric among contributions. The dream accounts submitted by readers (following Moritz's request to establish an empirical collection before submitting fundamental principles of an *Erfahrungsseelenkunde*, including a dream theory) were intended as a collective effort to help the authors as well as the *Magazin*'s dream commentators to decipher the inner forces and workings of human nature. Which mental processes were subject to will and which worked involuntarily? The answer was as urgent as it was important, for it set up the framework for conscious independent behavior and action in civil society.

One group of dream accounts gave immediate and very direct insight into the constellations of social and cultural relationships, tensions, and desires. For example, a physician dreamt of neglecting his professional duties and of intentionally making himself incapable of working in the hospital.[17] A "very upright and truth-loving man" dreamt of beating to death a man with whom he argued in a coffeehouse,[18] and a "learned man" admitted that at the moment of falling asleep "against my will and without any instigation" he was obliged to struggle "with the most alluring images of sensu-

---

15. Immanuel Kant, *Anthropologie in pragmatischer Hinsicht* (Stuttgart: Reclam, 1983), 29.

16. Salomon Maimon, "Über den Traum und über das Divinationsvermögen," *MzE*, vol. 9, p. 64.

17. "Merkwürdiger Gang der Phantasie in einem Delirium: Aus einem Briefe, von Herrn D. Dunker aus Klitschdorf bei Bunzlau in Schlesien," *MzE*, vol. 2, pp. 201–8.

18. Aaron Wolfssohn, "Erfahrungen über Träume," *MzE*, vol. 9, pp. 273–77.

ality" and sudden notions of "degrading appellations for the Godhead and things divine."[19]

Aside from these transgressions of social norms of behavior and moral boundaries, the medium of dreams also articulated fundamental cultural conflicts. Rahel Varnhagen von Ense, a Jewish writer who recorded dreams in her diary and in letters to friends, told of her experience of social marginalization in dreams as another form of reality. Thus she commented upon a dream in which she, having "departed this life," discussed with other women the sufferings of their past existence. She found comfort and purification but, in the end, had to bear alone the "disgrace" of Jewish birth: "and upon waking the burden still remained, for I truly do bear it; and if there really were people who could understand it completely, I would feel some relief."[20]

For contemporaries, the obvious meeting of the two worlds—the dream world and the real world—in these dreams raised the question of the sleepers' moral responsibility for their dreams' content. Although under the rule of the imagination the higher faculties of the soul acted only "mechanically," dream images were nevertheless—as Kant put it—"images produced by the dreamer himself."[21] This problem occupied the Enlightenment public beyond the psychological journals, as the Enlightenment theologian Johann Abegg's 1798 account of his journey through the German states in search of self-improvement illustrates. Abegg discussed the "psychological topic, whether dreams were moral?" with the philologist and educational reformer Carl Gotthold Lenz.

> In general, I thought, one could not say with certainty. One would need to know the individual. He alone could know this, a stranger only with difficulty. Lenz agreed with me, but believed nevertheless that, generally speaking, dreams could be imputed morally, for surely each human being was more or less guilty if dreams were not absolutely moral. Nonetheless strange phenomena do occur. Professor Weber in Jena, for example, recognized as an honest and wise man, struggled much with melancholy during his last years. In his brighter moments he wrote down the thoughts that occurred to him in his miserable periods, including his dreams. And this otherwise so exemplary man reported that despicable, completely immoral ideas often came to him, and he did not know how they did so.[22]

19. "Über den Einfluß der Finsterniß in unsere Vorstellungen und Empfindungen, nebst einigen Gedanken über die Träume," MzE, vol. 5, pp. 164–65.

20. "Im Schlaf bin ich wacher," in Die Träume der Rahel Levin Varnhagen, ed. Barbara Hahn (Frankfurt am Main: Luchterhand, 1990), 20–22, at 22.

21. Immanuel Kant, Träume eines Geistersehers, erläutert durch Träume der Metaphysik (Stuttgart: Reclam, 1982), 40.

22. Johann Friedrich Abegg, Reisetagebuch von 1798 (Frankfurt am Main: Insel, 1987), 45.

The educator Friedrich Pockels, one of the *Magazin's* editors, tried to answer this question:

> An absolute absence of shame, wild emotions, contempt for religious questions, blasphemies, and other abominable thoughts and sentiments, not troubling us when awake, are experienced by even the most excellent persons while dreaming . . . One either already had such notions during one's waking hours, or an association of contrasting notions leads us to them in a dream, or the emotions, in order to act all the more freely, instill images in the reflection, or—perhaps when awake one never, or seldom, acted upon religious principles, for then the dream is only a copy of waking life.[23]

Karl Philipp Moritz accused Pockels of positing a "mechanism of imagination," since he "himself proceeded mechanically," "without once considering that beyond the obvious surface there might well be something as yet unexamined by human thought." Again Moritz vigorously emphasized the healing powers of Enlightened self-knowledge: "At the point where our nature perfects itself, it truly must not shrink from itself; in its deepest recesses it holds firmly onto itself, and where it is recognized, all imagined horrors flee before its glowing clarity."[24] Moritz himself, however, found remembering his dreams "highly unpleasant."[25] He resolved the question of the dreams' origin and of their morality in the imperative to "obscure the ideas which we receive in dreams in an orderly fashion." Moritz saw the equilibrium of the faculties of the soul and the soundness of mind in direct relationship to this filtering ability. An "adequate number" of ideas that were constantly "flowing into the mind daily and hourly" had to be suppressed, because otherwise an "overabundance of ideas would arise, causing disorder and confusion." This intervention was more difficult to perform during dreams, because "in this state the self is only floating," as the philosopher Joseph Veit wrote in a debate with Salomon Maimon on dreams and delusions published in the *Magazin*.[26] There was the danger, Veit believed, that in dreams man would, "forget his own true self."

This observation, accompanied by terrible anxiety, recurred in a number of dream accounts in Moritz's *Magazin*. The dreamers experienced a total

---

23. Friedrich Pockels, "Psychologische Bemerkungen über Träume und Nachtwandler," *MzE*, vol. 6, pp. 238–39.

24. Moritz, "Revision über die Revisionen des Hrn. Pockels in diesem Magazin," *MzE*, vol. 7, pp. 198, 199.

25. Moritz, "Grundlinien zu einem ohngefähren Entwurf in Rücksicht auf die Seelenkrankheitskunde," *MzE*, vol. 1, p. 30, also for the two following quotations.

26. Veit, "Schreiben über Täuschung und besonders vom Traume," *MzE*, vol. 8, p. 200. The following quotation is from p. 204.

dissolution of the boundaries of the self.[27] Such nightmares had to be broken off, because this was the only way that the self could again cohere.[28] The fact that this often happened proved to the Enlightened self-observers the existence of a faculty of the soul that they considered the crucial sign of its soundness: willpower, i.e., the ability to "keep the imagination under control," ultimately even in states of "incomplete consciousness."[29] According to Maimon this faculty could be all the more effective "the more we sense our self, the more we regard this self as a source of our ideas, the more we are convinced that we are no mere suffering creatures simply receiving ideas, but in part produce them ourselves; and finally, the more we recognize the value of guiding our ideas, the more we will be encouraged to rule them."[30]

The ability to make a conscious distinction between dream world and real world, i.e., to interrupt a dream or to know that one was dreaming, testified to a strong and conscious self. It almost became a feature distinguishing sanity from madness.

What, however, had then caused the "strange crowding and confusion" in the head of Johann Joachim Spalding, Enlightenment thinker and Protestant theologian, dean of the Nikolaikirche in Berlin? He had been attacked by a "swarm of thrusting tangled images" one day in 1772, while making out interest receipts for the parish poor.[31] Among the "tumultuous disorder in a portion" of his "ideas," as he put it in his case history reported to the Magazin, the member of the Berlin Consistory was capable neither of writing nor of intelligible speech. Yet with another "part of his brain" he was "fully and firmly" conscious of the "familiar principles of religion and conscience." Moreover, he was able to reflect upon his loss of social communication and its consequences. Spalding was therefore reminded of the "probationer [for an ecclesiastical living] in the local lunatic asylum" who

27. For example, Carl Gotthold Lenz, "Auszug aus einem Briefe über Ahndungen und Feuerbesprechen," MzE, vol. 4, pp. 55, 56: "Everything was spinning around inside me like a disk, accompanied by creative ideas of eternal millennia and spaces I had to wander through, the thought of the impossibility of completing this journey, this vastness, which I saw always before me like an unending circle (and all of this is in an awakening state) aroused in me an extraordinary unease, in which I often could not stop myself from springing out of bed in a single leap . . . in order to escape that terror." Salomon Maimon wrote of the violation of sexual boundaries in dreams in "Revision der Erfahrungsseelenkunde," MzE, vol. 10, pp. 10–11.

28. S. Maimon, "Fortsetzung des Aufsatzes über Täuschung und besonders vom Traume," MzE, vol. 9, pp. 105–15; Joseph Veit, "Über die Anmerkungen des Herrn Maimon zu der Fortsetzung des Aufsatzes über Täuschung," MzE, vol. 10, pp. 76–98.

29. Maimon, "Fortsetzung," 110.

30. Ibid., 111.

31. Spalding, "Ein Brief an Sulzern über eine an sich selbst gemachte Erfahrung," MzE, vol. 1, pp. 117–21.

had "begun by speaking confusedly and incompletely," only to fall into a lasting silence. "Who knows, I thought, whether he does not have his own orderly thoughts just as I do mine, and only does not wish to speak because he knows and senses that he is not the master of his innermost organs of speech, and is thus reluctant to appear insane in speech, because he is not so in his thoughts."

Spalding's story of an involuntary "dissociation of the self" became a central case in the *Magazin zur Erfahrungsseelenkunde*, followed by numerous similar self-observations.[32] The authors of these case histories all shared the experience that personal freedom could be severely impaired by the power of involuntary ideas arising from physical causes, which could place the mind in a "state of slavery."[33]

This provided a first answer to the controversial and explosive, because potentially materialist, question of whether "there is something mechanical or, so to speak, physical in the workings of the soul?"[34] The experience of many observers confirmed a connection between disruptions to the "regular activity of the mind," especially in dreams, and disorder in the brain, which, as an organ of the body, was subject to the influence of "mechanical laws." Its disorder could also produce a disorder in the capacity of reason, considered the core or the highest faculty of the soul.[35] A teacher at the Graue Kloster in Berlin, a famous Enlightenment college, who described his nightmares in the *Magazin* concluded that if a simple wrong positioning of the head during sleep could unleash "terrifying brainless visions" and "ideas devoid of any human sense," setting off "a state of great alarm," nobody could really know where to draw the "borderline" to the "higher faculties" of the soul that would remain untouched by potential "horrifying disorders in the machine."[36]

If, however, one part of the self was capable of observing and describing the other part, "which of the two says 'I'?" the theologian Spalding asked.[37]

---

32. "Selbsterfahrung des Herrn Kirchenrath Stroth in Gotha," *MzE*, vol. 2, pp. 59–60; Ernestine Christiane Reiske, "Parallel zu der Selbstbeobachtung des Hr. O. C. R. Spalding im 2ten Stück des ersten Bandes," *MzE*, vol. 3, pp. 218–20; "Auszug aus einem Briefe, von Hrn. K. Gemeinheits-Commissarius Gädicke zu Cammin," *MzE*, vol. 4, pp. 207–8; "Anmerkungen und Berichtigungen zu dem Magazin zur Erfahrungsseelenkunde, von Herrn van Goens," *MzE*, vol. 8, pp. 239–40.

33. "Geschichte eines im frühesten Jünglingsalter intendirten Brudermords, von V . . . s. in Br——g," *MzE*, vol. 3, p. 41.

34. Fischer, "Stärke des Selbstbewußtseyns," *MzE*, vol. 1, p. 41.

35. For the discourse on the soul as a bodily organ, see Michael Hagner, *Homo cerebralis: Der Wandel vom Seelenorgan zum Gehirn* (Berlin: Berlin, 1997).

36. Fischer, "Stärke des Selbstbewußtseyns," 39, 41.

37. Spalding, "Ein Brief an Sulzern," 121 n. 31.

The prominent philosopher Moses Mendelssohn replied to him in the *Magazin zur Erfahrungsseelenkunde,* testifying to the significance of such questions. His answer also pointed to a typical characteristic of the *Magazin's* articles, namely, the blurring of genres between public discussion and scientific reflection on dreams. In his "Psychologische Beobachtungen auf Veranlassung einer von dem Herrn Oberkonsistorialrath Spalding an sich selbst gemachten Erfahrung"[38] (Psychological observations occasioned by Consistorial Councillor Spalding's own experience) Mendelssohn therefore reported on his own "nervous weakness," which he experienced as a "fit" after awaking from a troubled sleep. He lay in bed, fully conscious and capable of "following any sequence of thoughts I undertook with order and clarity," but incapable of movement.

> I felt as though something burning was trying to flow down my spine from my brain and was encountering resistance, or as if someone was whipping the back of my neck with burning switches. I thus had to keep perfectly still until an impression from without opened the sluices of my vital spirits, allowing them free reign, and in that very moment everything was suddenly restored, and I was once again master over my voluntary motions.[39]

Mendelssohn interpreted his and Spalding's experience by assuring the latter that "neither the location nor the purpose of his self had changed." It was only that "strange, inappropriate ideas had attained more influence than he had intended." According to Mendelssohn, human inner nature was organized like an Enlightened absolutist monarchy under the reign of the mind:

> Only it [the mind] does not rule absolutely in its kingdom, and not all of its orders are carried out unquestioningly . . . Sometimes an idea attains greater force, refuses obedience, and wishes to act on its own where it should not; it displaces an appropriate idea . . . which necessarily occasions disorder and interruption in public affairs. The ruler hurries to steer the disorder. It thus seeks to turn more of the attention, which it already possesses in part, to appropriate ideas, and thus make them more effective. It is understandable, however, that the unruly idea will not always give up straightaway, but rather may even win the first battle and produce an organic reaction, which the self's dominant part fails to recognize and finds inimical to its ultimate objective.[40]

38. Moses Mendelssohn, "Psychologische Beobachtungen auf Veranlassung einer von dem Herrn Oberkonsistorialrath Spalding an sich selbst gemachten Erfahrung," *MzE,* vol. 1, pp. 211–32.

39. Ibid., 227–28.

40. Ibid., 232.

2

How such a "tumultuous disorder of the ideas"—whether suffered in dreams or in madness—could come about, and which role "the self-controlling faculty of the soul" played in all this was one of the central and controversial themes of discussion in the different but related discourse of the developing scientific field of *Erfahrungsseelenkunde*. Dreams became an essential object there. Beginning at midcentury and gathering momentum in the last third of the eighteenth century, an increasing number of books appeared on dreams, and also on visions, presentiments, and sleepwalking. They were written by scholars, academic empirical psychologists *(Erfahrungsseelenkundler)*, who were largely trained as physicians, and also as philosophers and theologians.[41] They referred to the case histories on dreams in the *Magazin zur Erfahrungsseelenkunde* and in other publications of this genre as empirical sources.

Most of these empirical psychologists echoed the emotional response of participants in the broader Enlightenment discourse on self-knowledge. They associated reflections on dreams with experiences of fear. The professor of medicine Johann Christian Reil, for example, a leading figure in the nascent field of German psychiatry, gave the following disturbing description of a dreamer's state of mind:

> The self-consciousness wavers in all its relations. The fantasy ebbs and flows within itself, no sensory impression restrains it anymore. The dreamer has no idea whatsoever of his objectivity, and conceives of his subject wrongly. He believes his visions to be real objects and plays each alien role as his own . . . Tied neither to actual time nor place he exists now in the past, now in the future, among the living and the dead.[42]

Carl August Eschenmayer, professor of medicine and philosophy, noted that in dreams the persona became "diffuse and detached and often slipped into another."[43] His colleague Dieterich Tiedemann described in his *Handbuch der Psychologie* (Handbook of psychology) how, shortly after awaking, the dream's "imaginary reality" often could "not be reconciled at first with that reality experienced by the senses." One felt oneself

41. This indicates that the control of the soul is transferred to a new professional group. Textbooks on the science of the mind (psychology) always devoted much attention to the subject of dreams.

42. Johann Christian Reil, *Rhapsodieen über die Anwendung der psychischen Kurmethode auf Geisteszerrüttungen*, 2d ed. (Halle: Curt, 1818), 92.

43. Carl August Eschenmayer, *Psychologie* (1817; reprint, Frankfurt am Main: Ullstein, 1982), 226.

"doubled," a pathological experience one might also have "after a grave ill-ness."[44]

Academic psychologists, however, tried to dissolve and overcome such fears by the "objectivity" of scientific dream explanation.[45]

Three main approaches to a theory of dreaming emerged at the end of the eighteenth and the beginning of the nineteenth centuries. They point to the importance of dreams as a constitutive scientific object for the develop-ment of explanatory concepts of human inner nature. The different dream theories also already hint at the different future directions of this enter-prise in empirical psychology and psychiatry.

In quantitative terms, the most important group among the three ap-proaches to a dream theory were those empirical psychologists who fol-lowed the lines of the Enlightenment public's discussion of dreams. They based their reflections of dreams on the faculty-based model of the soul al-ready described in Mendelssohn's reply to the theologian Spalding. When dreaming, the equilibrium of the various faculties of the soul in the wak-ing state was destroyed in favor of the absolute rule of imagination. The soul thus turned into a "spectator" of its own actions.[46] The physician, philosopher, and experimental psychologist Johann Gottlob Krüger de-scribed the soul as "similar, in dreams, to a puppeteer who moves her own puppets, and does so without knowing that she does it."[47] The ideas an individual had "more or less consciously"[48] when dreaming were, after all, not connected through outward sensory impressions and feelings to "objectivity with its firm realities."[49] The powers of reason and will were active only to a limited extent and no longer capable of "reigning in" the "ideas and images that fantasy strings together by using the magic wand of the association of ideas."[50] The borderline between the internal and ex-ternal world was abolished, and dreamers took their inward pictures for outward reality. This experience was shared by dreamers and the insane alike, and psychiatrists in particular took up this theme and commented on

44. Dieterich Tiedemann, *Handbuch der Psychologie,* ed. LudwigWachler (Leipzig: Barth, 1804).

45. See Georges Devereux, *From Anxiety to Method in the Behavioral Sciences* (the Hague: Mouton, 1967).

46. Eschenmayer, *Psychologie,* 226.

47. Johann Gottlob Krüger, *Versuch einer Experimental-Seelenlehre* (Halle: Hemmerde, 1756), 197. See Gary Hatfield, "Remaking the Science of Mind: Psychology as Natural Sci-ence," in Fox, Porter, and Wokler, *Inventing Human Science,* 201–5, for Krüger's attempt to create an experimental science of the mind.

48. Wolf Davidson, *Versuch über den Schlaf* (Berlin: Belitz & Braun, 1799), 105.

49. Eschenmayer, *Psychologie,* 226.

50. Davidson, *Versuch,* 104.

the natural transition from dream states to madness. Probably in order to quiet fears, the powers of reason and judgment (repressed while dreaming), were brought back into play in the theoretical reflections at this point. These faculties were thought to be able to interrupt dreams or to make them recognizable as such by scanning the overflowing chaotic associations of ideas for logical conclusions. Thus, for example, the physician Wolf Davidson described a dream in which he was standing at an open window with his landlord when the latter leaned out and fell. Davidson thereupon rushed into the courtyard gripped "by the most awful fear that people might think I had pushed him out," but found nobody. This "made me doubt everything, I believed it a dream and awoke with the greatest feeling of joy."[51]

The origin of these dream sequences was interpreted variously as reminiscences of daytime occurrences and occupations and of far-off (childhood) experiences that the mind now recalled, undisturbed by outward impressions. Imagination, however, was also regarded as a productive activity, as a creative power. To the objection that nobody could dream of anything not experienced before, Johann Gottlob Krüger replied: "[T]hrough amalgamation of ideas imagination has a capacity to produce new ones. It is a creator like chemistry, which, through the mixing of those substances provided to it by nature, produces new ones that nature herself would not have produced." Imagination therefore could, alongside highly unpleasant nightmares, also bring forth very pleasant dreams. Krüger asked: "Do not say that these are mere imaginary pleasures, for what would remain of the real ones if we removed all belonging to imagination?"[52] Karl Philipp Moritz described the joy of immersing oneself in the world of "fantastical dreams" in the first German psychological novel, *Anton Reiser.* But he castigated such pleasures as negative antisocial behavior, because then "dreams and madness would be preferred to order, illumination, and truth."[53]

Following this line, the particular group of empirical psychologists introduced here stated in their dream reflections that to turn the internal world into an external world meant for waking and sleeping dreamers to experience isolation and loneliness. "When we are awake, we have a shared world; but when we sleep, each has his own."[54] In this world of one's own, unreflected by outward impressions and the control of others, a loss of moral principles occurred. It plunged the dreamer into a "wild chaos of in-

51. Ibid., 134.
52. Johann Gottlob Krüger, preface to *Träume,* for the preceding quotation as well.
53. See Moritz, *Anton Reiser,* 364.
54. Davidson, *Versuch,* 138.

cidents" and led him to doubt "morality, human dignity, the Creation, existence, and duration."[55]

All similarities to the argumentation of the Enlightenment discourse on self-knowledge notwithstanding, it is this more emphatic and almost universally negative interpretation of the dream event as an expression of the "limited autonomy of the higher faculties of the soul" that distinguished this approach to dream theory from the public Enlightenment dream discussion. Concerning the general judgment, this negative interpretation connects it to the fin de siècle physiological approach to dreaming, which paid little attention to dreams as objects of scientific research—because the autonomy of the higher faculties of the soul during dreams seemed, so to speak, not limited enough and dreams as manifestations of mental life too independent of demonstrable organic changes. This approach began in the mid–nineteenth century and was connected to the rise of a physiological approach in the life sciences. In the second theory of dreams emerging at the end of the eighteenth century this tendency and its underlying reason are already visible.

A fundamental change in thinking was ushered in by those dream theorists who no longer proceeded from the interplay of various mental faculties. They instead declared the nervous system—as the organ of the soul—to be the constituting factor of self-consciousness. The latter was therefore thought to be dependent upon the regular working of physical processes.[56] The dream, J. C. Reil explained, was "the product of a partial waking of the nervous system" without a "synthesis" with self-consciousness.[57] During both dreaming and sleepwalking a person might be "partially conscious of himself; he may act, observe himself, reflect upon himself, even consider whether he is doing all of this awake or asleep . . . We may carry out the most sublime operations of the higher faculties of soul consciously or unconsciously, as mere automata."[58] In his 1802 *Rapports du physique et du moral de l'homme*, the French professor of medicine Pierre-Jean-Georges Cabanis went much further than Reil in developing the consequences of this approach. In the face of a theory that regarded physical reactions to external stimuli and the motions of the inner organs as the causes of "disorders of the intellect and the will," it was no longer relevant to ask questions of individual responsibility and the

55. Ibid., 119.

56. See George Rousseau, "Cultural History in a New Key: Towards a Semiotics of the Nerve," in *Interpretation and Cultural History*, ed. Joan H. Pittock and Andrew Wear (New York: St. Martin's, 1991), 25–81, for the discourse on nerves in the eighteenth century.

57. Reil, *Rhapsodieen*, 90, 92.

58. Ibid., 96–97.

loss of moral principles, of the dissociation of the self and of inner conflict in dreams, and of the causes for certain associations of ideas which occurred in dreams.[59]

Cabanis had taken the "daring step" of "reducing all of anthropology to physiology," the German translator and editor Professor Ludwig Heinrich Jakob remarked critically in his preface. Moderating Cabanis's work and explaining it to German readers, he had also added his own treatise "Über die Grenzen der Physiologie in der philosophischen Anthropologie" (On the limits of physiology for philosophical anthropology) in order to encourage "some of our German physiologists who recently favor the same system in their writings . . . to consider their claims more carefully."[60] This worry seemed to be quite baseless, for Jakob's reflections on the tasks of a science of man written in opposition to Cabanis were also an accurate description of the theoretical level of German physiologically oriented dream theory. According to Jakob, Cabanis's main error was "not only his endeavor to explain all states of inner nature in terms of physical causes, but primarily that he considers them to be themselves physical conditions." While Jakob approved of the "maxim to avoid the introduction of a spiritual substance distinct from the body into science," he insisted that "physical processes and mental ideas belonged to two wholly different classes of sensory phenomena."[61] But a "causal connection" existed between them. Physiology as the "science of the system of physical processes and changes" was an "auxiliary science indispensable to anthropology." But the latter also required "empirical psychology, i.e., knowledge of the ultimate inner changes in the workings of human nature and of the system of ideas." Anthropology or the science of man should investigate the relationship between the fields of physiology and empirical psychology.[62]

The third approach to theories of the dream I would like to sketch here emphasized the therapeutic value of dreams and their significance for cur-

---

59. Pierre-Jean-Georges Cabanis, *Rapports du physique et du moral de l'homme*, 2 vols. (Paris 1802; trans. *Über die Verbindung des Physischen mit dem Moralischen*, Halle: Reinicke, 1804), 1:532. Also: "Thus for example cramps of the intestines and diaphragm and the entire epigastric region, the filling up of the vessels of the portal vein, or the fear of difficult digestion produce quite different images in the brain during sleep than during the waking state, and the means by which the sleeping state produces these images corresponds perfectly, as we shall see, to the means by which the crazed images of madness and delirium are produced in the diseased processes of the various internal organs": 1:567. On Cabanis, see Martin S. Staum, *Cabanis: Enlightenment and Medical Philosophy in the French Revolution* (Princeton: Princeton University Press, 1980).

60. Ludwig Heinrich Jakob, "Über die Grenzen der Physiologie in der philosophischen Anthropologie," in Cabanis, *Über die Verbindung*, v–vi.

61. Ibid., xxxvi, xlvii, and xl.

62. Ibid., li.

ing "maladies of the soul," rather than dream stimuli and dream sources from external and internal motions of the organs. The most prominent representative of this approach was the professor of philosophy and psychology Friedrich August Carus (1770–1807), an older relative of the famous Romantic physician and artist Carl Gustav Carus (1789–1869). In the chapter on dreams in his book *Psychologie* published in 1808, F. A. Carus defined the dream as an "involuntary uninterrupted continuous and often all the more powerfully productive or poetic activity of the faculties of the soul in the state of sleep."[63] There was, during the individual's lifetime, no "complete cessation of all mental activity." So the mind occupied itself during sleep, when the "senses were closed off," with the "stock of ideas resting within it." It revived the "images slumbering in its depth and the earlier notions much obscured during waking life."[64]

> What we did, felt, and thought, with outer and inner senses open, is not lost even if it was interrupted. Our inward drive takes up the thread once more and carries on . . . Even more, whatsoever we practiced in the past, even in our earliest childhood, to which we were accustomed and which we enjoyed at that time, it is with those things that we continue to occupy ourselves during the silent nights.[65]

F. A. Carus also incorporated nightmares into this pattern of interpretation. "Frightening dreams are thus also nothing more than a continuation of our feelings. But these are not intended to frighten us, but only to rouse us and bring us suddenly to a full consciousness of our self, even if it be a hideous one . . . The dream may be considered man's secret face; but here too he, as an independent and responsible individual, should be his own judge."[66] From this he derived a "law," namely "there is no dream without a relationship to the issues the dreamer dealt with in a waking state, however long ago." For this reason each dream contained "some truth," and the "essential feature of each dream" referred to the dreamer's particular character, his "ways of thinking, his inclinations, and his memories, however old they might be."

In analogy to his idea of different historical layers of experience embedded in a human being, F. A. Carus also considered human psychic dispositions to be the substratum of social evolution. In his book *Geschichte der Psychologie* (History of psychology) he therefore interpreted and pre-

63. Friedrich August Carus, *Psychologie,* in *Nachgelassene Werke* (Leipzig: Barth & Kummer, 1808), 2:181–82.
64. Ibid., 2:186.
65. Ibid., 2:189–90.
66. Ibid., 2:190–91.

sented a universal history of mankind—in fact the history of cognition and its limits—as the history of psychology.[67]

The experience of a dissociation of the self in dreams, usually into a second, worse person, was for him also an allusion to the dreamer's past, present, or future potential.[68] Carus confronted the assumption of the chance and chaotic nature of associations of ideas and images in dreams with the thesis of their coherence and causality based on the dreamer's personal history. "Objections may easily be raised here, as the content of dreams often appears too motley, too caricaturish, and as most jump from one object to the next. And yet here too the mind obeys the laws of causality and surely there is always a thread along which all are strung, even if it remains hidden from us."[69]

Ninety-two years later in his *Interpretation of Dreams*, Freud claimed to have discovered this thread and with it the secret of dreams:

> I will bring forward proof that there is a psychological technique which makes it possible to interpret dreams, and that, if that procedure is employed, every dream reveals itself as a psychical structure which has a meaning and which can be inserted at an assignable point in the mental activities of waking life. I shall further endeavor to elucidate the processes to which the strangeness and obscurity of dreams are due and to deduce from those processes the nature of the psychical forces by whose concurrent or mutually opposing action dreams are generated.[70]

Freud did not realize however, that much of his thinking on dreams had already been present in the Enlightenment discourse on dreams, particularly the use of dreams for healing mental and psychic diseases, the method of analyzing dream events and searching for laws of causality in a patient's personal history as well as the narrative presentation of a dream theory—based on the scientist's own experience. This discourse of the late Enlightenment—in fact several discourses—had been contradictory and fragmented and had left the future orientation of psychiatric and psychological research undetermined and with it the importance of dreams as its object.

At the end of the nineteenth century this question seemed to be settled. In his *Interpretation of Dreams* Freud noted the clear primacy of the physiological approach in contemporary psychiatry, which meant almost no attention to dreams as an object of research. According to him this low

67. Friedrich August Carus, *Geschichte der Psychologie* (Leipzig, 1808; reprint, Berlin: Springer, 1990).

68. Carus, *Psychologie*, 2:192–93.

69. Ibid., 2:195.

70. Freud, *Interpretation*, 57.

evaluation of dream-life was a result of the triumph of the specifically "scientific way of thinking" *(naturwissenschaftliche Denkweise)* that had entered psychological and psychiatric research in the second half of the nineteenth century.[71] Freud summarized:

> It is true that the dominance of the brain over the organism is asserted with apparent confidence. Nevertheless, anything that might indicate that mental life is in any way independent of demonstrable organic changes or that its manifestations are in any way spontaneous alarms the modern psychiatrist, as though a recognition of such things would inevitably bring back the days of the *Naturphilosophie,* and of the metaphysical view of the nature of mind. The suspicions of the psychiatrists have put the mind, as it were, under tutelage, and they now insist that none of its impulses shall be allowed to suggest that it has any means of its own.[72]

A case in point was for example the German neurologist Adolf Strümpell, who interpreted dreams as "an eclipse of all the logical operations of the mind which are based on relations and connections." He therefore judged them useless for scientific research on the brain.[73]

Almost one hundred years after the discussion on dreams as a threatening phenomenon of inner nature in the context of *Erfahrungsseelenkunde* Freud became the figure around whom the unsolved contradictory elements in the earlier discussion cohered. Though Freud also trusted in physiological and anatomical explanations of psychic disorders and mental diseases, he criticized the limited and—so to speak—mechanical understanding of the physical realm by his contemporaries. Both at the end of the eighteenth and at the end of the nineteenth centuries, there was a surprisingly similar constellation between collective and individual awareness of crisis within the middle class and among competing psychiatric and psychological attempts at an interpretation. The underlying reason was the search for a *bürgerliche* identity. For the fin de siècle, a period of social, cultural, and political crisis, saw the dissolution and destabilization of middle-class patterns of thought and behavior that had been established in the course of the nineteenth century. Again the workings of inner human nature became the main focus in the struggles over the redefinition of a *bürgerliche* identity. And again the question in psychiatry of whether dreams were or were not a significant object for explaining human nature came up.

In this article I have outlined how a threatening phenomenon of inner human nature—the dream—had been constituted by the Enlightenment

71. Ibid., 130.
72. Ibid., 105.
73. Quoted ibid., 122.

public and by scientific discussion as an object of observation, description, and empirical inquiry together with a common language and a narrative form, namely the case study. This coming into being of dreams as an object in *Erfahrungsseelenkunde* was linked to the everyday needs of the new middle-class strata, and emerged from the public sphere of the late Enlightenment at the end of the eighteenth century.

The public discussion of dreams directly influenced the theoretical considerations of the early psychologists and psychiatrists. Though they were as frightened by their own dreams as their middle-class correspondents were by theirs, they did not write primarily from the dreamer's point of view. The empirical collections of dreams were used for diagnosis and treatment of the insane, thereby lending scientific substance to the claim for exclusive expertise in the treatment of the disorders of the mind. And most importantly, because dreams promised to provide insight into the workings of human inner nature, reflections and theories on dreams became the starting point for different concepts of inner nature. The coming into being of dreams as an object of *Erfahrungsseelenkunde* and later of psychoanalysis point to the close correlation between the specific historical nature of consciousness and *mentalité* and of scientific developments.

Jan Goldstein

Mutations of the Self in Old Regime and Postrevolutionary France

FROM *AME* TO *MOI* TO *LE MOI*

MOI   It has been contended that this personal pronoun has the same meaning as the *je* or as the Latin *ego*. The *je* has been condemned by the word *egotism*, but that does not prevent it from being suitable on certain occasions. It follows still less that the *moi* cannot sometimes be sublime or admirably placed. Here are some examples . . .
—*Encyclopédie, ou Dictionnaire raisonné des sciences, des arts, et des métiers,* 1765

MOI   This is the name by which modern philosophers customarily designate the soul [*âme*] insofar as it has consciousness of itself and is familiar with its own operations, or is simultaneously the subject and object of its thought. When Descartes defined himself as a thinking substance, a *res cogitans,* or when he set forth the famous proposition "I think therefore I am," he truly put the *moi* in the place of the *âme.* And he was not content to found that substitution (or, to put it more exactly, that equation) on the very nature of things, he also made it pass into language . . . However, in his own usage and that of his disciples, the new expression never took on the rigorous and absolute meaning later attached to it. Descartes said, unmistakably and deliberately, *moi,* instead of saying *mon âme;* but he did not say *le moi* . . .
—*Dictionnaire des sciences philosophiques,* 1849

The French encyclopedic impulse, which flourished so luxuriantly during the *siècle des lumières*, continued through the nineteenth century, when it characteristically produced multivolume works taking the form not of encyclopedias properly so-called but of encyclopedic dictionaries, or "dictionaries of things and not of words," as the genre was sometimes described.[1] This constancy of intellectual aspiration and publishing trend affords a handy way to begin our investigation of the coming into being of the self as a scientific object in France, to map out the semantic field related to that event. Since there exist, on both sides of the Revolutionary divide, compilations of the most up-to-date knowledge on an exhaustive array of topics, we are in a good position to chart sea changes in this subtle and tricky area, which might otherwise be so difficult of access. For the eighteenth century the relevant text is, of course, the celebrated *Encyclopédie* of Diderot and d'Alembert. The early nineteenth-century text that I have consulted here, the *Dictionnaire des sciences philosophiques*, is less well known; and to establish its credentials as a source, I should say a bit about its genealogy.

The *Dictionnaire des sciences philosophiques* is in fact connected to the *Encyclopédie* by an unbroken lineage, one in which a third work, the *Dictionnaire des sciences médicales*, functions as the intermediary link. In 1768 an aggressive capitalist publisher named Charles-Joseph Panckoucke, having correctly perceived the potential market for less costly versions of the *Encyclopédie* than the first folio, bought the rights to all future editions.[2] By the early 1780s he had expanded his operations beyond reprinting the *Encyclopédie* in cheaper formats and had masterminded the project for the *Encyclopédie méthodique*, an updating of the original *Encyclopédie* and, more significantly, a division of it into forty specialized series, including agriculture, chemistry, jurisprudence, medicine, political economy.[3] Panckoucke's decision to introduce the *Encyclopédie méthodique* suggests that the fast-growing corpus of knowledge had, in his view, become too cumbersome to submit to alphabetical organization under a single title, and that a market for more selective slices of the whole could be tapped. In the opening years of the nineteenth century, Panckoucke fils, carrying on the family tradition, brought the *Méthodique* to a still higher

1. See "Prospectus," in *Dictionaires des sciences médicales*, 60 vols. (Paris: C.L.F. Panckoucke, 1812–22), 1:viii. See also "Le lexicographe et l'encyclopédiste,"in *Le siècle des dictionnaires*, ed. Nicole Savy and Georges Vigne, Les Dossiers du Musée d'Orsay, no. 10 (Paris: Editions de la réunion des musées nationaux, 1987), 26–28.

2. On Panckoucke and his entrepreneurial activities, see Robert Darnton, *The Business of Enlightenment: A Publishing History of the Encyclopédie, 1775–1800* (Cambridge: Harvard University Press, 1979).

3. *Encyclopédie méthodique, ou par ordre des matières*, 197 vols. (Paris: Panckoucke, 1782–1832).

level of specialization and conceived of the plan for an encyclopedic dictionary of medicine. Published between 1812 and 1822, the resulting *Dictionnaire des sciences médicales* grew to an imposing sixty volumes and assembled a large stable of authors that included the most prominent physicians of the day. That it also became something of a cultural icon—an emblem of the boiled-down sum total of medical knowledge—is seen vividly in a passage in Flaubert's *Madame Bovary* (1857). Enumerating the contents of the consulting room of Charles Bovary, licensed *officier de santé,* Flaubert observes: "Volumes of the 'Dictionary of Medical Science,' uncut, but the binding rather the worse for the successive sales through which they had gone, occupied almost alone the six shelves of a pinewood bookcase."[4]

The considerably smaller *Dictionnaire des sciences philosophiques* was constructed on the model of its medical forebear: not only was there a direct parallelism in the titles of the two works and in the representations of their collective authorship ("By a Society of Physicians and Surgeons," itself a variant on the phrase "By a Society of Men of Letters" used in the *Encyclopédie,* became "By a Society of Philosophy Professors"), but the firm of Panckoucke kept its hand in the enterprise, serving in this case not as publisher but as printer.[5]

The anonymous "authors' preface" says a good deal about the intellectual motivation behind this compendium. Most conspicuously, these spokesmen for philosophy betray a strong sense of embattlement. They allude darkly to the "abundant self-interested hatreds [that] rise up against [philosophy]" and to the widespread allegations that "after three thousand years, [philosophy] can still do no more than haltingly address frivolous questions, being condemned on more serious matters to the most shameful and incorrigible confusion." But while on the defensive, the philosophy professors also attempt to mount an offense, using the occasion of the publication of the first volume of the *Dictionnaire* to declare that the field of philosophy has been constituted as a science. Their opening sentence states proudly, "When, after much trial and error and many vicissitudes, by dint of struggles, conquests, and the vanquishing of prejudice, a science finally manages to constitute itself, it then faces an easier, more modest, but not less useful task: it must in some fashion conduct its own inventory." That "inventory" is, of course, the *Dictionnaire* itself. In keeping with the "ex-

4. On this point, see Lawrence Rothfield, *Vital Signs: Medical Realism in Nineteenth-Century Fiction* (Princeton: Princeton University Press, 1992), 17. The quotation is from *Madame Bovary,* trans. Paul de Man, Norton Critical Editions (New York: W. W. Norton, 1965), 22–23.

5. See the title page of volume 1, which gives the publisher as L. Hachette; the facing page indicates "Impr[imerie] Panckoucke."

ample given by the last century," it seeks not to reproduce the strenuous processes of reasoning by which philosophy arrived at its truths but to give a simple exposition of those truths for purposes of dissemination. It will "spread them out beneath everyone's eyes," inviting "each person, whether savant or man of the world, to draw from [the *Dictionary*], effortlessly and according to the needs or even whims of the moment." The time had come, they proclaim, for philosophy to cross the threshhold of the schoolroom and enter the public realm.[6]

Philosophy, with its tradition in the West going back at least to Plato, a newly constituted science in 1844? Philosophy, in the wake of the Enlightenment, just becoming matter for public consumption? Clearly, if implicitly, the authors of the preface are addressing the new, peculiarly nineteenth-century condition of French philosophy, when the prevailing definition of science had changed and the materialist trends associated with medicine and empiricist philosophy had threatened to subsume mental phenomena under the laws of biology and thus to put philosophy out of business altogether.[7] Under this protopositivist and, by the 1830s, bona fide positivist barrage, philosophy lost the high status and the currency in the world of public affairs that it had enjoyed during the Enlightenment, when to be a *philosophe* was an honored calling. Now, as the publication of the *Dictionnaire* indicated, philosophy was attempting to *re*constitute itself, not as a master science but simply as one specialized science among many. Whatever its own epistemological commitments, it had not failed to notice the prestige attached to such observational sciences as medicine, and it was sufficiently savvy and opportunistic to deck out its own *Dictionnaire* with all the formal trappings of the famous *Dictionnaire des sciences médicales*, thus tacitly asserting a full parity between philosophy and medicine.[8]

It is in the context of this scaling down of French philosophy first for purposes of survival and later for purposes of renewed expansion, that the "self" did not so much freshly emerge as a scientific object in France as it be-

6. "Preface des Auteurs," *Dictionnaire des sciences philosophiques,* 6 vols. (Paris: L. Hachette, 1844–52) 1:v–vi.

7. I discuss the beleaguered situation of early nineteenth-century French philosophy in *Console and Classify: The French Psychiatric Profession in the Nineteenth Century* (New York: Cambridge University Press, 1987), chap. 7.

8. The opening passages of the "Prospectus" for the *Dictionnaire des sciences médicales* focus on the definition of a science (a collection of facts given by Nature and a collection of the rules governing them, which are the discovery of the human intellect and are geared to intervention in the facts) and the problems characteristic of medicine as a science (its facts are so plentiful and unstable that it must multiply its rules, thus undercutting their certainty). The model of the *Dictionnaire des sciences médicales* was, in other words, one of a self-conscious effort to make medicine conform to what can be called the positivist ideal.

came for the first time a *salient* scientific object, much discussed and, in influential quarters, much insisted upon and even lionized. But what kind of scientific object is the self, anyway?

Just as I would agree with Marcel Mauss that "there has never existed a human being, who has not been aware not only of his body, but also at the same time of his individuality, both spiritual and physical,"[9] so I would hazard that the scrutiny of the contours of that awareness and the development of specialized and in some manner "scientific" vocabularies to describe it is also a ubiquitous phenomenon. I would also readily assent to Mauss's claim about the mutability of the self, its assumption of significantly different forms in different societies and time periods. But beyond this point of (to my mind) axiomatic clarity, the issue becomes murky. Many competing systems of classification, each arrayed along a temporal axis, have been proposed to trace the conceptual varieties of selfhood, personhood, subjectivity—terms that, moreover, may or may not be regarded as interchangeable by those who employ them.[10]

Rather than adopting one of these preexisting schemes, or recklessly ad-

9. Marcel Mauss, "A Category of the Human Mind: The Notion of the Person; the Notion of the Self" (1938), trans. W. D. Halls, in *The Category of the Person: Anthropology, Philosophy, History*, ed. Michael Carrithers, Steven Collins, and Steven Lukes (Cambridge: Cambridge University Press, 1985), 1–25, at 3.

10. Thus, for example, Mauss himself believed the originary form of the self, found alike among indigenous Australian and Northwest American tribes, to be the persona, role, or mask, a concept referring to its possessor's social function. According to Mauss's unabashedly progressive account, this primitive form evolved in ancient Rome into the self as a bearer of legal rights and obligations, was then enriched by the Stoics with a consciousness of good and evil and by the early Christians with a metaphysical aspect and finally, sometime during the eighteenth century, achieved its current form as a self-knowing psychological being. See Mauss, "A Category of Mind." Charles Taylor found the modern Western "self" or "identity" to be triply characterized by an inwardness, or sense of having inner depths, that began its career with Augustine; by an affirmation of the ordinary life of work and family as the arena for the realization of selfhood, a development that awaited the Protestant Reformation; and by a late eighteenth-century Romantic-inspired belief in the voice of nature as expressive of the authentic self. See Taylor, *Sources of the Self: The Making of the Modern Identity* (Cambridge: Harvard University Press, 1989). Michel Foucault offered yet another rendition of chronology and terminology. He distinguished between the "self," which had in his view existed as a category at least since classical antiquity, and the "subject," a distinctly modern invention. The former, fundamentally ethical and aesthetic in nature, was capable of obtaining truth only if well cared for by its owner. The latter, introduced by Descartes, could obtain truth by seeing what was evident and was thus functionally equivalent to all other subjects. With the mid seventeenth-century advent of the subject, in other words, evidence supplanted the vagaries of "care of the self" as the road to truth, and the enterprise of modern science was made possible. See Foucault, "On the Genealogy of Ethics: An Overview of Work in Progress," in *The Foucault Reader*, ed. Paul Rabinow (New York: Pantheon, 1984), 340–72, esp. 371–72. To judge only from the learned contributions of Mauss, Taylor, and Foucault, the possibilities for dating the coming into being of the self would appear myriad, perhaps endless.

vancing one of my own, I will for purposes of this essay embrace a minimalist theoretical attitude toward the self. I will regard it as a perennial scientific object whose form and degree of cultural salience are prone to extremely wide variation. What is noteworthy about the early nineteenth-century French moment with respect to the self, then, is not its absolute novelty but rather the heightened, almost obsessive attention paid to that object and the dramatic shift in the relevant vocabulary. The sense of localized everyday selfhood denoted by the humble vernacular *moi*—as opposed to the high-flown *âme*—came to be intensively theorized. The two quotations that begin this essay attest to the vast difference in the treatment accorded the *moi* in the *Encyclopédie*, where a few brief paragraphs suffice to cover a suspect term whose only meaning is grammatical, and in the *Dictionnaire des sciences philosophiques*, where the same entity has become the designated heir of the Cartesian *cogito* and Descartes himself is assigned a role in initiating the transformation. That difference in turn makes plain the vast conceptual distance that the *moi* has traveled in the space of less than a century.

The same point is brought home by tracing the evolution of the term *âme*, meaning in English "soul," "spirit" or "mind." The long and complicated article "Ame" in the *Encyclopédie* defines that traditional category as "a principle endowed with consciousness and feeling" and goes on to ponder, with reference to Western philosophy from the ancient Egyptians and Greeks forward, whether soul is a pure quality or a substance, how it is related to the divinity, and in what sorts of beings it resides. The article never even mentions the *moi* and certainly never suggests the workaday personal pronoun as a synonym for the *âme*. The same article in the *Dictionnaire des sciences philosophiques* is, by contrast, fixated on the *moi*. It starts with a basic distinction between modern philosophers like Descartes who, we are told, use the term *âme* to refer to the substance of the human self *(moi humain)*, and ancient and medieval philosophers, who used it in an extended and etymologically more correct sense to mean the principle of life and movement in organized bodies. It then goes on to fine-tune the "modernist" view, stipulating that while *âme* and *moi* are certainly overlapping categories, not entirely distinct from one another, they are not coterminous. The *moi*, characterized by reflexivity, self-consciousness, and generally expanded faculties, represents a decided development of the "spiritual principle," or *âme*, and occupies only a portion of its conceptual space.[11]

11. "Ame," *Encyclopédie*, 2:294–322; and "Ame," *Dictionnaire des sciences philosophiques*, vol. 1 (1844), 81–92. Hence, according to the article "Moi" in the *Dictionnaire des sciences philosophiques*, the critical, but ultimately insufficient move made by Descartes in

The early nineteenth-century French philosophers who assumed the related tasks of the disciplinary defense of an embattled philosophy and the foregrounding of the self were the Sorbonne *maître* Victor Cousin and the members of his carefully groomed school. The latter exclusively comprised the authorship of the *Dictionnaire des sciences philosophiques,* turning that work into the repository and codification of Cousinian orthodoxy. Since Cousin was a derivative philosopher but an academic entrepreneur of true genius, we can assume even at this early stage of our investigation that the coming into being of the self as a scientific object involved many extraintellectual considerations, especially those related to politics, both national and professional. The mere fact that the title page attributed the creation of the *Dictionnaire* to a society of philosophy professors indicates that philosophy production had acquired institutional moorings in the post-Revolutionary period, that its locus had shifted from the independent freelance Enlightenment *philosophe* to the salaried functionary of the new state educational system. The thoroughness with which the *moi* had invaded and overrun Cousinian philosophy, as well as the extent to which the term in its newly technical sense had penetrated the general culture, can be seen in the article "Moi" in the *Dictionnaire de la conversation et de la lecture,* an all-purpose reference work intended to meet the needs of the bourgeois household.[12] Prepared by a minor Cousinian philosopher, the article begins with a flourish: "That word [*moi*], which formerly belonged only to the domain of grammar and was nothing more than the most notable of pronouns, has become, after the word 'God,' the substantive noun par excellence. It now plays, and justly

---

the sixth meditation; see above, the second epigraph to this paper. Descartes wrote, "[O]n the one hand, I have a clear and distinct idea of myself (*moi-même*) insofar as I am simply a thinking, non-extended thing; and on the other hand I have a distinct idea of my body, insofar as this is simply an extended, non-thinking thing. And accordingly, it is certain that I, that is, my soul, by which I am what I am, is really distinct from my body and can exist without it." The English translation comes from Descartes: *Selected Philosophical Writings,* trans. John Cottingham, Robert Stoothoff, and Dugald Murdoch (Cambridge: Cambridge University Press, 1988), 115 n. 2. The note indicates that the phrase equating the *moi* and the *âme* was an addition to the Latin text made by Descartes in the French version—a fact that tends to support the Cousinian point that Descartes was interested in the linguistic innovation of bringing the term *moi* into technical, philosophical usage. Descartes's French reads: "[I]l est certain que moi, c'est-à-dire, mon âme, par laquelle je suis ce que je suis . . . "

12. It also prided itself on toeing no party line but instead giving voice through its different articles to controversy and divergent opinions. See the untitled preface to the *Dictionnaire de la conversation et de la lecture,* 52 vols. (Paris: Beilin-Mondar, 1832–39) 1:3. A glance at the list of principal collaborators on the page facing the title page confirms this claim. It includes such representatives of opposing camps as Victor Cousin and François Guizot, on the one side, and F.-J.-V. Broussais and Armand Marrast on the other.

so, a powerful role in philosophy. In fact, we could say without exaggeration that it epitomizes all of philosophy."[13]

IS THERE A SELF IN THIS MENTAL APPARATUS?

The article "Moi" in the *Dictionnaire des sciences philosophiques* suggested a temporal sequence in which Descartes's *moi* was directly metamorphosed into *le moi* of Cousin. But that capsule narrative omitted an intervening dialectical stage that in fact bore primary causal responsibility for the advent of the Cousinian self: the eighteenth-century vogue in France of Condillac's sensationalist psychology. While Condillac functioned in effect as the French Locke, significant differences separated the two philosophers. In the context of the present discussion, it is particularly noteworthy that, writing a half century before Condillac, Locke had explicitly posed and had wrestled at length with the problem of the implications of a sensationalist epistemology for personal identity, or the unity and coherence of the self.

If, Locke asked in a chapter written for the second edition of *An Essay on Human Understanding* at the express request of his friend William Molyneux,[14] we discard the Cartesian contention that the indivisibility of the self or thinking substance is a self-evident truth and postulate instead that all our mental contents are derived from discrete sensory impressions, then what is the ground of selfhood, of the "sameness of a rational Being" that persists through space and time? In a somewhat rambling argument, Locke located that ground in a combination of consciousness and memory. We cannot think, feel, sense, or will without being aware that we do so, he asserts, and this consciousness inevitably accompanying our mental processes "makes every one to be, what he calls *self;* and thereby distinguishes himself from all other thinking things, [and] in this alone consists *personal Identity.*" Still, in order to ensure this identity, memory must be added to consciousness because, as Locke readily concedes, consciousness is discontinuous, "being interrupted always by forgetfulness, there being no moment of our Lives wherein we have the whole train of our past Actions

13. A.-Jacques Matter, "Moi," Dictionnaire de la conversation, 38:259–61, at 259.

14. See Henry E. Allison, "Locke's Theory of Personal Identity: A Re-examination," in Locke on Human Understanding: Selected Essays, ed. I. C. Tipton (Oxford: Oxford University Press, 1977), 105–22, esp. 106 and 106 n. 3. In a letter of 2 March 1693, Molyneux, replying to Locke's request for "any new heads from logick or metaphysicks to be inserted," suggested that a discussion of the principium individuationis be included in the new edition. See The Correspondence of John Locke, 8 vols. (Oxford University Press, 1976–89), letter 1609, 4:647–51, esp. 650.

before our Eyes in one view." Sometimes sheer absorption in present thoughts momentarily obliterates our awareness of our past selves; once a day consciousness itself is suspended in sleep. Hence memory must be enlisted to fill in the gaps and restore that continuity of consciousness called self. [15]

But Locke probes further, posing as a more strenuous objection to the coherence of the concept of selfhood the possibility that certain portions of lived experience may be lost beyond retrieval—those, for example, that occur when an individual is drunk or in a state of somnambulism. Locke now salvages his basic contention about the persistence of the self through recourse to what he terms a "forensick" conception. A court of law, he says, lacks any sure means of assessing the authenticity of a plea that an accused should be found not guilty for reason of drunkenness or sleepwalking. Hence convention deems that the court avoid the issue entirely, ignoring any alleged gap in consciousness and, hence, any lapse in moral responsibility attendant upon it, and punishing the person in question on the purely factual basis of the crime committed by his hand. But this pragmatic arrangement lasts only as long as our temporal existence, becoming irrelevant on the Day of Judgment, "wherein the Secrets of all Hearts shall be laid open . . . [and] no one shall be made to answer for what he knows nothing of; but shall receive his Doom, his Conscience accusing or excusing him." Presumably, then, the postulation of the unity of the self is for Locke a necessary expedient to sustain the concept of moral responsibility in daily life in face of the ultimate imperfection of our terrestrial knowledge about other people's states of consciousness. [16]

While Locke recognized the magnitude of the problem on his hands and spun out a long and tortuous argument attempting to resolve it, his French successor dealt surprisingly casually with the unity of the self. In fact, his first psychological treatise, the *Essai sur l'origine des connaissances humaines* (1746), mentioned the self only in passing. When describing the

15. John Locke, *An Essay Concerning Human Understanding*, ed., Peter H. Nidditch (Oxford: Clarendon, 1975), 2.27.9–10, pp. 335–36.

16. Ibid., 2.27.20–26, at 344, 346. It should be pointed out that Locke does not assimilate madness to drunkenness and somnambulism. In his view madness qualifies both as a valid legal reason for exemption from responsibility for a criminal act and as an instance of "duplication" of the self. As Locke observes, "[I]f it be possible for the same Man to have distinct incommunicable consciousness at different times, it is past doubt the same Man would at different times make different Persons; which we see, is the Sense of Mankind . . ., Humane Laws not punishing the Mad Man for the Sober Man's Actions, nor the Sober Man for what the Mad Man did, thereby making them two Persons; which is somewhat explained by our way of speaking in English, when we say such an one *is not himself*, or is *besides himself.*" See pp. 342–43.

generation of the various mental operations from the primal capacity for sensation that, in his view, gives birth by stages to our whole mental apparatus, Condillac noted that the operation of reminiscence enables us to preserve the sequential linkage between perceptions that we have experienced at different moments in time; as such, he opined, reminiscence is a necessary condition for a persisting, unified self. "If this linkage were each night interrupted, I would so to speak begin a new life each day, and no one could convince me that today's *moi* was the *moi* of the day before." Condillac then went on to analyze two distinct aspects of reminiscence, one that "makes us recognize our own being," the other that "makes us recognize the perceptions that are there repeated." In other words, he tersely predicated selfhood on memory, making memory its sufficient condition; but he failed utterly to acknowledge the immensity of that claim.[17]

By the time of his second psychological treatise, the *Traité des sensations* (1754), Condillac was somewhat more deliberate in his treatment of the self, but he still disposed of that topic promptly and without obvious intellectual agony. Condillac's hypothetical case history of a statue gradually endowed with each of the five senses included, in book 1 (in which the statue's exclusive sensory organ is his nose), a succinct chapter entitled "Of the *Moi*, or of the Personality of a Man Limited to the Sense of Smell." Here once again, selfhood and memory are tightly bound. The statue, we are told, could not say "I" at the moment when it first experienced an odor. "Insofar as a being does not change, it exists without any folding back on itself *[retour sur lui-même]*. But insofar as it changes, it judges that it is still in some manner the same as it previously was, and it says *moi*." Condillac then recasts this point in a stunning definition of the self: the "*moi* is nothing but the collection of the sensations that [the statue or person] experiences and of those that memory recalls to it. In a word, it is the simultaneous consciousness of what [the statue or person] is and the memory of what it was."[18]

17. Condillac, *Essai sur l'origine des connoissances humaines,* 1.1.1.15, in *Oeuvres philosophiques,* 3 vols., ed. Georges Le Roy (Paris: Presses universitaires de France, 1956), 1:14. In equating Condillac's "reminiscence" with memory here, I am taking certain liberties with the subtlety of his categories. In the *Essai,* "mémoire" is generated after reminiscence (see 1.1.2) and is a more sophisticated mental operation; while reminiscence merely preserves past perceptions, memory processes past perceptions to which we have affixed linguistic signs and hence enables us to retrieve those perceptions whenever we wish.

18. Condillac, *Traité des sensations,* 1.6, in *Oeuvres philosophiques,* 1:238–39. It should be noted that while this chapter includes the word "*moi*" in its title, the *moi* is treated in the *Essai* in a chapter that, implicitly denying the importance of that concept, is called "De la perception, de la conscience, de l'attention, et de la reminiscence."

The brevity and nonchalance of Condillac's handling of the self is quite remarkable in view not only of Locke's extensive discussion but also in view of the controversial nature of the concept, the barrage of criticism that Locke's account had sustained from all sides since its first appearance in 1694. In Britain, Bishop Butler and Thomas Reid attacked Locke's argument as circular, presupposing what it allegedly proved by defining personal identity as consciousness of personal identity.[19] Noting the brouhaha surrounding personal identity (it had become, he wrote, "so great a question in philosophy, especially of late years in England, where all the abstruser sciences are study'd with a peculiar ardour and application"), David Hume offered in his *Treatise of Human Nature* (1739–40) a more elaborate critique of Locke—and one all the more devastating because, unlike Butler and Reid, he shared Locke's sensationalist epistemology. Postulating that there "must be some one impression, that gives rise to every real idea," Hume argued that no such single impression could possibly be found to undergird the idea of a self. Ordinary reflection revealed that far from being unitary, we are all "nothing but a bundle or collection of different perceptions, which succeed each other with an inconceivable rapidity, and are in a perpetual flux and movement." Hume therefore concluded that the self was a "fiction" or "artifice." As such, it was a construction of that most unreliable of human mental faculties, the imagination, here aided by our characteristically sloppy perceptual processes, which overlooked slight alterations and pronounced sameness where none existed.[20]

In France, the critique of the Lockean self was undertaken by Catholic Cartesians intent upon exposing the immoral and atheistic implications of empiricism.[21] Preaching to the converted, these critics tended to be more declamatory than argumentative. Their main point was that a sensationalist epistemology could never satisfactorily ground a self recognizable as such to a Catholic. Hence recourse to a philosophy that postulated a self given all at once as a spiritual substance, instead of being assembled serially from material sensations, was necessary. The Reverend Father Hayer asserted that while the alleged unity of *physical* bodies was "only an abstrac-

19. See Allison, "Locke's Theory of Personal Identity," 112.

20. David Hume, *A Treatise of Human Nature,* ed. L. A. Selby-Bigge and P. H. Nidditch, 2d ed. (Oxford: Clarendon, 1978), 1.4.6, "Of Personal Identity," esp. pp. 251, 252, 259, for the passages quoted. Hume very much favored the vocabulary of fiction as applied to the self; see also p. 259: "The identity, which we ascribe to the mind of man, is only a fictitious one"; and p. 262: "All the disputes concerning the identity of connected objects are merely verbal, except so far as the relation of parts gives rise to some fiction . . . of union."

21. On this point, see the excellent discussion in R. R. Palmer, *Catholics and Unbelievers in Eighteenth-Century France* (Princeton: Princeton University Press, 1947), chap. 6, "Soul and Mind."

tion of our minds," in the essentially spiritual creature that is a human being "we find a really and substantially indivisible center, where everything that interests man is brought back to unity." And how, one might ask, do we find this center, which Hayer called the *moi*? Hayer's answer is that the situation simply could not be otherwise: "If for this unique self [*moi*] we substituted a multitude of selves, what strange confusion would result!" The hypothetical multiple individual would be like an "anarchical society" composed of isolated, self-absorbed parts functioning as wholes, each part-whole in perfect ignorance of the needs of the others.[22]

Hayer went on to invoke other proofs of the unified and spiritual nature of the human *moi*, some of which relied—as was typical of this mode of Catholic-Cartesian apologetic—upon the self-evidence of introspective experience. ("Having retreated into a pleasant solitude, solely occupied with the desire of knowing myself, I begin to consider with the eyes of my *âme*, my *âme* itself. That is to say, my *moi*, folding back upon itself, . . . contemplates itself . . . ")[23] Introspection and the psychic reality to which it bears witness were also at the heart of the argument of the abbé de Lignac against Locke's theory of personal identity. In his preface, in which he also articulates his intention to write a book enlisting contemporary philosophy to vindicate the wisdom of the church fathers, Lignac explains and justifies his confident, declamatory tone. "Just as a witness ought to be firm when, before the court, he makes a deposition concerning what he has seen, . . . so ought I to refrain from weighing pros and cons or appearing to have the slightest doubt about the verities I discover."[24] Lignac gave his book a title consonant with that motif—he called it "testimony of the *sens intime*"— and he proceeded accordingly:

> By the *sens intime* of existence, I have always understood, Monseigneur [the cleric to whom the book is dedicated], the consciousness of the identity of our *âme* at all times, under an infinite variety of different ways of being, which the substance of our *âme* sheds without thereby ceasing to be the same person, the same *moi*. This consciousness of my identity I find at the bottom of all my thoughts, sensations, emotions. I sense myself perceiving while perceiving something, as Locke says. *But what Locke does not say is that, when perceiving the letters that I am now tracing, I sense myself as the same being*

22. Hubert Hayer, *Le spiritualité et l'immortalité de l'Ame*, 2 vols. (Paris: Chaubert, 1757), 2:1–3.

23. Ibid., 2:6–7. Hayer also offers a spiritualist response to Locke's argument about the discontinuity in the self introduced by deep sleep; see pp. 2:13–18.

24. Le Large de Lignac, *Le témoignage du sens intime et de l'expérience opposé à la foi profane et ridicule des Fatalistes modernes* (Auxerre: Fournier, 1760), preface, 1:n.p.

who received his first writing lessons so many years ago. If this experience
. . . is common to all men . . . I am correct in insisting that our *âme* is a substance and correct in defining it as the consciousness of identity. For substance is that which remains the same whatever form it assumes, whatever
modification it is subjected to.[25]

But, despite the aggressive tone of these Catholic critics, Condillac did
not engage them in a debate about personal identity. Nor did he engage
Hume, whose *Treatise* was never translated into French during the eighteenth century.[26] With respect to the former, he seems to have shied away
from polemics on religious matters. His reply to Lignac's critique of his
Traité des animaux, for example, counsels the Catholic apologist simply to
accept or reject a philosophical argument on its internal merits, bracketing
its doctrinal consequences. A valid argument, he promises, will never harbor danger for religion because "Truth cannot be contrary to truth."[27]

With respect to Hume, whose *Treatise* he probably never read, Condillac
is in the odd, almost perverse position of appearing to side with Locke about
the cogency of a self founded on sensations while sounding a great deal like
Locke's Scottish detractor. The very same image of the mind as a "collection" of fleeting sensations and perceptions, which Hume deliberately employs to damn Locke's theory of personal identity, is employed by Condillac
in a completely neutral register, simply to describe the *moi* as Condillac
believes it is, without commentary on the cogency or absurdity of the concept. For Hume, the presumed fact that the self is nothing but an arbitrary
collection of sensations and their by-products reveals the scandalous back-
ruptcy, the fictive nature of Locke's claims about personal identity. But for
Condillac the self as an empty space, as the theatrical stage (to use Hume's
metaphor)[28] where a succession of sensory events are momentarily enacted, seems all the self that he could ever envision. Condillac evinces no
discomfort, certainly no horror, with the flimsiness and lack of grandeur of
such a self. Not inclined to dwell on the self in the first place, he seems obliv-

25. Ibid., 1:392–93, my italics.
26. In the early nineteenth century, Victor Cousin would regret that it had taken so long
for Hume's corrosive argument against the sensationalist self to reach and be appreciated in
France; in 1816, there was still no French translation of Hume's *Treatise.* See Cousin, *Premiers
essais de philosophie,* 3d ed. (Paris: Librairie Nouvelle, 1855), 57–58. Cousin's comments were
made as part of his 1816 course at the Sorbonne.
27. "Lettre de M. l'Abbé de Condillac à l'Auteur des Lettres à un Amériquain," reprinted
from the *Mercure de France,* April 1756, and bound with the Bibliothèque Nationale, Paris,
copy of Condillac, *Traité des animaux* (Amsterdam, 1755), p. 10.
28. See Hume, *Treatise of Human Nature,* 1.6, p. 253: "The mind is a kind of theatre, where
several perceptions successively make their appearance; pass, re-pass, glide away, and mingle
in an infinite variety of postures and situations."

ious to the controversy swirling around his formulations—much like the intrepid truth seeker, the philosopher committed to exploring the limits of human knowledge, that he would later depict in his reply to Lignac.[29]

It remained for Cousin and his school to inject the (in their view) requisite note of horror, to reveal the mental apparatus of sensationalist psychology as shamefully lacking in a proper self. They would do for Condillac what the eighteenth-century Catholic apologists had done for Locke. But, while they would rehabilitate many of the apologists' old keywords, like substance and *sens intime*, they would meet with notably greater success. Before turning to the pivotal conceptual move of the Cousinians, let me examine the terminology of the eighteenth-century phase of the story: the transition from *âme* to *moi*.

As the citations from the eighteenth-century French texts, both Cartesian and sensationalist, suggest, the term *moi* was used with some frequency before the Cousinians swept the philosophical field. It was not, however, used with any systematicity. On the one hand, Condillac's 1746 *Essai* implicitly defines it as that aspect of the *âme* that has cognizance of its persisting sameness and is the locus of personal identity; the chapter in which Condillac introduces the *moi* as a corollary of the mental operation of reminiscence is included in a section of the book called "The Analysis and Generation of the Operations of the *Ame*," and this organizational device appears to designate the *moi* as a subset of the *âme*.[30] On the other hand, both Hayer and Lignac, as quoted above, use *moi* and *âme* as synonyms and seem simply to equate the spiritual substance with the sense of personal identity. And at least one eighteenth-century figure scrupulously avoided using the term *moi* to mean enduring personal identity: Locke's first French translator, the Huguenot emigré to England, Pierre Coste. In a fascinating footnote to his translation of the chapter "Of Identity and Di-

29. That is, I think, the point of Condillac's long footnote to his discussion of the *moi* in the *Traité des sensations*. He cites a passage from Pascal that poses the question of whether we love other persons for their particular mental and physical qualities or for the abstract conception of the substance of their soul. Pascal insists that human love is confined to the former, and Condillac quotes him as saying, "We never love the person, then, but only the qualities; or, if we love the person, then it is the assemblage of qualities that makes the person." Commenting on Pascal's text in the same footnote, Condillac then denies that this "assemblage of qualities" is really what Pascal takes a person to be. He concludes, "In Pascal's meaning of the term [person or *moi*], only God can say *moi*." *Traité des sensations*, 1.6, p. 1:239 n. 1. In other words, Condillac is fully aware that his own definition of the *moi* as a "collection of sensations" does not exhaust all the meanings of that term in the language of his day or, perhaps, even capture the most desirable meanings. But in keeping with the principle of epistemological modesty that undergirds his work, he contends that the *moi*-collection is the only self knowable by human intelligence.

30. Condillac, *Essai*, title of section 2 of part 1.

versity" of Locke's *Essay*, Coste explained why he used the terms *le soy* and *soy-même* to translate Locke's "self." Part of the reason for his choice was the indelible coloration that Pascal had, in Coste's view at least, imparted to the term *le moi*; the other part was Locke's own alleged neologizing in English:

> The *moi* of Monsieur Pascal in some manner authorizes me to make use of the words *soy, soy-même*, to express the sentiment that each one has within himself that he is the same. Or, better put, I was obliged to do so by an indispensable necessity, for I would not know how otherwise to express the meaning of my author, who has taken a parallel liberty in his language. The roundabout terms I would have to employ on this occasion would clutter the prose and perhaps render it completely unintelligible.[31]

In some famous passages in his *Pensées*, Pascal used the noun *moi* to refer to the fallen self that had not yet found God. "The *moi* is hateful *(haïssable)*," he declared bluntly. Its hatefulness derived from its exclusive self-love ("it makes itself the center of everything") and from its desire to rule tyrannically over others. One version of the *Pensées* had Pascal pronouncing the rhetorical rule, similar to the one later disputed in the *Encyclopédie* article "Moi," that an *"honnête homme* ought to avoid . . . using the words 'je' and 'moi.'" In Pascal's theological scheme, conversion to the love of God would bring about not merely a forgetfulness of the *moi* but a total annihilation of it.[32] The term *moi* was so thoroughly imbued with these Pascalian associations for Coste that he regarded it as inappropriate to signify the respectable entity, the bearer of moral responsibility, that was the Lockean self. But by 1839 when a new French edition of Locke's *Essay* appeared, a "revised [and] corrected" version of the Coste translation, Locke's "self" was routinely

---

31. See Locke, *Essai philosophique concernant l'entendement humain*, trans. by Pierre Coste from the 4th ed. (Amsterdam: Henri Schlete, 1700), 403 n*. I do not know whether Coste is correct in attributing to Locke the coinage of the noun "self." In any case, Coste engaged in other neologistic gestures in French, for example, translating Locke's "consciousness" as the hyphenated *con-science*, instead of the ordinary *conscience*, in order to stress the Latin etymology of the term and thus make Locke's meaning clearer; see 404 n*.

32. For the well-known passages about the *moi* in the *Pensées*, see Pascal, *Oeuvres complètes*, ed. Jacques Chevalier (Paris: Gallimard/Pléiade, 1954), para. 130, p. 1123 ("Le nature de l'amour-propre et de ce *moi* humain est de n'aimer que soi et de ne considérer que soi . . ."); para. 136, pp. 1126–27 ("Le *moi* est haïssable . . ."); para. 443, p. 1211 ("le *moi* consiste dans ma pensée . . ."). The rhetorical rule attributed to Pascal is found in Victor Cousin, *Des Pensées de Pascal, rapport à l'Académie francaise sur la nécessité d'une nouvelle édition de cette ouvrage* (Paris: Ladrange, 1843), 45, which also quotes Pascal as saying "Christian piety annihilates the human *moi*" and "human civility hides and suppresses it." For a discussion of the annihilation of the *moi* in Pascal's theology, see Henri Gouhier, *Blaise Pascal: Conversion et apologétique* (Paris: Vrin, 1986), 49–53.

rendered as the *moi*. This change no doubt owed a good deal to the Cousinian philosophical revolution of the intervening decades.[33]

THE COUSINIAN PHILOSOPHICAL OFFENSIVE

As he often noted, Victor Cousin articulated his brand of philosophy at a critical moment in the history of France. Influenced by Hegel, whom he had visited in Germany, and passionately convinced that all philosophical practice was historically situated, he never intended his own philosophy as mere intellectual tinkering. Rather he conceived of it as a vehicle for the reconfiguration of French society and politics in the wake of the upheavals of the 1789 Revolution, a revolution whose liberal principles he basically affirmed but whose episodes of disorder he deplored. To the famous lament about the origins of the Revolution, that it was "the fault of Voltaire, the fault of Rousseau," Cousin would probably have made the emendation that it was really the fault of Condillac and of the sensationalists in general. As he wrote in 1826 of "that sad philosophy": "It is an incontestable fact that, in eighteenth-century England and France, Locke and Condillac replaced the great antecedent schools and that they have reigned supreme until today. Instead of being irritated by that fact, we must try to understand it."[34]

In Cousin's view, the unchallenged ascendancy of sensationalism had the disastrous effect of eroding the moral verities that must, if society is to remain stable, serve as a brake on human impulse. The precise source of that erosion was the failure of sensationalism to ground a durable, unified self— one that would bear moral responsibility both as a duty in this life and because its immortality would entail eternal punishment if it strayed. In the Cousinian scheme of things, repairing the self by philosophical means was therefore the linchpin in the project of the post-Revolutionary stabilization of France.

Cousin's *moi*-centered philosophy was linked to his politics not only as means to end; the two also shared formal analogies. After the demise of Napoleon in 1815, Cousin belonged to a group of politician-philosophers, including the future prime minister François Guizot, who believed that the stability of France required ending the country's ideological polarization and forging a deliberately middle-of-the-road path—a *juste milieu*—between

33. *Oeuvres de Locke et de Leibnitz contenant l'essai sur l'entendement humain*, revised, corrected, and annotated by M. F. Thurot (Paris: Firmin Didot, 1839). In bk. 2, chap. 27, compare the different versions of, e.g., para. 23, last sentence (Thurot p. 203, Coste p. 418) and para. 24, first sentence (Thurot p. 203, Coste p. 418).

34. Cousin, "Préface à la premiére édition," *Fragmens philosophiques*, 2d ed. (Paris: Ladrange, 1833), 1–50, at 3, 5.

the egalitarianism of the radical Revolution and the traditional hierarchies of the Old Regime. In politics, this middlingness translated into a peculiarly cautious and conservative brand of liberalism, one that was not only unabashedly antidemocratic but was as much concerned with using authority to prevent excesses of liberty as with safeguarding liberty to begin with. As was noted by Adolphe Franck (the first Jew to become a philosophy professor in France and the faithful disciple of Cousin who undertook the editing of the *Dictionnaire des sciences philosophiques* and penned its article "Moi"), the "mere name of democracy never reached [Cousin's] ears without causing him obvious displeasure."[35] The counterpart of this *juste-milieu* political position was a philosophical position called rational spiritualism or, more usually, eclecticism. It aimed at harmonizing sensationalist philosophy in the manner of Locke and Condillac, and especially its reliance on observation and experience, with a rationalist philosophy that would restore the legitimacy of ontology and metaphysics and thus reinsert human beings into a world of stable, transcendent meanings. Both the political and the epistemological prongs of this *juste-milieu* conception gained hegemonic status under the July Monarchy (1830–48), the constitutional regime with a non-Bourbon king and high property qualifications for voting that emblematized the will to achieve durability through the reconciliation of opposites.

What is significant for our purposes is that, almost immediately upon assuming a public role, Cousin began to hammer out his message about the grandeur of the human *moi* and the inability of a sensationalist philosophy to provide a foundation for that indispensable entity. Indeed, his fixation on the *moi* surfaced from the moment he emerged as a public philosopher and, simultaneously, as a charismatic professor who enraptured the new student generation.[36] He made his entrance onto the public scene in December 1815, when he began a five-year stint as Royer-Collard's substitute as professor of the history of modern philosophy at the Sorbonne.[37] His inau-

35. Adolphe Franck, "Victor Cousin," in *Moralistes et philosophes* (Paris: Didier, 1872), 291–321, esp. 304. All of the unsigned articles in the *Dictionnaire des sciences philosophiques*, of which the article "Moi" was one, were written by Franck as editor of the compendium.

36. See Alan B. Spitzer, *The French Generation of 1820* (Princeton: Princeton University Press, 1987), chap. 3, "Victor Cousin: The Professor as Guru." As one of Cousin's detractors described his lecture style, "Monsieur Cousin . . . speaks like a high priest; his rich intonation, his mobile features, his weighty and cadenced diction, the painful childbirth of a thought that seems to have gestated in his gut—everything he does favors the impression that he makes on his audience." See Armand Marrast, *Examen critique du cours de philosophie de M. Cousin (Leçon par Leçon)* (Paris: Corréard Jeune, 1828–29), 7.

37. See Cousin, *Premiers essais de philosophie*, 3d ed. (Paris: Librairie Nouvelle, 1855). This volume provides a version of the text of Cousin's Sorbonne lectures of 1815–16 as well as course outline fragments for those of 1816–17. First published many years after the lectures

gural lecture had announced a course on the perception of external phenomena. But after several classes on this subject Cousin quickly—and, no
doubt, all the more memorably—reversed direction and devoted the entire
year to the philosophical issue he regarded as "first both chronologically
and in importance, that of the *moi* and of personal existence."[38]

The maiden course treated the history of philosophy exclusively
through this lens. Thus when Cousin turned to Condillac, he warned his
audience not to "expect any general consideration" of that philosopher or
of the eighteenth century. "I will limit myself . . . to all the passages about
the *moi* that can be found in [Condillac's] writings." His criticism was
harsh. Condillac, he charged, "departs from a radically defective hypothesis" and subsequently "gets lost in nihilism." He is able to make the self dependent on something as flimsy and feeble as memory only by confusing it
with another concept, that of the self-identical self. After all, Cousin
pointed out, "memory returns to that which was; if there is nothing prior, it
is mute. It can say 'still me' only after an initial act of intelligence mingled
with consciousness has said 'me.'" Cousin thus corrected Condillac by insisting that the *moi* exists a priori, that it is already there at the first sensation and the first glimmerings of consciousness. Only its utter priority, its
foundational nature, enables a more highly elaborated entity, the self-identical *moi* that endures through space and time, to come into existence at the
moment of the second sensation.[39]

The *moi* is a substance, Cousin furthermore insists, controverting
Condillac; but, pace Spinoza, it is not a substance that can be defined ax-
iomatically. "Metaphysics," Cousin states, "is not a part of mathematics. It
is instead a science of observation, like physics or the natural sciences."
Hence, Cousin's eclecticism will not be a simple reprise of the seventeenth-
century systems; it will return to metaphysics while inflecting it with the
modality of observation emphasized by the eighteenth-century empiricists. Cousin's insistence on self-as-substance is also marshalled against
Condillac's definition of the self as a collection of sensations, a collection
that, Cousin says, paraphrasing Condillac, is located in an indeterminate
"somewhere" and is nothing but a "logical and grammatical subject," a
"sign" affixed to an assemblage of floating qualities that is "imagined" as a
subject. By contrast, Cousin's substantial self is no nebulous, jerry-built

---

were delivered, these materials were drawn, Cousin tells us, both from his own yellowing and
barely legible notes and from the notes taken by students; see "Avertissement de la 2e édition
de 1846," 1–2.

38. Cousin, *Premiers essais,* 24.

39. Ibid., 128–29, 132, 134, 138.

somewhere; as "common sense and the entire human race" attest, it is a "real subject."[40]

The core of Cousin's observationally based metaphysics, of that "so sought-after alliance of metaphysical and physical science," was what he called the "psychological method." Descriptions of that method lard his teaching like incantations. Here is a description from 1826:

> The psychological method consists in isolating oneself from any other world but that of consciousness in order to establish and orient oneself there, where everything is real but where the reality is exceedingly diverse and delicate. The talent for the psychological consists in voluntarily inserting oneself into this entirely interior world, in giving oneself to oneself as spectacle, and in reproducing freely and distinctly all the phenomena [faits] that, in the circumstances of ordinary life, are thrown up only in an accidental and confused manner.

The rule of this interior observation was totality: undertaken without bias, observation "must be complete, must exhaust its object, and can be allowed to stop only when there are no phenomena [faits] left to observe." No wonder, then, that there were "many different levels of depth in the psychological method," as years of practice had taught Cousin.[41] But the method had led to at least one discovery of capital importance. In direct refutation of the doctrine of the *sensualistes*, it had revealed that sensation and its derivatives constituted only one of the categories of the so-called real contents of consciousness. Distinct from and "impossible to confound with" sensation, but equally incontestably real, were two additional components: volition (which often existed in combat with sensation) and reason (which alone was capable of supplying such concepts as substance, cause, time, space, and unity).[42] The three components operated as a seamless ensemble, but they could be teased apart by means of analysis. They were, Cousin noted in a phrase heavy-handed in its neologizing as well as in its religious reference, "a triplicity that resolves itself into a unity and a unity that develops into a triplicity."[43]

The psychological method was for Cousin the key to the philosophic enterprise not only because its supposed scientific rigor as an observational

---

40. Ibid., 134–36.

41. Cousin, "Préface à la première èdition," *Fragmens philosophiques,* 2d ed., 11–12.

42. Ibid., 13–14.

43. Cousin, *Introduction à l'histoire de la philosophie* (Paris: Pichon & Didier, 1828), lesson 5, p. 15 (each of the lessons is separately paginated in this edition). He said much the same thing two years earlier; see "Préface à la première édition," *Fragmens philosophiques,* 2d ed., 38.

practice lent it credibility and cachet in an early nineteenth-century intellectual environment. The method, as the *maître* construed it, also extended well beyond itself, functioning (in another of his favorite phrases) as the "vestibule" to ontology and metaphysics.[44] Indeed, Cousin contended, any fledgling philosopher who began by scrutinizing his own consciousness would soon learn that its so-called triplicity provided a map of the very structure of the universe. "Ontology is given to us in its entirety at the same time as psychology." The three internal elements of consciousness had their external counterparts, voluntary activity translating into mankind, sensibility into nature, and reason into God.[45] Delving deeply inside himself, then, the student of Cousinian philosophy would be quickly propelled outward and upward, arriving at an intimate conviction of time-honored verities.

But how was the *moi* related to the three distinct elements of consciousness? While Cousin did not pose the question so baldly in his earliest teachings,[46] by the mid-1820s he made it clear that volition was the stuff of selfhood. "The will alone is the person or the *moi*," he announced in 1826, only to reiterate a few sentences later, "Our personality is the will and nothing more."[47] In contrast to those "movements that external agents determine in us, despite ourselves, we have the power to initiate a different kind of movement," one that "in the eyes of consciousness, assumes a new character." We "impute [such a movement] to ourselves," consider ourselves as its cause; indeed it serves for us as the very origin of the concept of cause.[48] Voluntary activity, be it spontaneous or reflective, is that element of consciousness that we perceive as our own; it belongs to us. It exists in a "foreign world, amidst two orders of phenomena [the sensible and the rational] that do not belong to us, that we can perceive only on the condition of separating them from ourselves."[49]

The motif of belonging and not belonging was one that Cousin mined

---

44. For the vestibule metaphor, see, e.g., Cousin, *Introduction à l'histoire de la philosophie*, lesson 13, p. 14. Two years earlier, Cousin employed a slightly different version of the metaphor: "Psychology is thus the condition for and, as it were, the vestibule of philosophy"; see *Fragmens philosophiques*, "Préface à la première édition," 12.

45. Cousin, "Préface à la première édition," *Fragmens philosophiques*, 2d ed., 39.

46. See "Du fait de conscience," a brief excerpt from Cousin's 1817 Sorbonne course, reprinted in his *Fragmens philosophiques*, 2d ed., 242–52. This text discusses the three components of consciousness but does not explicitly locate the term *moi* with respect to them. Cousin does, however, indulge his predilection for the Fichtean term *non-moi* as a synonym for the external world as known through sensation.

47. "Préface à la première édition," *Fragmens philosophiques*, 2d ed., 17.

48. Ibid., 25.

49. Ibid., 17.

extensively. It formed the basis of his celebrated concept of "impersonal reason."[50] As he instructed the students in his 1828 course, "Your intelligence is not free. . . . You do not constitute your reason, and it *does not belong to you.*" There was nothing "less individual" or "less personal" than reason. If the products of reason were merely personal, imposing them on others "would be an exaggerated form of despotism." Instead, the "universal and absolute nature" of reason obliged everyone to bend to its dictates. We are entirely within our rights when "we declare entirely crazy *(en délire)* those who do not accept the truths of arithmetic or the difference between beauty and ugliness, justice and injustice."[51] In other words, Cousin deployed the claim that reason does not "belong to us" (but rather to God) as a powerful argument in favor of common standards and values and against the kind of social and political contestation that bred instability and revolution.

Conversely, the claim that voluntary activity, or the *moi,* did indeed "belong to us" provided Cousin with an equally powerful argument in favor of private property. It is worth noting that both Locke and Cousin provided a philosophical blueprint of the self with reference to its legal implications. Locke's forensic reference was, as discussed above, the criminal law. He emphasized the pragmatic need to postulate a continuous self-identical self to ensure the just punishment of wrongdoers. Cousin's forensic reference was, on the other hand, the civil law. He stressed the inextricable intertwinement of the theory of the self with the right to private property.

As Cousin declared in his 1818–19 Sorbonne lectures on the history of moral philosophy, the "first and most intimate development of the free *moi* is thought; all thought, considered within the bounds of the individual sphere, is sacred." Its quality of inviolability derived from the transfer to it of an essential quality of the self—that it belongs to us. Thus, in keeping with the principle that "our original property is ourselves, our *moi,*" Cousin asserts that the "first act of free, personal thought is the first act of property." This rhetorical move then enabled him to make sweeping assertions about property in general. Property was not, he assured his student audience, based upon mere convention; after all, conventions could be annulled by the parties who had agreed to them. Rather it was founded on a "superior principle—that of the sanctity of liberty." Property consisted in the "free imposition of the personality," that is, of volitional activity, "on

50. "Reason is impersonal by its nature," Cousin declared; see ibid., 18. One of Cousin's students, who wrote an entire book on impersonal reason, regarded it as among Cousin's "glories" as well as the "link uniting the whole eclectic school"; see Francisque Bouillier, *Théorie de la raison impersonnelle* (Paris: Joubert, 1844), ii, iv.

51. Cousin, *Introduction à l'histoire de la philosophie,* lesson 5, pp. 9–10; emphasis added.

things." Once acquired, those things "participate in some manner in my personality." They obtain rights by this relationship or, what is the same thing, "I have rights in them," so that "by augmenting [my] property, [I] extend the circle of [my] rights." The natural right to property thus rested on the principle of human liberty, and that natural right in turn became institutionalized as a right protected by civil law.[52]

To sum up, private property was for Cousin not only rooted in and protected by the homologous structures of the human psyche and the universe but was also an arena of distinctly human self-development: nowhere more than in its gloss on property do eclectic philosophy and its conception of the *moi* appear so clearly as a justification for the bourgeois order. The urgent intensity of the eclectic need to shore up property can be seen by comparing Cousin's argument with those of Locke and of Destutt de Tracy, the foremost representative the post-Revolutionary brand of sensationalism called Idéologie.

In his late seventeenth-century *Second Treatise of Government*, Locke also derived the right of private property from the contention that "every Man has a Property in his own Person." But his understanding of the person in this context was a corporeal rather than a spiritual or psychological one ("The Labour of his Body, and the Work of his Hands, we may say, are property"), and he cast his whole discussion in terms of man's natural right to physical self-preservation. While an unambiguous advocate of private property, Locke did not attempt to depict ownership as a spiritual desideratum; nor did he return to the issue in his *Essay Concerning Human Understanding* in order to treat it from a psychological angle.[53]

In an argument quite similar to and nearly contemporaneous with Cousin's, Destutt de Tracy did derive private property from human psychology and, in particular, from the will. But that will was itself a rather more contingent affair than it was in the eclectic system, being itself a consequence of the prior capacity for sensation. Destutt insisted that the capacity for sensation was originary. It was "that beyond which we cannot go," and as such it was "the same thing as *us*," the "existence of the *moi* and the sensitivity of the *moi*" being simply identical. Hence, to regard "one's will as the equivalent of oneself is to take the part for the whole." Destutt even entertained the possibility that a being endowed with sensitivity but lacking a will could have individuality or personality; but such a being, he

---

52. Cousin, *Cours d'histoire de la philosophie morale au 18e siècle, professé à la Faculté des Lettres en 1819 et 1820*, ed. E. Vacherot (Paris: Ladrange, 1841), 11–13.

53. John Locke, *Two Treatises of Government*, 5th ed. (London, 1728), 159–61, 172 (chap. 5, paras. 25–27, 44, of the *Second Treatise*).

opined, could never come up with the idea of property. On the other hand, once the sensory capacity had generated a will, the idea of property would be born "necessarily and inevitably in all its plenitude."[54] In other words, Idéologues like Destutt were entirely committed to the idea of private property, but they thought that idea sufficiently hardy that it did not need the fortresslike protection of an a priori *moi* nor of a tripartite division of consciousness replicated in the structure of the universe as a whole.

Let me conclude this section on a rather speculative note. During his frequent sessions of interior observation, Cousin had perceived voluntary activity, or the *moi*, inevitably present in consciousness as its center. (According to the typically metaphor-laden formula that he offered, "The sensibility is the external condition of consciousness, the will is its center, and the reason is the source of light.")[55] He thus believed that he had empirical evidence for rejecting Condillac's conception of consciousness as an empty space awaiting the random entrance of sensory experience. "Consciousness," he asserted, "is not a deserted stage where the events of the intellectual life occur while someone in the pit *[parterre]* contemplates them." Because of the continual presence of the *moi*, "the audience is, so to speak, onstage."[56]

But what would consciousness be like in the absence of the controlling center provided by this a priori *moi?* What, in other words, if a strictly sensationalist model of consciousness obtained? Such a decentered consciousness would be a shifting series of disconnected images and sensory traces, a kind of phantasmagoria. In fact, the term phantasmagoria, so formally consistent with the sensationalist construction of mental experience but lending a touch of horror to it, was a neologism of this period. It was coined in French by a Belgian physicist, student of optics, and showman, Etienne-Gaspard Robertson, who presented his first *fantasmagorie* in Paris in 1796. The term referred to a lugubrious and terrifying form of entertainment, a public exhibition of optical illusions that were produced chiefly by means of a magic lantern and billed to the audience as spectral apparitions of the dead. So overwhelmingly popular did phantasmagorias become that they soon sprang up everywhere in Paris.[57] And, to judge from Stendhal's account,

54. Antoine-Louis-Claude Destutt de Tracy, *Traité de la volonté et de ses effets,* 2d ed. (Paris: Courcier, 1818; reprint, Geneva and Paris: Slatkine Reprints, 1984), 49, 60–63, 66–67.
55. Cousin, "Préface à la première édition," *Fragmens philosophiques,* 2d ed., 18.
56. Cousin, "Du fait de conscience," 245.
57. On the history of the phantasmagoria, see Terry Castle, "Phantasmagoria and the Metaphorics of Modern Reverie," in *The Female Thermometer: Eighteenth-Century Culture and the Invention of the Uncanny* (New York: Oxford University Press, 1995), 140–67, esp. 144–55. Castle's interest in the phantasmagoria as evidence of the "displaced supernaturalism" of the post-Enlightenment period is quite different from my interest here; but she

they played an important role in the provinces as well.[58] The public fascination with the phantasmagoria might have been, at least in part, an unreflec-
tive version of Cousin's philosophical recoil from sensationalism. That is to
say, the lantern show might have been on some level perceived as a concrete
representation of the empiricist model of the mind; but in this rendition,
the string of projections on an empty screen was far from reassuring. It produced in the public a frisson of horror, conjuring up a world of hallucina-
tions and feverish delusions, a world gone out of control. Cousin did not, to
my knowledge, use the term "phantasmagoria" to describe and express his
distate for the phenomenology of the Condillacian mind, but he did use that
term to mean the opposite of his supremely ordered eclectic system. Speaking of the sensationalist philosophy of the eighteenth century, which,
though misguided, still had its place in the larger scheme of things, Cousin
commented, "The apologia for a century is in its existence, for its existence
is a decree and a judgment of God, or else history is nothing but an insignificant phantasmagoria."[59]

INSTITUTIONALIZING THE COUSINIAN SELF

If Cousin had merely articulated his doctrine of the *moi* in books and lectures, it would not be entirely clear why he rather than, say, Condillac
would deserve the credit for bringing the self-as-*moi* into being as a scientific object. There are, I think, two main reasons for Cousin's unambiguous
salience in this endeavor. The first is the pivotal place of the *moi* in his philosophy as compared with its decidedly minor role in Condillac's. (If any
further proof of the latter contention is needed, it should be noted that
Condillac's *Dictionnaire des synonymes* includes an entry for "âme" but
none for "moi.")[60] The second and by far the more important reason is the
currency acquired by the Cousinian *moi* as a result of its creator's combined vocation for philosophy and bureaucracy. The *maître* succeeded in
institutionalizing his grandiose version of the self in the curriculum of the
state educational system, an achievement never matched by the sensa-

nonetheless notes (p. 144) the formal correspondence between the phantasmagoria and the
empiricist model of the mind.

58. See *Mémoires d'un touriste* (1838) in *Oeuvres complètes* (Geneva: Cercle du Biblio-
phile, 1968), 15:44–45, where Stendhal describes a locally celebrated episode in a small town
in the Nivernais that took place between 1815 and 1820 at a soirée at which Robertson presented his phantasmagoria.

59. Cousin, "Préface à la première édition," *Fragmens philosophiques*, 4.

60. Le Roy, ed., *Oeuvres philosophiques de Condillac*, vol. 3. The *Dictionnaire* was not
published during Condillac's lifetime but was found in manuscript among his papers.

tionalists and their modest, pared-down self. To be sure, the sensationalist Idéologues had their moment in the sun under the Directory (1795–99), when they were put in charge of formulating a curriculum for the new central schools, designed as the republican replacement for the Jesuit collèges of the Old Regime. For this purpose they created a master course in "general grammar" that was really a course in Condillac's epistemology and theory of language. The voluminous archive generated by the revolutionary government in its effort to ascertain the degree to which pedagogical practice in the central schools corresponded to the founders' ideals indicates that at least one professor of general grammar—the one stationed in the remote region of the Basses Pyrénées—included Condillac's conception of the self in his lesson plans.[61] But whether or not this Idéologue pedagogical experiment gave much attention to the sensationalist concept of the *moi*, the experiment itself was of notably short duration. The central schools fell with the Directory, to be supplanted by lycées under Napoleon.

The lycées, of course, proved a far more lasting educational innovation, and it was Cousin's great administrative feat to have acquired enough control over them at a formative phase that he could firmly install his eclectic philosophy centered on the *moi*. In fact, during the period of the constitutional monarchy, Cousin set up a formidable educational machine. At the Ecole Normale and the Sorbonne, he trained a "philosophical regiment" that obtained academic employment and carried his message ("our cause") throughout the provinces. His position as president of the national jury of the philosophy *agrégation* further strengthened his gatekeeper role in philosophy teaching in France. And perhaps most significantly, as a member of the Conseil royal de l'instruction publique, he authored an 1832 decree that added a subject called "psychology" to the standard lycée course in philosophy, giving it pride of place as the first substantive section.[62] Henceforth the youth of France, or more precisely the male bourgeois youth who alone attended the lycées, would be instructed about the dynamic, action-initiating *moi* that was always already there and would learn the meticulous in-

61. See Archives Nationales, Paris, F17 1344/3, reply of Germain Baradère, professor of general grammar at the Central School of the Department of the Basses-Pyrénées, to the ministerial circular of the Year 7. The reply includes his manuscript, "Cours de Grammaire Générale: 1ère année," which presents (ms. p. 48) an account of the self drawn from Condillac's *Essai* and *Traité*. While I did not make an exhaustive study of these archives, my impression is that the *moi* was an infrequent item in the lesson plans submitted and that Baradère was more the exception than the rule.

62. Archives Nationales, Paris, F17* 1795, "Procès-verbaux des délibérations du Conseil royal de l'instruction publique," 28 September 1832, fols. 434–36.

trospective techniques required to explore it directly. They would thus presumably come to identify themselves as possessors of such a *moi* or, in another standard Cousinian locution, as bearers of the "sentiment of personality."

Moreover, the hegemonic power of Cousinianism over the philosophy curriculum of the lycées proved extraordinarily durable, persisting at least until the end of the nineteenth century. A reform of the curriculum in 1880 might have been expected to unseat the Cousinian *moi* from the instructional program, given the strong positivist affiliations of the newly republicanized Third Republic and its anticlerical distaste for metaphysics. But, through a combination of institutional inertia and the "fit" between eclecticism and generic French bourgeois values, the *moi* survived the change in political climate and retained its old curricular centrality.[63]

Among Cousin's entourage, the *moi* passed readily from psychology and metaphysics into other discursive contexts, adding to the domains in which it functioned as a scientific object. In his history lectures at the Sorbonne in the 1820s, later to become the best-selling *History of Civilization in Europe*, François Guizot ascribed the "sentiment of personality" to those individuals whose actions he deemed decisive in moving along the metanarrative of liberal progress. Guizot's rhetoric thus posited a strong *moi* as one of the explanatory factors in the unfolding of history.[64]

Similarly, Cousin's student Théodore Jouffroy marshalled the sentiment of personality in support of an aesthetic theory. Objects move us aesthetically, he asserted, by "their invisible element," by the force in them similar to and therefore able to address the "force that animates us—that is to say, [human consciousness] endowed with the three principal attributes of sensibility, intelligence, and freedom." It is by reference to this trio that we distinguish the merely agreeable from the beautiful and the beautiful from the sublime. For example, the spontaneous movements of a woman

63. The brunt of the 1880 reform of the philosophy curriculum was variously interpreted at the time; according to some accounts, the *moi* may have lost some of its privileged status as an a priori entity. See my discussion of this matter in "Saying 'I': Victor Cousin, Caroline Angebert, and the Politics of Selfhood in Nineteenth-Century France," in *Rediscovering History: Culture, Politics, and the Psyche*, ed. Michael S. Roth (Stanford: Stanford University Press, 1994), 321–35. As I stress there, the *moi* survived in the lycées for boys but was deliberately omitted from the curriculum of the newly created lycées for girls.

64. François Guizot, *Histoire de la civilisation en Europe* (Paris: Didier, 1846), 58–59. Here Guizot credits the German barbarian invaders with having introduced one of the "fundamental elements" of modern European civilization: "le plaisir de se sentir homme, le sentiment de la personnalité, de la spontanéité humaine dans son libre développement." By the twelfth century the burghers in the towns will have picked up this sentiment ("la volonté individuelle se déployant dans toute son énergie") from the feudal seigneurs; see 195–96.

are childlike and obey the impulsions of the passions, but they "do not give us the idea of a free force that understands its goal and heads toward it." Such movements therefore strike us as agreeable and nothing more. To acquire either beauty or sublimity, movements must express psychological attributes other than mere sensibility. "Only in face of the spectacle of a man who develops himself with intelligence and freedom, who pursues with his freedom the goal that he identifies with his intelligence, . . . can the beautiful and the sublime appear." The fundamental difference between the latter two lies in their relationship to struggle and, hence, in the quality of the sentiment of personality that they disclose. Sublimity attaches to the "idea of a free, intelligent force struggling against obstacles that impede its development," beauty to the idea of that same force "arriving at its goal easily and without effort." In other words, Jouffroy continues, what we label sublime evokes its characteristically intense aesthetic response because it provides an especially pure, strong, and concentrated expression of the sentiment of personality. By contrast, "there is in the development of a force operating with ease"—and that we consequently experience only as beautiful—"a self-forgetfulness [oubli de soi-même] entirely contrary to the sentiment of personality that dominates us when we develop ourselves painfully." In the hierarchy of aesthetic responses according to Jouffroy, the peak is attainable only in the presence of a distilled manifestation of the moi.[65]

THE COMING INTO BEING OF A SCIENTIFIC OBJECT: SOME REFLECTIONS

The Cousinian moi was a very particular and almost paradoxical kind of self, one capable of sublimity yet at the same time carefully circumscribed. Defined as an entirely individual will and a personal principle of activity that could impose itself on inanimate matter, its options for titanic self-making were nonetheless severely limited by the ontology to which it was attached. Radically free and capable of profound introspection, its life's journey would be one of quasi-comic deflation. For the grandiose moi was destined to be thoroughly unoriginal, to rediscover and take as its guide the eternal verities about The True, the Beautiful, and the Good described in Cousin's lectures of that title, which became the official philosophy text-

65. Jouffroy, Cours d'esthétique, ed. Ph. Damiron (Paris: Hachette, 1843), lesson 14, esp. pp. 315–18. As noted in Damiron's preface, these lectures were given by Jouffroy as a private course to some twenty to twenty-five young people on the rue du Four in 1826—that is, during the period when the Restoration monarchy had banned the eclectics from public instruction.

book in France for most of the nineteenth century.[66] In short, the Cousinian combination of "personal will" and "impersonal reason" flattered the possessor of the *moi* that he enjoyed a thrilling degree of individuality and efficacy yet at the same time guaranteed that he would not rock the boat.

A *moi* replete with such contradictions clearly corresponded in manifold ways to its historical moment, which I will take for these purposes to be the aftermath of the 1789 Revolution, the emergence of the bourgeoisie as the socially and politically dominant class, and its deep fear of renewed revolution. The bourgeois, unaccustomed to his new leadership role and anxious about his capabilities, needed a "sense of self." In part, this was furnished by a bevy of social practices that constituted him as an object of deference, a man to be reckoned with. But the equally important linguistic aspect of that sense of self—that is, a precise vocabulary of robust selfhood—was supplied by the Cousinian discourse on the *moi* that he imbibed in adolescence at the lycée.

Still, a bourgeoisie fearful of renewed revolution and eager to restore consensus could not afford to produce a race of willful heroes, even among its own membership. Hence it was appropriate that the Cousinian *moi* come into the world already anchored in an ontology and foreordained to embrace the blandest of value systems. Religious politics also helped to shape this self. The project of reestablishing social stability required that the old principles attached to the Catholic soul—moral responsibility, immortality, and eternal punishment for serious transgression—be revived and that the sensationalist version of the self, smacking of materialism and moral unaccountability, be eradicated. But since the Church was a pillar of the pre-1789 order, the religious roots of those reassuring principles had also to be played down. Thus Cousinian discourse spoke garrulously of the *moi*, a thoroughly secular term, and painstakingly distinguished that entity from its predecessor, the *âme*, which was awash in religious connotations.

As part of the process by which the Cousinian *moi* gained discursive currency in nineteenth-century France, a new social division of labor took place. The cleric continued as the caretaker of the *âme*. The biomedical scientist took firm charge of physiological researches. The professor of philosophy emerged as a new social type.[67] Entrusted with inculcating the secular but resolutely nonmaterial *moi*, he also claimed the right to monitor the

66. On Cousin's textbook *Du vrai, du beau et du bien*, see Theodore Zeldin, *France, 1848– 1945*, 2 vols. (Oxford: Oxford University Press, 1973–77) 2:409.

67. On the role of Cousin in bringing about the professionalization of philosophy in France by making possible the secure, full-time employment of philosophy specialists in the lycées and arts faculties, see R. R. Bolgar, "Victor Cousin and Nineteenth-Century Education," *Cambridge Journal* 2 (1949); 357–68, esp. 358–59.

biomedical sciences in order to keep them from overstepping their limits and attempting to reduce mind to body. A new professional man staffing the national educational system created by the Revolution, the philosophy professor thus embodied the middlingness of all eclecticism, in this case by representing and trying to keep ascendant the middle term between religion and biomedical science.

Both by its internal intellectual features and by the mode of its institutionalization, the Cousinian *moi* also participated in creating critical distinctions *within* the new bourgeois order, especially distinctions of gender and class. The reader will probably have already noted the gender implications of Cousin's psychological theory, which are writ large in the passage from Jouffroy's aesthetics discussed above. Within the parameters of the tripartite Cousinian consciousnesss, women are assigned to the realm of sensibility; their movements can therefore be at best no more than agreeable. Men, inhabiting the realm of reason and will, are by contrast capable of movements that reveal a sentiment of personality and can therefore strike the beholder as beautiful or even sublime. This gender bias was to some extent built into eclectic psychology through Cousin's reliance on the binary opposition between activity (the quality par excellence of the will) and passivity (the quality par excellence of sensitive matter, which awaits a form-imposing will) and through the presumed operationalization of that binary opposition in the mechanics of human sexual reproduction. The *maître* thus implicitly divided the world into the sectors of man-activity-rationality-culture and woman-passivity-feeling-nature, additionally positing as an axiom the superiority of the former sector over the latter. This strategy, a nineteenth-century intellectual commonplace, dovetailed nicely with the domestic ideology—that is, the relegation of women to home and family, the reservation to men of the public spheres of work and politics—that everywhere in Europe and America accompanied the ascension of the bourgeoisie.[68]

At the same time that Cousinian eclecticism reinforced and rationalized a hierarchical relationship between male and female, it reinforced and rationalized a distinction between the working class and the bourgeoisie befitting its obdurate opposition to democracy. Here Cousin's typology of the principle of personal mental activity was key: that principle, he stipulated,

---

68. But as Caroline Angebert, an autodidact female admirer and critic of Cousin, aptly pointed out to the *maître* in the late 1820s, the eclectic philosophy, which zealously affirmed the mind-body distinction against sensationalists and physiologists, should therefore have affirmed the disembodied nature of mental attributes; it was by no means *logically* wedded to the principle of the intellectual inferiority of women. Angebert's correspondence with and criticism of Cousin is discussed at length in my essay, "Saying 'I.'"

could be either "spontaneous" or "reflective."[69] Hence everyone had a *moi* in principle and, to some degree, in practice; but every *moi* was not equivalent to every other *moi*. Only in an elite minority would the *moi* be detached by reflection from the "primitive synthesis" of elements of consciousness in which it was ordinarily fused.[70] In the vast majority of people, on the other hand, the mental activity that was the spiritual endowment of all human beings would remain in an aggregated, inarticulate, and "spontaneous" condition.[71] It was in fact along this axis of spontaneity or reflection, fusion or delineation of the elements of consciousness, that Cousin located "the sole difference possible" among people.[72]

Evidently (although Cousin left this implicit), the reflective possessors of a strong sentiment of personality became the ruling class, the spontaneously inspired masses became the ruled. But whether or not one was a bourgeois male who had gone to the lycée to learn philosophy and to hone his *moi*, one was intimately bound by the moral truths that the psychological method and its resultant ontology revealed. As Cousin pointed out in a pamphlet called *Popular Philosophy* that he wrote immediately after the bloody working-class insurrection of June 1848, there were two sorts of philosophy, "one artificial and learned, reserved for the few, the other natural and human, for everyone's use." Although the latter, which springs from "the spontaneous suggestions of consciousness" rather than the exercise of reflection, lacked the specialized scientific vocabulary of the former, the basic contents of the two were identical: the distinction between mind and body, the moral freedom of human beings to choose between good and evil, the existence of God.[73] The working class might, in other words, have fewer capacities than the bourgeoisie, but there could be no doubt that it operated within the same constraints.

The Cousinian *moi* thus corresponded neatly to a quite detailed sociopolitical agenda—that of a conservative brand of nineteenth-century liberal-

69. See, e.g., Cousin, "Préface à la première édition," *Fragmens philosophiques*, 2d ed., pp. 27–29, a passage that concludes, "Search as hard as we might, we will find no other modes of action. All the real forms of activity are covered by reflection and spontaneity."

70. Cousin, *Introduction à l'histoire de la philosophie*, lesson 5, pp. 39–40.

71. Cousin, "Préface à la première édition," *Fragmens philosophiques*, 2d ed., 45. The passage reads: "Now in my view, the mass of humanity is spontaneous and not reflective; humanity is inspired. The divine breath that is always and everywhere in it reveals all truths to it in one or another form . . . Spontaneity is the genius of human nature, reflection is the genius of certain men."

72. Ibid. Cousin made the same point in *Introduction à l'histoire de philosophie*, lesson 5, pp. 39–40.

73. Victor Cousin, *Philosophie populaire, suivie de la première partie de la profession de foi du vicaire savoyard . . .* (Paris: Pagnerre, 1848), 1–14, quote at 2.

ism dedicated to the empowerment of the male bourgeoisie and the protection of its property. This *moi* came into being as a scientific object charged with multiple extrascientific roles, including the demarcation of a "self-possessed" ruling class from the "unselved" masses of workers and women. The use of the ordinary personal pronoun to designate this self secularized it without democratizing it. In order to constitute what Michel Foucault felicitously named the "everyday individuality of everybody," the nineteenth century relied not on elite, state-sponsored instruction in philosophy, but on another aspect of the so-called disciplinary regimen: the individual, data-filled dossiers that resulted from the sustained, scientific observation of the occupants of hospitals, asylums, primary schools, and prisons.[74]

74. Michel Foucault, *Discipline and Punish: The Birth of the Prison,* trans. Alan Sheridan (New York: Pantheon, 1977), 191–93, quote at 191. The phrase is in fact rather more felicitous in Sheridan's English translation than in Foucault's original: "l'individualité quelconque—celle d'en bas et de tout le monde." See *Surveiller et punir: Naissance de la prison* (Paris: Gallimard, 1975), 193.

5   Gérard Jorland

The Coming into Being and Passing Away of Value Theories in Economics (1776–1976)

History of science gives a biased picture of science since it deals with what Bachelard called *"science sanctionnée"*—science vindicated—rather than science in progress. It aims at understanding how we ultimately came to, say, Newtonian mechanics, Maxwellian electromagnetism, Darwinian natural history, Pasteurian microbiology, Walrasian mathematical economics, etc., rather than showing how, at every historical moment (but looked upon prospectively) scientists are at a crossroads and might follow one of several paths.[1] If one wished to get a less distorted picture of the eighteenth century, at least its first half, one would have to deal with both the attraction and the vortex explanations. Scientists not only had to choose between them; they also had to argue against the one they rejected and account for the shortcomings of the one they would pursue. It suffices to recall that Euler was in favor of the vortex theory and on good grounds. Similarly in the 1950s, elementary particle physicists had to choose between bootstrap theory—which turned out to be a dead end but had the appeal of intertwining quantum mechanics and relativity theory—and quark theory, which eventually took hold.

However, as Hegel put it, one cannot jump over his time, one cannot help knowing the outcome, viz., which explanation has been vindicated and which rivals have been eliminated. Therefore, in order to get an unbiased picture of sciences, by which I mean of sciences in progress, one has to raise

---

1. For a case study of what I have in mind, see Berno Müller-Hill, *The lac Operon: A Short History of a Genetic Paradigm* (Berlin: Walter de Gruyter, 1996).

117

the issue of the coming into being of scientific theories and their passing away. The question is not so much how scientific theories are vindicated or fail, but rather what it means to be vindicated or to fail. Do scientists switch from one theory to the other like fads? After all, they are so deeply committed to innovation that it could be so. Would it not be the case that to be vindicated, a theory would have to be identified with by a network of scientists who brandish it as a banner in their career game? And it would therefore be eliminated whenever this network collapses, if only because its endorsement did not help win the hoped-for positions.

These questions are relevant to a forward history, a history written from the point of view of the working scientist. However, they might lead to shortsighted answers, or to conceptions of history as a series of snapshots without any continuity—in short, no history at all. But there is no alternative; one cannot proceed the other way round, for then we would no longer be at a crossroads but following a definite path. We would be back to science vindicated when we were looking for science in progress. And so we are reduced to a dilemma: either we try to get a true picture of science at every moment of time although it does not make a history, or we hold to the continuity of the historical process, but are then left with a distorted picture of science as judged from the viewpoint of the working scientist. One way out would be to trace an intergenerational solidarity among scientists, every generation leaving to the following genuinely new problems to solve so that individual scientists belonging to the next generation can compete with one another in order to make a name for themselves.

In what follows, I will endeavor to show that a scientific theory comes into being to solve an emerging problem, and remains in being so long as it helps to solve that problem, provided other problems that it itself raises are solved in turn. And it passes away under the burden of unsolved problems that it raises whenever there is a simpler way to solve the starting problem, which may mean not trying to solve it any longer or not even posing it at all. Pursuing it would mean ruining one's career prospects. Since my argument is closely tied to my example—value theory in economics—I can claim to offer no more than a case study.

THE PHYSIOCRATIC PARADIGM

Value theory has always been around, at least ever since Aristotle. As long as Aristotelianism was paradigmatic, namely up to the eighteenth century, the Aristotelian theory of value had remained unchallenged. So the only question we need to answer is why Aristotle had any need for a theory of value and why precisely this one.

In order to understand Aristotle's economics, one has to unfold his re-gional ontology.[2] Economics finds its locus as a result of a twofold differ-entiation, first between becoming by internal agency, as is the case for natural entities, and becoming by external agency;[3] and then, within the latter category, between action and production, which duplicates the former differentiation since, once again, the telos, or end, is inside action whereas it is outside production. However, production is only one means of securing riches; there is another one, through exchange. Now, the telos of riches is the fulfillment of our needs, i.e., its utility. In this scheme,[4] one can easily understand Aristotle's condemnation of chrematistics—or the accumula-tion of riches—that money makes possible: in that case, the end is no longer outside but inside.

Thus, the value of economic goods is determined by their utility since their very nature is to fulfil a need. The value of any commodity is a func-tion of its utility and its utility is a function of the need that the commodity fulfils. But it is also a function of the rarity of the commodity. A rarity is something that somebody needs but does not own; that thing is the rarest that is owned by the fewest and needed by the most people. Now, if some-body owns something in excess of his own needs, that excess will be useless and thus valueless; if through exchange, the trader can get something he needs but does not own for something he owns but does not need, he increases his overall utility. He trades plenty for rarity. Therefore, the ex-change value of economic goods is a function of utility on the demand side.

Aristotle's formula for exchange value follows the pattern of distribu-tive justice: good *a* must be to trader *B*, in terms of utility, as good *b* to trader *A*. Money, having utility for everyone, is the universal standard of needs: Aristotle can state that, since nobody should value money for itself, there is no utility of money per se. So, supposing that *b* is money and that *A* trades his commodity *a* for a certain amount of money *b*, the formula of exchange would read: *a* must be to *B* as *b*, money, is to *A*, which is as good a definition of nominal price as one can get. But what is the meaning of "as money is to *A*"? It can only be: as *A* is in need of money to secure for himself the neces-sities of life. Thus *A*'s need for money is a function of the utility of these necessities to him.

2. Edmund Husserl, *Ideen zur einer reinen Phaenomenologie und phaenomenologischen Philosophie*, 3d ed. (Halle: M. Niemeyer, 1928), § 9.

3. This difference accounts for two kinds of becoming, like becoming a grown-up and be-coming an educated person.

4. On Aristotelian economics, see B. J. Gordon, "Aristotle and the Development of Value Theory," *Quarterly Journal of Economics* 78 (1964): 115–28; J. Soudek, "Aristotle's Theory of Exchange," *Proceedings of the American Philosophical Society* 96 (1952): 45–75.

The Aristotelian theory of value is a theory of price determination, a theory of exchange, and it is required because economic goods cannot have prices on their own since their telos lies outside of them. Thus Aristotle came to his specific theory of value—utility theory—because of his principle of the telos operating either within or without.

After the physiocrats in the eighteenth century, the theory of value becomes more than a mere theory of price determination. We will not go here into the details of the origin of the *tableau économique* since it is a different story that would take us too far afield.[5] Suffice it to say, in contemporary wording, that it was the first macroeconomic model describing the exchange relationships between three classes—and hence a three-sector model—at equilibrium. The farmers—exclusive producers of wealth—grow products worth, say, five million. Of these five million, they must keep a certain amount, worth, say, two million as seeds for the next crop; they must pay, out of the rest, say, another two million as rents to the landowners; and they use the remaining one million to obtain implements, tools, and all sorts of conveniences from the craftsmen. Now, the landowners buy foods from the farmers (say, worth one million) and luxuries from the craftsmen (worth another one million). In turn, the craftsmen buy raw products (worth, say, the one million they received from the landowner) and food (worth the one million they received from the farmers). At the end of the exchange process, the five million are back in the hands of the farmers for another economic round.

I will only add that this model had two purposes.[6] The first was to show that the landowners and not the farmers should be taxed, contrary to the rule prevailing under the ancien régime in France; otherwise there is a leak in the circuit. And second, in order for the circuit to close on itself, the landowners must spend all they get; they must not hoard, since that would create another leak. Hence the *tableau économique* implies a theory of exchange of commodities, a theory of income distribution, and a theory of resource allocation.

In a certain sense, the history of economics is a tale of only one paradigm: the physiocratic paradigm. Of course, its implementation has required many changes, some of them dramatic, so much so that some have even spoken from time to time of "revolutions," as for example "the marginalist revolution" or "the Keynesian revolution," although they did not break

5. Cf. *François Quesnay et la physiocratie*, 2 vols. (Paris: Ined, 1958); Ronald L. Meek, *The Economics of Physiocracy* (Cambridge: Harvard University Press, 1963); and the communications at the Colloque international François Quesnay (1694–1774), Versailles, 1–4 June 1994.

6. François Quesnay, "Maximes générales" (1767), in *François Quesnay et la Physiocratie*, 2:949–57, in particular maxims 5, 7, and 31.

with the main physiocratic paradigm but rather with the previous way of handling it. Both static and dynamic economics—equilibrium theory and growth theory—stem from the *tableau économique*. I am convinced that we have not yet exhausted all of its potentialities. And yet the physiocrats had no theory of price whatsoever and their *tableau* was simply a balance, as Norton Wise thoughtfully pointed at. Why has it given rise to a theory that singled out prices as the equating variable between supply and demand? Boiling down to a balance, and thus intrinsically static, how did it give rise to dynamic economics? The answer to both questions is that there is, within the *tableau*, an equilibrium of exchanges and a multisector model of a self-reproducing economy.

The *tableau économique* is a model of an agricultural economy, of a *Royaume agricole*, as the physiocrats expressed it. This feature enabled them to conceive of the economy as an automaton. They considered the agricultural products as a whole, as if these products were simply modes of the same substance. Moreover, they considered manufactured goods not as specific products in their own right, but as mere transformations of agricultural raw materials that did not alter their substance. Therefore, there was no heterogeneity among economic goods, which derived from only one physical substance. And since agricultural products have the ability to reproduce themselves, the physiocrats could conceive the whole economy, made of one and the same substance, a kind of naturalized Cartesian *res extensa*, as self-reproducing, as an automaton based on the principle of *causa sui*. And since they were dealing with one homogeneous substance, no problem of valuation arose, since prices were immediate expressions of quantities exchanged. Of course, the physiocrats' strong Cartesian claim was immediately challenged by Smith and Condillac among others. For it led them, as a corollary, to claim that only agriculture was productive and that manufacturing trades were sterile. Agriculture produced more of the same thing that it consumed, whereas manufacture transformed raw materials without apparently adding to them in quantity.

It is within this context that modern value theory emerged. Its role was not only to regulate exchanges but to regulate the whole *tableau économique*, that is to say, the whole economy. In other words, theory of value was no longer a mere theory of exchange but also (and simultaneously) a theory of income distribution and a theory of resource allocation. It must determine prices at which the ensuing income distribution will result in the optimum allocation of resources, i.e., the allocation compatible with the reproduction of the economy. Already Quesnay had defined a "prix fondamental des marchandises [qui] est établi par les dépenses, ou les frais qu'il faut faire pour leurs productions, ou pour leurs préparations" [fundamental price of merchandise, which is established by expenses, or the cost necessary for their production or preparation];[7] and, more explicitly: "il faut entendre dans le prix fondamental, les impositions et le fermage des terres" [one must understand (as included) in the fundamental price, the assessments and rents of land], as well as "la subsistance des ouvriers" [the subsistence of the workers]. Thus, "les frais de production" [costs of production] include "les gains des habitants des campagnes, les revenus des propriétaires des biens-fonds, et les revenus du Roi" [the profits of the rural inhabitants, the revenues of the proprietors of real property, and the royal revenues].[8] However, for lack of a proper theory of value, Quesnay could not explain why the determination of prices should entail a certain distribution of income. That is precisely what Adam Smith was to do.

CLASSICAL THEORY OF VALUE

But in order to do so, he first had to break with the paradigmatic Aristotelian utility concept of value, for it could not assume these new regulatory functions. As long as economics was not a science in its own right but part of a theory of justice, it did not matter that the *ultima ratio* of price determination was the needs of the trading parties, that is to say, was relevant to another science, viz., biology. Needs are determined by human nature, not by economic causality. The Aristotelian utility concept of value could lead only to the medieval doctrine of just price, not to the actual price charged and paid. Under the physiocratic paradigm, which is self-contained and therefore can delineate a new scientific domain, all relevant variables must be economic. Enlightenment political economists, like Turgot or Condillac, who tried to apply the Aristotelian utility concept of value to the *tableau économique*, simply failed.[9] It was only thanks to Cournot's demand function, over half a century later—according to which demand is a function of price and not price a function of demand, itself a function of need—that utility theory returned to the field as a potent contestant. But if it did not vanish altogether, it was because the alternative concept of value was not without problems of its own.[10]

Adam Smith could account for price determination as well as for income

7. François Quesnay, "Hommes" (1757), in *François Quesnay et la Physiocratie*, 2:529.

8. Ibid., 555.

9. Gérard Jorland, "Le problème Adam Smith," *Annales: Economies, Sociétés, Civilisations*, no. 4 (July–August 1984): 831–48, at 841–42.

10. Gérard Jorland, "Position historique de l'oeuvre économique de Cournot," in *A. Cournot: Etudes pour le centenaire de sa mort (1877–1977)*, ed. Jean Brun and André Robinet (Paris: Economica, 1978), 12–22.

distribution and resource allocation with his labor theory of value. Two background points are worth noting: first, he started his economic studies after traveling to France, where he had extensive discussions with the physiocrats; and second, he turned to a labor theory of value after having rejected the utility theory on two grounds.[11] The first was ethical. In contrast to his friend Hume, he did not make utility the basis of human behavior, neither ethical nor economical. The second was theoretical, already mentioned above: utility was not relevant to economics but to biology. That is what he meant with his famous paradox of utility: some goods are of great value but little utility, like diamonds, others are of little value but great utility, like water. Of course, the standard response was that once one had taken quantity into account the paradox vanishes. Aristotle had already said that utility was a function of scarcity. But if that is so, why should economists bother with needs, which are not relevant to their field? Why not start immediately with quantities and find out whether they can also be dependent variables? And that is indeed what Adam Smith did: quantities are for him functions of the division of labor, itself a function of capital and labor. Henceforth, for over two centuries, economists would have to face the choice between two competing theories of value, both of which must also constitute a theory of economic regulation. The labor value theory came into being because it initially seemed to succeed at the job where the venerable utility theory failed. Let us see how.

Since labor "is the real measure of the exchangeable value of all commodities,"[12] "in that early and rude state of society which precedes both the accumulation of stock and the appropriation of land . . . the whole produce of labour belongs to the labourer."[13] The embodied labor theory of value is equivalent to the labor command theory: "the quantity of labour commonly employed in acquiring or producing any commodity, is the only circumstance which can regulate the quantity of labour which it ought commonly to purchase, command or exchange for."[14] After capital "has accumulated in the hands of particular persons," and the land "has all become private property," "the whole produce of labour does not always belong to

11. Jorland, "Le problème Adam Smith," 837–39 and 843, where, relying on the Glasgow edition of Adam Smith's works and correspondence, *The Theory of Moral Sentiments* (1759; reprint, Oxford: Clarendon Press, 1976), pt. 4, "Of the Effect of Utility upon the Sentiment of Approbation," and *An Inquiry into the Nature and Causes of the Wealth of Nations* (1776; reprint, Oxford: Clarendon Press, 1976), bk. 1, chap. 11, sec. c, paras. 3–7, I state that, for Adam Smith, utility is a concept relevant to aesthetics, not to ethics nor to economics.

12. Smith, *The Wealth of Nations*, 1.5.1.

13. Ibid., 1.6.1, 1.6.4.

14. Ibid., 1.6.4.

the labourer," "he must in most cases share it with the owner of the stock which employs him," and "he must give up to the landlord a portion of what his labour either collects or produces."[15] Labor incorporated is no longer equivalent to labor commanded, since one has to add profits and rent, which are both measured in terms of labor commanded. The quantity of labor embodied in a commodity can command more or less labor according to the share of capital and land; or, in other words, the lower profit and rent are, the higher real wage is. So Adam Smith defined the *natural price* as the price that "is sufficient to pay the rent of the land, the wages of the labour, and the profits of the stock."[16] Profit and rent are defined as deduction from the produce of labor that is employed by the capital upon the land.[17] Hence this theory of value is also a combined theory of price and of distribution because it is a theory of allocation of resources. Stock or capital can get a share of the product because it employs labor. The quantity of labor performed in an economy, and thus the quantity of wealth it creates, is a function of the stock of capital at its disposal: first, because capital employs labor, and second, because it raises its productivity.

But the matter is different for the rent. It is a monopoly price[18] and it comes from the natural fertility of the land, i.e., the fact that land produces more than it consumes under the form of seeds as well as of laborer consumption. Therefore, there is an asymmetry: "High or low wages and profit, are the causes of high or low price; high or low rent is the effect of it."[19] In fact, Adam Smith's theory of value cannot fully account for rent, for which the physiocratic analysis in physical terms still holds.

Adam Smith's theory of value required a standard measure. Suppose that the value of commodities has changed: it might be accounted for by a modification of the conditions of production of some of them, i.e., by the quantity of labor that they might command. It is all-important to know which commodities have experienced such a change because the income distribution and the resource allocation are affected. If corn is exchanged for a different amount of a manufactured commodity, one has to know whether it is the value of the former or the value of the latter that has changed. In the first case, wages and rents will move in the same direction and profits in the opposite one, whereas in the second case, they remain unaffected. If profits were lower, not only would stock not accumulate but it might even disaccumulate; the quantity of labor employed would not rise

15. Ibid., 1.6.5, 1.6.7, 1.6.8.
16. Ibid., 1.7.4.
17. Cf. ibid., 1.8.6, 1.8.7.
18. Cf. ibid., 1.9.a.5.
19. Ibid., 1.9.a.8.

or might even diminish; and the rate of growth of the economy might be lowered or even become negative.

But if modern value theory is a theory of economic regulation encompassing price formation, income distribution, and resources allocation,[20] it faces a problem that Ricardo first uncovered and that economists have tried to overcome ever since: prices must be independent of distribution. A rise in wages must be balanced out by a lowering of profit, not by a rise in prices. Otherwise, as is well known, wage hikes would trigger off an inflationary spiral.

Keynes wrote that the friendship between Ricardo and Malthus "will live in history on account of its having given rise to the most important literary correspondence in the whole development of Political Economy."[21] Indeed, one can read there the minute account of how the shortcomings of the modern theory of value were uncovered. At the outset, the banker and the churchman split the Smithian labor theory of value between them. Malthus supported a strict command labor theory of value whereas Ricardo developed an embodied labor theory of value. Malthus the clergyman's motivation for such a theoretical choice was plain. Dreading above all overpopulation, it suited him that a rise in wages might be checked by a rise in prices, otherwise it would spur workers to increase their offspring, thus worsening their condition. Ricardo the banker's motivation was of quite another kind. Fearful of a lasting drop of the profit rate, he tried to understand why it does happen in order to ward it off. If Malthus was right in his tracing out the consequences of a rise in wages, he nonetheless stopped short of the proper conclusion. For an increase in population would imply an increase of demand for food, and since land is limited, an increase in the price of food and therefore of rent since, as we have already seen, after Smith rent is high as a consequence of high prices of food. The last step was for Ricardo to show that, as consequences of high rents and high wages, profits must be low, or income distribution must be independent of price formation. If this is so, then the only means of offsetting the increase of rent would be the free trade of corn, for then land would no longer be a scarce resource. To yield these consequences in sequence was the purpose of Ricardo's embodied labor theory of value.

Hence after modern value theory arose out of a conceptual need to regulate the *tableau économique*, it unfolded at first in order to fit other requisites, which one can label "political" without leaving economics, which was

---

20. Gérard Jorland, *Les paradoxes du capital* (Paris: Odile Jacob, 1995), 40–47.

21. J. M. Keynes, *Essays in Biography* (1933), in *The Collected Writings of John Maynard Keynes* (London: Macmillan, 1972), 10:96.

called at that time "political economy." Both overpopulation and free trade were relevant to applied economics or political economy. In other words, when a theory emerges, it does not do so in a vacuum but in a theoretical context that gives rise to it in order to fulfil a definite function, provided that it fits in other respects. To check that it does is left to the next generation.

Thus Ricardo wanted to disprove Adam Smith's contention that wages, profit, and rent were constituent parts of prices. He was convinced that any change in wage rate was exactly compensated by an equivalent but opposite change in profit rate, so much so that the overall effect on price of a change in distribution was neutral. As for rent, it is not a cause but an effect of price. It is because lands are of unequal yields and there can be only one market price that covers the costs on the less productive land that there is a rent on the most fertile ones. Thus it is because food prices are high that rent is high, and food prices are high because demand is strong, and demand is strong because of population pressure. Now, if food prices are high, the wage rate tends to be high and the profit rate low, so that the price level of manufactured commodities remains unaffected.

However, Malthus pointed out to Ricardo that this rule was not valid, for in some cases a rise in wage rate might result in an increase of price whereas in other cases it might cause a lowering of price. The reason is that not all industries have the same organic composition of capital, as Marx would say, or the same capital intensity, some requiring more labor and less capital than the others. A commodity produced with more labor and less capital than the average will have its (relative) price raised by a rise in the wage rate and undergo the correlative fall in the profit rate, since the latter cannot compensate for the former. In contrast, a commodity produced with less labor and more capital than the average will have its (relative) price lowered by a rise in the wage rate and raised by a lowering of the wage rate and the correlative rise in the profit rate. More precisely, every commodity produced with more labor and less fixed capital than the standard commodity will have its price raised by a rise in the wage rate and, vice versa, decreased by a decrease in wage rate, whereas the opposite will hold for any commodity produced with less labor and more fixed capital than the standard commodity. What is paradoxical about these relationships between distribution changes and price changes is that not only do changes in the distributive shares affect the formation of prices, contrary to what Ricardo had claimed, but that, contrary to any sound intuition in general this time, a rise in the wage rate might result in a lowering of the relative price of some commodities.

The Ricardo paradox was the vexing puzzle of Ricardian economics throughout the nineteenth century. The embodied labor theory of value was prone to the same shortcomings as the labor command theory, namely the nonindependence of price formation from income distribution. Marx's tentative solution is known as the transformation of values into production prices.[22] The century-old discussion of Marx's transformation problem led to the acknowledgment that the labor theory of value led to a theory of price formation dependent on income distribution: one could get a theory of price formation and a theory of income distribution, i.e., a theory of production prices and a theory of exploitation, independently of any theory of value whatsoever.[23] At the end of the 1970s, there was no room left for a labor theory of value since it led to insuperable paradoxes and one could in any case do without it.

Meanwhile, the Ricardo paradox has also been the internal reason for the substitution of marginal utility for embodied labor as the cause and substance of value. The so-called marginalist revolution of the 1870s had several motivations. One was the mathematization of economics advanced by Cournot, which led to the rediscovery of the Bernoullian law of diminishing marginal utility.[24] In 1732, Daniel Bernoulli had claimed in a completely different context—a puzzle known in probability theory as the Saint Petersburg paradox—that the same amount of money did not have the same utility for everybody; it had less for the rich and more for the poor. Under the hypothesis that the utility of money was a logarithmic function of its quantity, he arrived at the first law of diminishing marginal utility. Now, the reason why this law has not been introduced in economics, in spite of the fact that utility theory was still the dominant theory of value, is that it was of no help in determining price formation since demand was not an economic variable. Thanks to Cournot, demand entered the field of economics as a function of price and, killing two birds with one stone, the mathematical form of his demand function could be deduced from the law of diminishing marginal utility.[25]

Here we see how the context in which working scientists use their intellectual apparatus, and, more broadly, theoretical concepts, can be understood as the set of inherited constraints that bear upon the relevance or the

22. Gilles Dostaler, *Valeur et Prix: Histoire d'un débat* (Montreal: Presses Universitaires du Québec, 1978); Jorland, *Les paradoxes du capital*.

23. Jorland, *Les paradoxes du capital*, 343–53.

24. Jorland, "Position historique de l'oeuvre économique de Cournot," 14–16; and Gérard Jorland, "The Saint-Petersburg Paradox, 1713–1937," in *The Probabilistic Revolution*, ed. Lorenz Krüger, Lorraine J. Daston, and Michael Heidelberger (Cambridge: MIT Press, 1987), 1:157–90.

25. Gérard Jorland, "Cournot et l'avénement de la théorie de la valeur-utilité," *Revue de synthèse* C1 (1980): 221–50.

effectiveness of apparatus or concepts. Concepts must fit in the context. However, the context is not a *social* context but an *intellectual* context, a paradigm. This way of looking at history of science in progress and not in retrospect entails a realist stance, whereas considering the social context as the main explanatory grounds entails a nominalist stance. Of course, ideas do not exist on their own, they have to be thought by humans socially contextualized. However, ideas are objects that one cannot manipulate at will; they have properties that remain for a long while unknown and unfold as ideas are pondered. And so, when I say "thanks to Cournot," I do not wish merely to acknowledge his personal merits; I want to stress the good fortune of a man who succeeded in devising the proper intellectual context within which a longstanding problem could be solved and, as he said, open a door for others to enter the field. And I say "good fortune" since it required an interdisciplinary mix that had previously failed, as is often the case.

Another motivation for the substitution of marginal utility for embodied labor was the necessity to counter the socialist Ricardians, among whom Marx had appeared already about 1860 as the leading figure, although less known than Proudhon in France or Lassalle and Rodbertus in Germany. If labor could be considered as the main determinant of value, and thus of the process of economic regulation, the whole product of labor should belong to the laborers since, as a class, they were ruling the economy if not yet society. Liberal economists had thus to refute the right of labor to the whole product by eradicating the underlying labor theory of value.[26] This political consequence of the labor theory of value motivated the return to the utility theory. Utility theory also profited from the Ricardo paradox, which rejected labor as a sound standard of value in any theory of regulation. Moreover, Cournot's demand theory had made the return to utility efficient. Together, these factors made for a sufficient (Cournot's demand curve) and a necessary (Ricardo's paradox) condition for the coming into being of marginal utility theory. However, the aim remained the same: to regulate the *tableau économique*.

Marginal utility theory explained prices through the determination of the shape of the Cournotian demand curve. Cournot had assumed that demand was a decreasing function of prices: the marginalists could give a *rationale* for that hypothesis and turn it into a corollary.[27] Since utility was a decreasing function of quantities as well, prices were simply indices of util-

26. For example, Emile de Laveleye, "Le socialisme contemporain en Allemagne," *Revue des deux mondes* 17 (1876):121–49, at 145–47; Jean Bourdeau, "Le parti de la démocratie sociale en Allemagne," *Revue des deux mondes* C4 (1891): 168–203; and 907–44, at 912; Maurice Block, *Karl Marx: Fictions et paradoxes* (Paris: Giard & Brière, 1900), 9–13.

27. Jorland, "Cournot et l'avénement de la théorie de la valeur-utilité."

ities. Moreover, marginal utility theory explained distributive shares through the principle of remuneration of productive factors (labor and capital) at their marginal productivity. In a production function where the quantity of output depends on the inputs of labor and capital, the first derivative of that function in each variable in turn gives the increase of output produced by one more unit of each input, and thus measures its productivity. Now, if wages are equal to the marginal productivity of labor and profits to the marginal productivity of capital, then Euler's theorem of homogeneous functions ensures that the distribution of income is exhaustive, i.e., there is no rent or free gift.[28] Finally, the Pareto optimality principle stated the condition at which resource allocation was optimal.

However, two blows were inflicted upon the (marginal) utility theory of value. The first, very early, was that the integrability conditions that marginal utilities, expressed in the form of partial differential equations, must fulfill in order to yield the demand curves, proved to be very restrictive indeed: either the order of consumption had to be prescribed (Pareto) or, more recently, the revealed preferences had to be transitive (Samuelson).[29] After the war, von Neumann and Morgenstern's attempt to build a standard of utility out of a lottery ticket tumbled after Maurice Allais showed that the independence axiom—the assumption that there is no utility of gambling, i.e., that no one gambles for the sheer excitement of gambling whatever the outcome, which would make for nonlinear utility functions—on which the construction rested was unsustainable.[30] As a result, demand curves are not determined by utility functions, except under stringent conditions. Marginal utility is not a satisfactory explanation and most economists prefer now to start at once with demand curves. Staunch utilitarians are left with ordinal utility and the transitivity axiom of revealed preference to account for well-behaved demand surfaces.

28. If one writes the production function $Q = F(L,K) - vL - \rho K$, where Q is the net product, v the rate of wages, and ρ the rate of profit, Q is a maximum provided $\partial Q / \partial L = F_L - v = 0$ and $\partial Q / \partial K = F_K - \rho = 0$; and so, the net product is maximum if the production factors are rewarded at their marginal productivity. If, moreover, the function F is linear and homogeneous, this repartition of the net product is exhaustive, i.e., there is no rent. Since F is homogeneous, it can be written as a function of only one variable, viz., $Q = f(k)$, where $k = K/L$ is the capital-labor ratio. Thus, $\partial Q / \partial L = f(k) - kf'(k)$ and $\partial Q / \partial K = f'(k)$, and so $L\partial Q / \partial L + K\partial Q / \partial K = Lf(k) = F(L,K) = Q$.

29. Vito Volterra, "L'economia matematica ed il nuovo manuale del Prof. Pareto," *Giornale degli economisti* 32 (April 1906): 296–301; Vilfredo Pareto, "L'ofelimita nei cicli non chiusi," *Giornale degli economisti* 33 (November 1906): 424–40; Paul Anthony Samuelson, "Consumption Theory in Terms of Revealed Preference," *Economica* 15 (1948): 243–53; H. S. Houthakker, "Revealed Preference and the Utility Function," *Economica* 17 (1950): 159–74.

30. Maurice Allais and Ole Hagen, eds., *Expected Utility Hypothesis and the Allais Paradox* (Dordrecht: D. Reidel, 1979); Jorland, "The Saint-Petersburg Paradox," 1:176–82.

The second blow came along with the Cambridge controversies in the 1950s and 1960s.[31] This time, it was the marginalist theory of distribution that came under fire. The whole controversy turned around the Wicksell effect. Wicksell had shown that the rate of interest, which determines the rate of profit, falls short of the marginal productivity of capital because, the function of capital being to employ labor, an increase in its stock is partly absorbed by an increase in the wage rate. Joan Robinson gave the Wicksell effect its full weight in a landmark 1953 paper that started the twenty years of bitter controversies between economists in Cambridge, Massachusetts, and Cambridge, Cambridgeshire.

The reversion of capital and reswitching of techniques paradoxes, which can be explained in terms of Wicksell effects, have falsified in a Lakatosian sense—i.e., not empirically, but theoretically—the neoclassical theory of resource allocation. According to that theory, stable economic growth requires the substitution of more capital-intensive techniques to less intensive ones as the rate of interest falls and the rate of wages increases. However, it has been shown that, depending on the form of the production functions, it might very well happen that a less capital-intensive technique follows one that is more so (capital reversing), and that a most profitable technique within a certain range of interest rates becomes such again within another, not-connected range after having been superseded by another technique (reswitching of technique). Hence the marginalist utility theory of value is a satisfactory theory neither of prices nor of distribution nor of allocation.

Why do these paradoxes occur? The labor theory of value is true if and only if there is but one sector in the economy: thus, commodities exchange at their value, and the transformation of the surplus value rate into a general profit rate introduces no divergence between price and value. Similarly, the neoclassical capital theory is true if and only if there is but one sector in the economy: thus the factor-prices frontier, which represents a family of techniques, is well-behaved so that no technique can switch back nor can capital intensity reverse. Both cases point at the aggregation conditions of economic magnitudes. A theorem by Leontief establishes that production functions of several variables can be aggregated only if any change of one variable leads to the same change in the microeconomic function and in the aggregated macroeconomic function. But that is possible only if it leaves the other variables unaffected. In other words, aggregating production functions of several variables is legitimate only if functions of one variable

31. Geoffrey Colin Harcourt, *Some Cambridge Controversies in the Theory of Capital* (Cambridge: Cambridge University Press, 1972); Jorland, *Les paradoxes du capital*, 355–464.

are added, which requires that the variables be separable. It has been shown that this condition implies that all production functions have the same proportion of factors or that all economic sectors have the same organic composition of capital or the same capital intensity: in short, that there is only one sector.

These mathematical-economic paradoxes as aggregation paradoxes echo the first of that kind based on the fact that the aggregation operator does not have the property of closure. For instance, Cournot had shown that if one considers a set of right triangles and wants to build a triangle such that each side has the mean length of those of the triangles of the set, it might well happen that the resulting triangle is not a right one or does not belong to the set.

The most famous paradox, because it has had a recent version, is Condorcet's. The aggregation of rational individual choices, rational if only because of their transitivity, does not lead necessarily to a collective choice that is also rational in the same sense. Arrow made an impossibility theorem out of it: whatever the rule of the vote, there is no way to dispel Condorcet's paradox.

The end result has been once again to forsake the idea of a comprehensive theory of value as an all-in-one theory of economic regulation. Economists resolved to do so only under the pressure of the many paradoxes of value theories, of whatever kind, that have resisted two centuries of effort to overcome them. Here, anomalies did not lead economists to a revolution but rather to a loss of faith in their grand design. Instead, they have splintered the field and buried their heads in the sand as they tried to take refuge in specialization.

The passing away of a scientific theory, in that case, comes from exhaustion. Most scientists survive lost hopes; some do not. As for the historian of science, he can observe the paradigm switch occurring since the 1970s. The vanishing of a theory is once again accompanied by the coming into being of another theory. A working scientist does not quit a theory except for another one, otherwise he could no longer think at all. This switch has substituted for the neo-neoclassical paradigm (a mixture of neoclassic economics and Keynesianism) the monetarist paradigm that inspires most of the economic policies all around the world, that brings to its students the highest academic rewards such as Nobel Prizes, and that has left its mark on heterodox economics as well.[32]

32. Michel Aglietta and André Orléan, *La violence de la monnaie* (Paris: Presses Universitaires de France, 1983), and Michel Aglietta and André Orléan, *La monnaie souveraine* (Paris: Odile Jacob, 1998).

6 Peter Wagner

"An Entirely New Object of Consciousness, of Volition, of Thought"

THE COMING INTO BEING AND (ALMOST) PASSING AWAY OF "SOCIETY" AS A SCIENTIFIC OBJECT

An entity called "society" became an object of scientific study during the nineteenth century. Its emergence, or its discovery, gave rise to what was then seen as new sciences, variously called "social science," "sociology" (a term coined by Auguste Comte), or directly "science of society" (or in German: *Gesellschaftswissenschaft*). While the study of the gregariousness of human life can be traced to almost any point in intellectual history, there is nevertheless some validity to the claim of novelty on the part of these sciences, a validity that hinges to a considerable extent on the existence of the new object "society." Whether there was such an object at all or whether it was of such novelty that a new science was required for its analysis, however, was contested from the beginnings and remains so up to the present day. The purpose of this essay is not to review exhaustively these debates—an objective impossible to achieve in the space of one chapter. Rather, I shall try to identify some basic problematics that were at the roots of the coming into being of "society," and shall attempt to demonstrate how such problematics shaped the form of this object of inquiry. My attempt will focus on the middle of the nineteenth century, the time when "society" had its strongest presence as an object and when debates about it had acquired considerable momentum. Moving from there briefly backward to the late eighteenth century, and then forward through the twentieth century to the present, will allow me, even though only cursorily, to relate the question of the existence of "society" to historical transformations of social configurations in the West.

THE COMING INTO BEING OF "SOCIETY"

Rupture and Continuity

From an etymological point of view, the term "social"—and its correlates in other European languages—refers to the connectedness of a human being to others. We could say that it enables us to talk about situations in which human beings create relations to each other. In this sense, we can regard as "social sciences" all those theories that reflect upon why and how human beings link up to each other, such as conceptualizations of passions and interests, of individualism and collectivism, of rational, expressive and other orientations of action toward others, Immanuel Kant's thoughts on the "unsocial sociability" of human beings, and many other theorems originating in the seventeenth and eighteenth centuries.[1] However, a very specific way of talking about connections among human beings was introduced into those discourses with the term "society"; and it is significant that the term "social sciences" emerged only in the eighteenth century for those modes of thinking that until then were referred to as "moral and political sciences" or as "state sciences."

From the mid–eighteenth century onward, the term "society" came to be used in the moral and political sciences, in particular within French and Scottish debates, and it became the denomination for the key object of sociopolitical life there. Originally, in combinations such as "political society" and "civil society," it referred to nothing else but the state, but from a point of view of contract theory, namely as the aggregation of human beings that have come together for a purpose. But in some late eighteenth-century theories, "civil society" came to be seen as a phenomenon that was different from the state—but different from the individual households as well. And it is here that the story of "society" as a scientific object starts.

Up until then, in everyday language, "society" used to refer to phenomena that existed in the interstices between the private and the public. In

1. For a recent analysis of this issue in "early social science," see some of the contributions to Johan Heilbron, Lars Magnusson, and Björn Wittrock, eds., *The Rise of the Social Sciences and the Formation of Modernity*, Sociology of the Sciences Yearbook, vol. 20 (Dordrecht: Kluwer, 1998).

France up until the beginning of the eighteenth century, the "society" denoted "small social units that belonged neither to the realm of 'the state' nor to that of the family or household."[2] These units were basically either social circles or legally defined associations, i.e., exactly human aggregations for a purpose. Such usage of the term continues today, as in "high society" or in "Society of Engineers," alongside the sociological meaning.

The introduction of the term into the moral and political sciences—and the change in the semantic position that accompanied it—can, on the one hand, be seen as a reflection of the growth and multiplication of these "small units." Rather than an arbitrary array of phenomena whose only commonality it was to be neither part of the state nor of any particular household, "society" may have gained in importance and coherence such that it could no longer be ignored in moral and political philosophy.[3] The broadening of the meaning of "society" is then a response to an observable change in the structure of social relations, i.e., in the ways the lives of human beings are connected to each other.

On the other hand, the specific new position of this term reveals its initial dependence on another discourse—a discourse indeed, against which the talk about society was directed. The new object society inherited the status of being neither state nor household. The new language thus affirmed that a moral-political entity consisted essentially of (a multitude of) households and a (single) state. It merely added a third category of phenomena; and in the way it did so, it also posited that this third category consisted of a single member rather than a multitude, though the oneness of society was of a different nature than that of the state.[4]

The new threefold division of the moral-political order has therefore to be understood against the background of the earlier twofold division. The latter stemmed, to stretch the point a bit, from some basic continuity from

2. Johan Heilbron, *The Rise of Social Theory* (Cambridge: Polity, 1995), 87.

3. Johan Heilbron argues rightly that "society" in the early social sciences allowed one to relate concerns of moral philosophy, dealing with manners, to political philosophy proper. My own argument could be read as saying that the creation of this relation also entailed some degree of conflation of concerns.

4. To avoid some of the epistemological issues related to attempts to describe an entity that (allegedly) comes into being before it exists or at times when its existence is in doubt (issues to which Bruno Latour refers in his contribution to this volume), I shall use the terms "structure of social relations" as well as "moral-political order" to denote what often is called "society." The former of these terms places the emphasis on the extent, form, and nature of connections between human beings. It tries to be less presupposition-rich than related terms (on the theoretical and methodological issues related to such choice of terminology, see Peter Wagner, "Dispute, Uncertainty, and Institution in Recent French Debates," *Journal of Political Philosophy* 2, no. 3 (1994): 270–89). The latter refers to the central concern of the "moral and political sciences," often regarded as the predecessor of the social sciences.

the Aristotelian conception of the *polis* and the *oikos* through to the political philosophy of liberalism, which made a fundamental distinction between individuals and the state. The "classical liberals" of the seventeenth and eighteenth centuries continued to assume that the free and responsibly acting "individual" citizens were owners of property, which included women, children, and servants, i.e., that they were heads of households rather than single human beings. Their relation to the members of their property was one of mastery; it was at the same time private and of little interest to political thought. Between citizens, however, whether Aristotelian or liberal ones, there was nothing but action and speech, essential freedom. This perspective, thus, gave a very clear-cut view of the structure of the moral-political order in relation to the ways human beings connect with each other: needs determined the private linkages in the house, and the public linkages between men were free. Admittedly, the image I give here comes close to a caricature, but it is important to recognize that the sociological discourse maintained an often implicit reference to this earlier view on the background of which it was modeled. To assess the relation of rupture and continuity in this intellectual transformation, we need to take a look at the historical context.

The Case for Society's Existence

Even though the coming into being of "society" can be traced to the mid–eighteenth century, the historical event that accelerated the intellectual transformation was the French—and to some extent also the American—Revolution.[5] The issue of a new structure of the social world was of crucial importance after the old order had been torn down and claims were made that a new one could be consciously built. To a large extent, the social sciences, as the scientific study of that new "society," owe their forms and contents to the transformation of political issues related to what one may call the onset of political modernity.[6] However, rather than providing the radical rupture that it had announced, the revolution ushered in a rather

5. Keith Michael Baker, in particular, has emphasized the changes of political language that took place before the revolution and, in his terminology, contributed to "inventing" it; see his *Inventing the French Revolution* (Cambridge: Cambridge University Press, 1990). Nevertheless, it was the event of the revolution that made some intellectual positions almost untenable and thus brought about a considerable shift in the discursive balance. See on this broad topic the works of Michel Foucault and, more recently, François Furet in France; of the Cambridge intellectual historians around Quentin Skinner in England; and the works on "history of concepts" around Reinhart Koselleck in Germany.

6. For more detail on this point, see Peter Wagner, "Certainty and Order, Liberty and Contingency: The Birth of Social Science as Empirical Political Philosophy," in Heilbron, Magnusson, and Wittrock, *The Rise of the Social Sciences*, 239–61.

gradual, and not at all linear, transformation of political and social life. "Society" had a slow coming into being, and it was only by the middle of the nineteenth century that some observers could call "the invention of the social" an accomplished fact, particularly for France, whereas others, notably Germans, remained doubtful that any important change had occurred at all.[7]

During this half-century, German observers had made persistent efforts to assess the importance of the events west of the Rhine and to decide whether and how far they would or should have to follow their neighbors' example in the German lands. In the 1850s, when the dust from revolutions and wars had settled, but France seemed to remain somewhat unstable, "state scientists" Robert von Mohl and Heinrich von Treitschke led a debate about the need for a recasting of the political sciences due to the transformations of the social world.[8] The existence of "society" as a scientific object was at the heart of their dispute, which, somewhat off the center of political events, is particularly elucidating for our purposes. It provides something like an anchor halfway down the stream. The (international) debate about the new object "society," and about the need for social sciences to analyze it, is fully developed at this point, and these two authors attempt a systematic assessment from a somewhat distant point of observation.

Mohl opened the debate in 1851, stating that "for about fifty years"— the reference to the French Revolution is evident—"something entirely new" has come into being, the "particular being" of society (6). Consequently, the political sciences, hitherto occupied only with the "individual" and the "entirety," should recognize that there is "between the two, and well distinct from either, a whole, wide area, which similarly has laws that accordingly demand research and ordering." The task was to look at the forms through which "human beings unite, not through the state and its commands, but by way of an accord of their immediate needs, through single but sufficiently powerful interests." These forms vary among peoples, but they "do not fail to appear in any bounded number of humans, i.e., in any people" (12–13). A few years later, Treitschke reviewed Mohl's argument and concluded that "it has not been proven that society was a particu-

7. Jacques Donzelot traced the long-term developments in France in his essay under the suggestive title *L'invention du social: Essai sur le déclin des passions politiques* (Paris: Fayard, 1984). I should note that "the social" is synonymous to "society," when, as is often the case, it is conceptualized as a realm between "the private" and "the political." Other understandings of the "social," often a result of further differentiations within this discourse, will be dealt with below.

8. Robert von Mohl, "Gesellschafts-Wissenschaften und Staats-Wissenschaften," *Zeitschrift für die gesammte Staatswissenschaft* 2 (1851): 3–71; Heinrich von Treitschke, *Die Gesellschaftswissenschaft: Ein kritischer Versuch* (1859; reprint, Halle: Niemeyer, 1927). Further references to Mohl and Treitschke are to these works.

lar element of human conviviality" (Treitschke, 58–59); there was no need to abandon or even complement the state sciences.

Mohl had tried to make a very general argument that took into account all the theorizing on society that had been offered during the preceding fifty years or so. To evaluate claims as to the existence of the new phenomenon "society," it was important to determine its nature and the change in the structure of social relations that it allegedly entailed. By the time of Mohl's writing, one main line of debate on these new kinds of connections between human beings described them as being of a commercial character. Need and work had left the sphere of the household to which they used to be confined and had been exposed to public light. Markets and the division of social labor became the basis of "society" in the tradition of political economy from Adam Smith onward. The argument was taken up and—twice—modified by Hegel in his integration of the division of labor into civil society in the *Elements of a Philosophy of Right* and by Marx in his critique of Hegel's conception.[9] By implication, this movement of needs-related activities into the public sphere entailed that the latter no longer consisted of speech and action alone.

A second line of debate focused on associations as the basis of society and regarded those associations, at least partly, as a response to the effects of the commercialization of life. Along those lines, Lorenz von Stein, for instance, reported in Germany about the "social movements" in France that announced a major change in the social order. Alexis de Tocqueville studied associative life as the basis of a democratic polity in America. And later Marx again gave a central place to the "working class," as a newly formed social phenomenon, in his social theory and philosophy of history.

All these approaches have one feature in common about which Mohl is very explicit. They all claim that major elements of the social world cannot, or can no longer, be grasped through the mere distinction between polity and household or, in the modern, liberal form, between polity and individual. In this sense, these "sociologists" break with an earlier representation of the moral-political order on grounds, as they claim, of a transformed empirical reality. Doing so, however, they used the earlier representation as a resource with which to model the new one; i.e., they added one key element to an existing discourse rather than developing an entirely new representation. This choice, I shall argue, was motivated by the fact that these authors

9. For a recent overview see Manfred Riedel's entry on "Gesellschaft, Gemeinschaft," in *Geschichtliche Grundbegriffe*, ed. Otto Brunner, Werner Conze, Reinhard Koselleck (Stuttgart: Klett-Cotta, 1975), 2:836–37; see also Riedel's entry on "Gesellschaft, bürgerliche," at 719–800.

retained an interest in theorizing the form and feasibility of moral-political order that had informed both Greek and classical liberal thought. "Society" was investigated *because of* the change in political reasoning its existence might require, not because, say, the production of pins or the associative life in America was found inherently interesting. The discovery-invention of "society" in early sociology was also an event in political philosophy, and sociology then can be regarded as a transformed, empirical version of political philosophy.[10]

As such, the introduction of "society" into the moral and political sciences created a very specific linkage of empirical-historical observation and normative-conceptual investment. Such linkages are rather common in the social sciences. In this case, it can even be considered as constitutive for the sociological debate on "society," as I will try to demonstrate. At the same time, however, it creates a basic tension between two perspectives on the "object." The main thrust of the remainder of the chapter will be to discuss how these two aspects have historically been constructed and have been related to each other (or, at times, conflated) in various ways.

CHARACTERISTICS OF "SOCIETY"

Mohl's writings have particular significance in this context. Though he shares the background in "state sciences" with other authors, he refuses to subordinate the "sciences of society" *(Gesellschafts-Wissenschaften)* he calls for to the exigencies of political theory. The phenomena of his "society" are not ontologized. They are many and manifold, and they emerge from crystallizations of more fluid interactions. "How many and what kind of interests are sufficiently big, persistent, and general enough to turn into the core of such a crystallization can neither be determined on general grounds nor on the basis of experience" (Mohl, 50). He does not open any path toward relating them directly to political issues, as Hegel did with his "civil society" or Marx with his "class." At the same time, his conceptual investigation, which combines programmatic ambition with skepticism as to strong assertions, explores all the three discursive possibilities that the introduction of "society" between the polity and the individual entailed, namely the counterposition of society to the polity (the state), the relation of society to the individual, and the argument that "society" could have causal effects.

10. As is reflected, for instance, in the title of Hegel's "Philosophy of Right," a term, incidentally, that was still used in Germany in the early twentieth century for quasi-sociological undertakings in the study of "society."

Society and the State

First, Mohl denies that societal phenomena are necessarily related to the state: "With regard to their extent, [they] do not at all orient themselves towards political boundaries" (44). This proposition may at first glance look innocent and fairly unproblematic. Unless reasons or mechanisms are given, it should not appear as evident why the phenomena belonging to "society" should acquire polity-wide extent. Indeed, straightforward observation should show that that was rather unlikely to occur. Most relations between human beings in the Western Europe of about 1850 were within a local community, such as many family relations. Some others, like production for a world market as described in classical political economy, acquired magnitudes beyond state boundaries, with great effort given available technologies of production and transport. In between the two, there certainly were social phenomena that could best be characterized as statewide, but nothing endowed them with special importance from the point of view of a "science of society." Any arbitrarily chosen multiple set of social relations should normally be expected to show an incoherent variety of magnitudes. The question then imposes itself, however, why Mohl should emphasize this fairly obvious fact.

In fact, though, it was this statement that made explicit that Mohl was arguing against a long tradition of "state sciences" in Germany and even against some of the newer social philosophies that had been proposed during the preceding half-century. He broke with the assumption that the emerging social structures of "society" would show a *coherence* and *boundedness* in a specific way, namely as being coextensive with the boundaries of existing polities.[11] Such an assumption, translated into the terms used here, means that chains of linkages extend from any given member of a "society," so that the web of all these chains forms a bounded whole that has a certain stability, *and* that the magnitude of this web coincides with that of the polity. The predominance of this—very strong—assumption in much of the debate on "society" as well as, indeed, in the very emergence of "society" has to be explained by the intention to analyze historical transformations with a view to issues of political philosophy and as a contribution to political problem solving.

11. In response to Mohl as well as to other authors who separate state and society, Treitschke ponders why this "erroneous political theory" of the "separation of state and society" should have emerged at this time and place, the European nineteenth century, and he finds some reason in the unnatural situation, as in the Germany of the 1850s, where state and society do not match (88). Significantly, he uses here a sociological mode of explanation (though a rather crude one), by deriving an intellectual state of affairs from a sociopolitical one.

Mohl, in contrast, took the autonomy of societal phenomena much more seriously. In a forceful criticism of German state-oriented sciences and philosophies, he tried to show that such phenomena were taken into consideration there only if they could be subordinated to the requirements of political order, their autonomous existence being denied. Treitschke, his great opponent, misunderstood the move Mohl made and showed himself to be part of the state-oriented German intellectual tradition when he claimed, after discussing all the empirical phenomena Mohl introduced, that there was no independent existence of society. In his view, all these relations and organisms had constitutive links to the state, the latter indeed being "organized society," society under the aspect of the state's organization (Treitschke, 73, 68).

It should be granted, though, that Treitschke could have been empirically right where he conceptually erred. Indeed, in the German states at mid–nineteenth century, very few social phenomena could be understood without taking their relations to the state into account. That this should remain so, that order needed to be safeguarded through state-led organization, was Treitschke's normative stand. With this emphasis on exigencies for order, he, the antisociologist, was closer to an important stream of the sociological tradition than Mohl, who tried to argue for sociology in Germany in an intellectually and politically hostile environment.

Without following Treitschke's political preferences, we see that his intervention revealed the necessity of a careful assessment of possible grounds for tying the shape of "society" to a polity, and for underlining its coherent and bounded nature rather than allowing for a more indeterminate pluralistic and open-ended entity. Among the arguments that were brought forward in this context during the nineteenth century, we can roughly distinguish among politico-historical, cultural, biological, and statistical variants.

In *politico-historical* terms, as has often been observed, sociology tended to conflate the idea of the nation-state with "society"; sociology's society was indeed "national society."[12] But why such an assumption was made is not entirely incomprehensible. The French Revolution posited the nation as the bearer of the liberal polity, and under its influence other such national projects of societal organization emerged in Europe. Thus, even if there is nothing inherent in "society" that focuses practices so as to create coherence and boundedness, such focusing could be a historical attempt pursued by nation-building elites. One could envisage historical actors trying to ex-

---

12. See most recently Neil Smelser, *Problematics of Sociology* (Berkeley and Los Angeles: University of California Press, 1997).

tend "smaller" practices to national dimensions, such as abolishing communal sick relief in favor of a compulsory national health insurance, and cutting off (or at least monitoring) those "longer" linkages that stretched across its boundaries, such as protectionist measures to reduce cross-border trade. Once regarded this way, it can indeed be shown that significant attempts of this kind were being made during the period of social-policy innovations and increasingly aggressive nationalism toward the end of the nineteenth century, with the state being a main organizer of these national social practices.[13] If this is the case, however, the autonomy of "society," and its justification as a fundamental, not merely historical, concept, is very much in question—as indeed Heinrich von Treitschke pointed out, though with his particular normative agenda in mind.

In *cultural* terms, it is possible to argue that the existence of "society" is insufficiently explained by recourse to connections between human beings through social practices. For such connections to come into being and to develop, there must be other, foundational principles for a collectivity, such as shared values and norms and commonality of language. This argument featured prominently in German romanticism, and it was in part consciously developed against an Enlightenment conception of individual human beings getting together on reasonable grounds. But it also inaugurated a tradition of social thought that emphasized preindividual sociality of humans and attempted to locate this sociality in observable and analyzable social phenomena. An understanding of society as "collective representations" can be found in Durkheim's early *Rules of Sociological Method*. In his later writings, it appears to gain dominance over the more material concept of societal integration through the division of social labor. Weber, too, is more ready to give conceptual status to "culture," as shared meanings, than to society, as coherent practices. A full elaboration of a social science that reflects on language as the starting point for conceptualizing the shared—and thus social—nature of human practices was more recently offered by Peter Winch in his Wittgensteinian *Idea of a Social Science* of 1958. Such thinking retreated from the full-fledged concept of coherent and bounded social practices across all—or at least major—realms of human action to a more limited range of phenomena that could more justifiably be considered as shared, bounded, and coherent.[14]

13. This "nationalization" of social practices is the focus of a recent study on France and Germany: see Bénédicte Zimmermann, Claude Didry, and Peter Wagner, eds., *Le travail et la nation: Histoire croisée de la France et l'Allemagne* (Paris: Editions de la Maison des Sciences de l'Homme, 1999).

14. It should be said that Mohl was not entirely free of this thinking, though he took a highly individualistic stance compared to most of his German contemporaries.

A similar form of argument, though based on a different justification, can be found in affirmations of the existence of "society" that used *biological* terms. Throughout the nineteenth century (and up to the present), it was contested whether the sciences of human social life operated exactly like the natural sciences, whether they were similarly scientific but devoted themselves to a distinct part of reality that demanded different concepts, or whether they had to develop a different "scientificity" owing to the essentially different nature of the social world as compared to the natural one.[15] The former two of these conceptions allowed conceptual borrowing from the natural sciences, and many social theorists of the nineteenth century likened "society" to an organism. An organism does not consist of individuals but of parts that together form a functional whole. The coherence and boundedness of an organism are not a matter of empirical observation, but are rather the goal of "society," for which its parts are functionally predisposed and constantly mobilized. Versions of such thinking, though mostly without explicit reference to biology, have persisted in functionalism and systems theory. These forms of theorizing fail, however, to provide underpinnings for the strong claim that "society" is capable of self-organization (and of organizing itself as it is), a claim that goes much beyond the one made for its mere existence prior to individuals in cultural thinking.

Finally, there was also an argument for the existence of "society" that appeared to operate without strong presuppositions but on the basis of empirical techniques alone. *Statistical* and demographic research was said to reveal some solid and lawlike features in the characteristics and movements of a population and thus to underpin the idea of the existence of "society." However, these kinds of research produced the unity and coherence as much as they revealed it, by performing on units assumed as given, normally by administrative boundaries. Statistics, as its name indicates, is historically the state science par excellence—and not a science of "society," though its results were often taken to give a proof of the latter's existence.[16]

Mohl was sober enough not to accept any of these arguments in their strong versions. Observation minded as he was, he rejected the biological one. He was too strongly rooted in the liberal tradition, which emphasized

15. See Johan Heilbron, "Natural Philosophy and Social Science," forthcoming in Theodore Porter and Dorothy Ross eds., *The Social and Behavioral Sciences,* Cambridge History of Science (New York: Cambridge University Press).

16. See Alain Desrosières, *La politique des grands nombres: Histoire de la raison statistique* (Paris: La Découverte, 1993); Theodore M. Porter, *Trust in Numbers: The Pursuit of Objectivity in Science and Public Life* (Princeton: Princeton University Press, 1995), 37–38; and Theodore M. Porter, *The Rise of Statistical Thinking, 1820–1900* (Princeton: Princeton University Press, 1986), chap. 2.

individual autonomy, to adhere straightforwardly to the cultural argument. However, this did not make him a classical liberal for whom there was nothing between the individual and the state. He could no longer follow the politico-historical argument, because postrevolutionary events had made social life much more complex than this would allow. And while he indeed turned the extent of societal phenomena into an issue of empirical investigation, he did not believe that statistical research had shown any unitary nature of the social body.

Society and the Individual

Second, if the debate on the relation between society and the state was shaped by concerns of political philosophy at least as much as by the empirical-analytical interest in knowing the "particular being" of society, the attempts to conceptualize the relation of society to individuals were not free from sociophilosophical assumptions either. The very counterposition of state and society made sense, in the traditional way, only if society was somehow on the side of the individuals. In Hegel's *Philosophy of Right*, for instance, civil society was the realm of the particular, where subjectivity was expressed and where interests governed action. Similarly, the economy, the relatively independent realm as it was portrayed in political economy, was a part of society, and distinct from the state, exactly because individuals expressed their particularity there.[17]

If this was the emerging conception in the late eighteenth and the early nineteenth centuries, it was to lose its persuasiveness later. Romanticism was an early reaction, as discussed above, but one that was often—often incorrectly—regarded as looking backward to a harmonious prerevolutionary order. Mohl's writings again mark a transitional point in the sense that, after him, society came to be seen as a sui generis reality that, while being composed of, or "crystallized" from, social relations among individuals, was more than the individuals themselves. If the romantic view of "society" presupposed, so to speak, preindividual connections between human beings that constituted this entity, it was the "modern" connections through the division of social labor or through individual will that brought society into existence in much of later sociology. This was, for instance, the case in Emile Durkheim's early works and in Ferdinand Toennies' writings on "community" and "society" that inaugurated a distinction between a traditional and a modern sense of belonging and of social unity. This view-

17. I have briefly discussed the discursive duality of "state" and "society" as well as the triangle that emerges with the "economy" being added, in *A Sociology of Modernity: Liberty and Discipline* (London: Routledge, 1994), 181–83.

point could receive a strong theoretically anti-individualist bent, opposing sociology as the study of collective phenomena to economics as the study of individuals and their rationalities.[18] The earlier social philosophy here split into two parts, diametrically opposed as to their ontological foundations.

Society and Causality

Third, such a society then was an entity that could have causal effects on individuals' actions. What a human being thought and did was determined by her or his position in a society, in a social structure. Actions and events could thus be explained, and under conditions of adequate knowledge even predicted. Mohl's text offers fine examples of how the argument for the relevance of societal phenomena can be linked to a form of social determinism, though a cautious one in his case.

In his lengthy identification of nonstate, nonindividual phenomena, Mohl described, for instance, aristocracy as a "widely diffused, lasting condition, that entails, for its adherents, a similar feeling, willing, and acting." On the clergy, he notes similarly that it is "a community of thought, of interest, and consequently of willing and acting" (Mohl, 36). To find yourself in a certain lasting social situation has an impact on your volition and action; this is how Mohl puts what has become known as the sociological mode of reasoning and explanation. Even though social-deterministic reasoning and explanation have always remained contested among sociologists, and even though they have been held in a variety of—weaker and stronger—versions, they certainly are defining characteristics of the sociological tradition. Their theoretical viability depends on the possibility of claiming the existence of causally efficacious social entities such as "society."[19]

Mohl himself moved from these observations to a general argument that both justified sociology and demonstrated its necessity. The societal phenomena he identified can be said to be, first, persistent; second, of major

18. The group around Emile Durkheim pursued a strategy of turning "society" into a respectable scientific object, not least on grounds of gaining intellectual reputation for the new discipline of sociology. Comparatively successful as this attempt was in France, it may hamper the current rethinking of the concept exactly because the fate of the discipline as a whole now appears too much tied to it.

19. The very broad debate about forms of determinism need not be taken up here; see, e.g., Ian Hacking, *The Taming of Chance* (Cambridge: Cambridge University Press, 1990). Recently, the link between the emergence of "society" and determinist reasoning has been analyzed, for instance, by Pierre Manent, *La cité de l'homme* (Paris: Fayard, 1994); see also my "Sociology and Contingency: Historicizing Epistemology," *Social Science Information* 34, no. 2 (1995): 179–204.

importance because of underlying interests; and third, more generally diffused (Mohl, 41–43). If that is the case, a science is needed to order these phenomena and identify the laws of their existence. "As soon as it has been stated that society is a specific human relation, the possibility of a special scientific perception of it has been proven" (52).

Mohl's advocacy of social science is a combination of boldness with regard to its necessity and caution with regard to its foundations and rules. Society exists, it acts causally, but it is autonomous, not coherent, not bounded, has no unequivocal relation either to the individuals of which it is composed or to the state. For a long time, this view remained unpersuasive, both in Germany, where the "state" continued to predominate intellectually over "society" until well into the twentieth century,[20] and in France, where the inverse was the case but society, in the Durkheimian tradition, was turned into a higher entity endowed with metaphysical properties. Or in other words, after about a century of sociological debate, the unfortunate dilemma persisted, forcing a choice between, on the one hand, denying the relevance of those "societal" phenomena that exist between the individual and the polity and conceptualizing the polity basically in the same way as before and, on the other, claiming their overarching importance and subordinating the political problematic to the structure of "society." Such a situation was unsatisfactory, both in intellectual and in political terms. The debate on "society" was to continue.[21]

PERSISTENCE OF "SOCIETY" AS AN OBJECT

From the preceding discussion we could provisionally conclude that, when its meaning broadened in the eighteenth century, "society" was an intellectual construction that served some needs of political philosophy—in confirmation of liberal political theory or as a counterargument to it, depending on the author. But that did not necessarily mean that it was completely devoid of empirical content; on the contrary.[22] To say that there was

20. For half a century, if not longer, Treitschke has been the winner of this dispute in Germany. A new edition of his *Gesellschaftswissenschaft* in 1927—in "the era of sociology"— carries a foreword by Erich Rothacker (incidentally, one of Jürgen Habermas's teachers), who claims Treitschke for a German tradition of the scientific study of societal life that should be preferred to French and English biologism (Erich Rothacker, "Zur Einfuehrung," in Treitschke, vii–viii).

21. A useful overview of such later sociological positions on the existence of "society," as will be discussed below, is David Frisby and Derek Sayer, *Society* (Chichester: Horwood; London: Tavistock, 1986).

22. Hans Medick, *Naturzustand und Naturgeschichte der bürgerlichen Gesellschaft: Die Ursprünge der bürgerlichen Sozialtheorie als Geschichtsphilosophie und Sozialwissenschaft*

an increasing number of phenomena in the social world that could be adequately subsumed neither under the category of the individual nor under that of the polity appeared perfectly plausible. However, there were few compelling reasons to assume that this "society" should show a strong degree of boundedness and coherence. The most important of those reasons was the existence of a nation-state apparatus that effectively focused social practices (along the lines of the politico-historical and statistical arguments presented above). No social theory, however, provided for inherent mechanisms as to why commercial exchanges, associative life, and division of labor should hold together in the absence of such an apparatus that did not itself emerge from societal exchanges. At the beginning of the twentieth century, a line of reasoning emerged that attempted to relax such strong theorizing on "society" without abandoning the entire project of a social science, broadly in the Mohlian tradition.

Interim Doubts

This proposal was to regard "society" no longer as an entity but in terms of (spatially and temporally open-ended) processes and relations. Max Weber directed his methodological considerations against what he perceived as a predominant inclination to analyze alleged collective phenomena with concepts that did not have solid foundations. In the German intellectual context and in his reworking of the methodology of the so-called Historical School, this admonition focused on terms like "people" or "spirit of the people," but by implication "society" was named in the same indictment. His own historical sociology started out from human action and its meanings and employed processual concepts instead, of which "rationalization" is certainly the best-known. Such concepts are also problematic, in particular because of their evolutionist leanings, but in this way Weber succeeded in avoiding any undue reification of social phenomena without giving up on the analysis of the social world. His *Protestant Ethic*, regardless of the validity of the particular argument, carefully—even if, as one may want to argue, ultimately not successfully—tries to maintain the objective of sociological explanation without succumbing to simple determinisms.

Georg Simmel, a contemporary of Weber, accepted the problematic or, as he called it, the "riddle" of society as indeed the fundamental "problem of sociology." To solve the riddle, however, exactly meant to ground sociology not on "society" as a "unified being" but rather on the ways human beings

---

bei Samuel Pufendorf, John Locke und Adam Smith (Göttingen: Vandenhoeck & Ruprecht, 1973), e.g., 23–24.

relate to each other, their forms of sociation—his terminological way of turning "society" into a relation rather than an entity—and interaction. The concept of the mutual constitution, in interaction, of self and other as well as of the totality in which interactions take place foreshadows the sociological perspectives that would later become known as symbolic interactionism and social constructionism.

"Society" as the Key Object of Sociological Study

With Weber and Simmel we come to the twentieth century and thus to the so-called classical period of sociology, which proved to be in many respects formative of the institutionalized discipline. In striking contrast to Weber's and Simmel's reasoning, Emile Durkheim then developed the strongest notion of society as an object that had existed hitherto, a notion that through its adaptation by Talcott Parsons had a lasting impact on one major current of sociological theorizing during the twentieth century. Durkheim offered a representation of society in which the elements of the social order were defined according to their position in the division of social labor and their relations regarded as interlocking in the form of "organic solidarity." Parsons modernized the vocabulary by introducing the concepts of "social system" and "system integration," and he enlarged it through elaborate ideas on role systems and mechanisms of functional adaptation.

Parsons and his associates had a strong influence on sociological theorizing between the 1940s and the early 1970s, first in the United States, later in Europe and Latin America as well. At various points, they advertised their approach in key statements about the social sciences, such as the two international encyclopedias that appeared in 1934 and 1968.[23] In both cases, the

23. A useful first step to determining whether a scientific object is said to exist is obviously a look at codified statements on what the science in question is about, i.e., handbooks and dictionaries. Sociology became somewhat codified and consolidated only after the beginning of the twentieth century, and such publications emerge from the 1930s onward, with a second wave of grand attempts being pursued during the expansion of the discipline at universities in the 1960s. Since then, markets seem to have been big enough for a somewhat steady flow of new works and new editions of old works. The closing decades of the nineteenth century already abounded with publications on "the foundations of sociology" and the like. However, these are the proposals and projects of individual authors, trying to assert their own versions of sociology, rather than attempts at a comprehensive representation of a consolidated discipline. It would be an interesting study in itself, not to be pursued here, to trace the changes in the characterization of "society" in these publications over time, across languages, and— given the continued and sometimes deliberate personal imprint of the author(s) in some such works—among authors. The two works that are discussed below are the only two international encyclopedias of the social sciences up to the present; work on a new, third one is under way.

presentation centers on the postulation of "society" as the main object of sociological study, and each time the key problems described above resurface—despite the intentions of the contributors to these works to conceal them.

In the entries on "society" and on "sociology" in the 1934 *Encyclopedia of the Social Sciences*,[24] Talcott Parsons and Robert MacIver respectively give their work a far-reaching genealogy by reading "society" back into earlier, pre-1750 writings in political philosophy. They go back to Plato and Aristotle, whose language did not yet have an "actual equivalent of the English word society" (Parsons, 225) and whose "thought on society never takes specific sociological form" (MacIver, 233), and gradually move toward the period when society is finally "thought of as an independent focus of theoretic interest and of scientific study" (MacIver, 235), and "as possessing in some sense independent reality" (Parsons, 229).

This strategy has the justificatory advantage of both endowing sociology with an object of eternal duration and demonstrating that the very emergence of sociology marked intellectual progress, namely the discovery of this object. However, it underestimates the importance of changes in social practices that go along with terminological shifts. If there was neither "word" nor "thought" in the proper form, could it be that "society" did not really exist before sociology?[25] In the ancient Greek view as well as in much of later political thought, there was no "social" world; and its postulation was not just a discovery but transformed key issues of political philosophy.[26] Parsons and MacIver disregard the alternative justification of the emerging "social" sciences, namely that they should be regarded as *new* sciences for a *new* phenomenon. As I have shown, however, this is how they were indeed presented by some proponents. But the historicization of the object would threaten its ontological status, and it would have opened the door to those kinds of questions Mohl and others struggled with but which Parsons and his colleagues hoped to bypass.

After having established the existence of "society," MacIver goes on to

24. Talcott Parsons, "Society," and Robert M. MacIver, "Sociology," in *Encyclopedia of the Social Sciences*, ed. E. Seligman (New York: McMillan, 1934), 225–32 and 232–47 respectively. Further references to Parsons and MacIver are to these works.

25. This is not necessarily exactly the same as saying that microbes did not exist before their discovery/invention by Louis Pasteur, as Bruno Latour claims (see, for instance, his contribution to this volume). The existence of "society" has sometimes been made explicitly dependent on human knowledge of it by sociologists. Latour provocatively extends such a viewpoint to the "natural sciences." But even in the social sciences, the more conventional approaches insisted on a knowledge-independent existence of scientific objects.

26. For a fundamental critique of the invention of "society," using ancient Greek philosophy, see Hannah Arendt, *The Human Condition* (Chicago: University of Chicago Press, 1958).

emphasize that it was exactly the insight that it was "distinct from the state" (MacIver, 244) that made society the object of a new discipline, sociology. Parsons also hailed the achievement of the separation of the social realm from the political. However, that separation was mostly conceived in terms of sociophilosophical presuppositions that did not give "society" the potentially autonomous existence that Mohl thought was required. Separation implied that human social life had its particularities and subjectivities that were to some extent independent of the laws of the state. But these sociologists of separation did not really think of social relations developing independently from state exigencies; the concern about the "social contract," now sociologized as the need for "integration," remained predominant. As Talcott Parsons put it, "[W]ithout a system common to the members of a community social order itself cannot be accounted for" (230–31).[27]

Again, however, it is difficult historically to confirm any such view of "society." And, as argued above, other sociologists such as Weber and Simmel were well aware of the flaws in such a conceptualization. If "society" was nevertheless increasingly (re)emphasized during the early decades of the twentieth century, the basic reason lies in the perception that social practices did in fact *not* cohere or remain within a bounded order, *and* in that this situation was seen as politically problematic. By that time sociology had more or less successfully presented itself as a scientific approach to contemporary society and its problems. Trying to show how a new coherence could emerge after the turmoils of industrialization, workers' struggles, urbanization, in short "the social question," was at least as much a political action as a sociological proposition. To some extent, sociologists participated in constructing "imagined communities," in Benedict Anderson's phrase for solving problems of social coherence and boundedness.[28] It is to their credit that many of them, and in particular those whom we have come to call "classical" sociologists, persistently used more sober terms than many of

27. Without specification of the term "social order," which Parsons accepted as a problem inherited from Hobbes through all of the history of social philosophy, this sentence reads as a tautology. It was left to American sociological approaches inspired by Simmel and pragmatism to disentangle what social order is and how it comes about. See most recently Dennis Strong, *The Problem of Order: What Unites and Divides Society* (Cambridge: Harvard University Press, 1994).

28. I have dealt with aspects of these developments in two other essays: "Crises of Modernity: Political Sociology in Its Historical Contexts," in *Sociology and Social Theory*, ed. Stephen P. Turner (Oxford: Blackwell, 1996), 97–115; and "Science of Society Lost: On the Failure to Establish Sociology in Europe during the 'Classical' Period," in *Discourses on Society: The Shaping of the Social Science Disciplines*, ed. Peter Wagner, Björn Wittrock, and Richard Whitley (Dordrecht: Kluwer, 1991), 219–45.

their contemporaries who tried to provide foundational readings for a new community.[29]

The concept of the social system then provided a kind of antifoundational foundation for social coherence and integration. It is the latest incarnation of "society" that attempts to maintain the double orientation toward the exigencies of political philosophy as well as to the empirical analysis of the structure of social relations. However, the tensions in the conceptual construction could not be overlooked. If they could not be resolved, they had to be glossed over. The entry on "Society" in the *International Encyclopedia of the Social Sciences*, published in 1968 at the likely historical zenith of the cultural importance of sociology, confirms this suspicion.

Through the voice of Leon Mayhew, then one of Parsons's doctoral students, this authoritative source defines society as "a relatively independent or self-sufficient population characterized by internal organization, territoriality, cultural distinctiveness, and sexual recruitment."[30] This definition sounds highly empirical, but it is also loaded with presuppositions. It gives

29. Along those lines, one might even try to rewrite the history of early twentieth-century sociology in terms of sociologists' concern about, and commitment to, strong notions of society. Their concern about such collective concepts would reveal something about the political self-understanding of their work. It would show, as in Weber's case, to what extent they maintained the aspect of political philosophy that had characterized sociological discourse from its beginnings. Those who, unlike Weber, simultaneously committed themselves to such collective concepts revealed an uneasiness about the sustainability of a liberal-individualist solution to issues of social coherence. Political preferences might not always correspond with intellectual ones, but there certainly were affinities between the two.

30. Leon H. Mayhew, "Society," in *International Encyclopedia of the Social Sciences*, ed. David L. Sills (London: Macmillan, 1968), 577–86, at 577. Further references to Mayhew are to this work. (I owe the information about Mayhew's position at that time to a personal communication from Neil Smelser.) Cf. Harry M. Johnson, *Sociology: A Systematic Introduction* (London: Routledge & Kegan Paul, 1961), 10, where society is characterized by "(1) definite territory, (2) sexual reproduction, (3) comprehensive culture, and (4) independence." Or: "A society exists to the degree that a territorially bounded population maintains ties of association and interdependence and enjoys autonomy." (Gerhard Lenski, *Human Societies: A Macrolevel Introduction to Sociology* [New York: McGraw-Hill, 1970], 9, as quoted in Paul B. Horton and Chester L. Hunt, *Sociology*, 3d ed. [Tokyo: McGraw-Hill Kogakusha, 1972], 49). Or: "The most complex macrostructure is a *society*, a comprehensive grouping of people who share the same territory and participate in a common culture" (Donald Light Jr. and Suzanne Keller, *Sociology*, 4th ed. [New York: Knopf, 1985], 93). Other encyclopedic works consulted include: William F. Ogburn and Meyer F. Nimkoff, *A Handbook of Sociology* (London: Routledge & Kegan Paul, 1947); G. Duncan Mitchell, *A Dictionary of Sociology* (Chicago: Aldine, 1968); Patricia M. Lengermann, *Definitions of Sociology: A Historical Approach* (Columbus, Oh.: Merrill, 1974); Theodor Geiger, "Gesellschaft," in *Handwörterbuch der Soziologie*, ed. Alfred Vierkandt (1931; reprint, Stuttgart: Enke, 1959), 201–11; Dankmar Ambros, "Gesellschaft," in *Handwörterbuch der Sozialwissenschaften* (Stuttgart: Fischer, Mohr, Vandenhoeck & Ruprecht, 1965), 427–33; Günter Endruweit and Gisela Trommsdorf, eds., *Wörterbuch der Soziologie* (Stuttgart: Enke, 1989); Gerd Reinold, *Soziologie-Lexikon* (Mu-

no fewer than four criteria, at first glance unrelated, that need to be fulfilled at the same time, and it turns them into strong criteria by demanding that the resulting phenomena be "independent and self-sufficient." A naive reader might readily arrive at the conclusion that "society" must be a rare occurrence in the social world. Without good reasons to the contrary being given, one might well assume that cultural practices and patterns of sexual partner choice may have different spatial dimensions, and that both could well be unrelated to other aspects of social organization—the boundaries of the polity, for instance. Why, first of all, should we think that human beings dwell in "independent and self-sufficient" groups rather than have a variety of links to other groups, and possibly not form any identifiable collective whole at all? In line with the tradition outlined above, our sociologist might answer that it turns out they do, as a matter of fact, and that this is not least the discovery of sociology. However, Leon Mayhew is well aware of the complexity of the issue. He takes a number of steps to qualify his assertion.

First of all, Mayhew lightens the burden of the criterion of self-sufficiency. It should not be understood as a situation of isolation and autarchy, but rather one of "controlled relations with an environment" (Mayhew, 584). People do interact across the boundaries of societies, but societies can control these boundary crossings. A different type of complication is evidently created here, since societies are now being endowed with agential capacities rather reminiscent of a state agency such as a customs office. However, the strong, and hardly tenable, criterion of self-sufficiency is abandoned.

Once this avenue is opened, further skeptical thoughts can no longer be avoided. Indeed, the different kinds of human practices could have different spatial and demographic extensions. Only when such spaces overlap, "when a relatively broad range of such systems cohere around a common population, we may speak of a society" (Mayhew, 583). Such rethinking may even go so far as to demand an alternative approach: "The emergence of a bounded, unified social system is no longer assumed but becomes an object of inquiry" (584), the consequence being that "the concept of a *society* with exclusive boundaries may be obsolete" (583). Sociologists could analyze the extensions of modes of internal organization, cultural practices, and sexual recruitment as well as of other social practices and could then try to characterize specific configurations that might emerge from such an analysis. The existence of society, as defined above, would become a purely empirical issue.

nich: Oldenbourg, 1992); Werner Fuchs-Heinritz et al., *Lexikon zur Soziologie* (Wiesbaden: Westdeutscher Verlag, 1994).

However, the reflection does not stop here. Mayhew assumes that there are other, apparently nonempirical reasons for sticking to the idea of society.[31] "Social theorists have found in 'society' a convenient foundation for relating their specific problems to a larger context" (578). Since Mayhew does not provide any further detail, this phrase remains enigmatic, to say the least. What are the specific problems of the social theorist? And why should it be useful to relate them to a larger context (instead of, for example, solving them)? What implicitly happens here is that he acknowledges the most important issue of "society" as a sociological object to be unresolved. The very specific linkage of empirical-historical observation and normative-conceptual problematic that was invested into this object created tensions that kept breaking it up.

The Passing Away of "Society"?

Since the late 1960s, the time when Mayhew wrote, the acknowledgment of this tension has become more widespread within sociological debates, and the insight that the idea of "society" as an object may need to be given up has gained ground.[32] The normative-political problematic is still with us. There are again good reasons to assume that the structures of social linkages do not cohere and show no strongly overlapping boundaries. And again, there is much talk about the need for new coherence, now often straightforwardly under the title of "community"—as well as renewed debates about the nation as the allegedly natural container of shared practices. In contrast, hardly anybody outside systems theory seems any longer convinced that "system integration"—or any other concept for "society"— could suffice to deal with the time-honored political problematic of relating the strivings of a multitude of individuals to the requirements of sustaining a polity. Significantly, sociologists are not even the key participants in these

31. Harry M. Johnson's remark that "the concept *society*, although unrealistic, might have as great scientific interest as, let us say, the concept of perfect competition in economics" (Johnson, *Sociology*, 13) is amazingly blunt about the problematic relation between concepts and experience. In his view "concepts" seems to refer to some overarching guides for social analysis and/or social life, but—unlike Weber's ideal types, for instance, which are no real sociohistorical phenomena either, but whose validity is measured against empirical findings—their relation to reality is not exactly an issue. The concept of perfect competition in economics has at least had the advantage of having acquired strong discourse-organizing power, which cannot to the same degree be said about "society" in sociology. Cf. also the chapter on "value" by Gérard Jorland in this volume.

32. See, for instance, the perspectives developed by Alain Touraine (from his early work *La sociologie d'action* [Paris: Seuil, 1965] onward) and Michael Mann (*The Sources of Social Power*, vol. 1 [Cambridge: Cambridge University Press, 1986], chap. 1).

debates (though some schools are present), which are led by political philosophers.

In this most recent development we may find indications that sociological analysis of the structure of social linkages might ultimately fully separate from the politico-philosophical concern about the sustainability of the polity. The tension in the concept of "society" would then be abolished and, as a consequence, it appears very likely that the concept could no longer be sustained; the interest in it would subside. If this were the case, one would have to conclude that it was only the ambivalent relation of sociology to political philosophy that upheld the idea of "society" as an existing object. By way of conclusion, I shall try to review the findings of my brief history of the concept in the light of the current possibility of such an imminent passing away.

THE NEED FOR SOME NEW LANGUAGE

In the preceding observations I have tried to identify the places "society" occupies in the various discursive formations that make the claim of the existence of this object. Without being explicit about it, I have gradually introduced a language for speaking about "society" that is not itself dependent on any specific one of those discourses under study. Thus, I tried to characterize the broadest possible space of a sociology, i.e., to offer a conceptualization that allows for the possibility that "society" exists but does not presuppose its existence. The purpose of this construction was to show how historically existing approaches to sociology have moved in a broader, common space and which relative positions they have taken with regard to each other in this space.

When the terms "social" and "society" were appropriated as key concepts of the emerging "social sciences," they were simultaneously considered as a rather general and as a very specific way of talking about connections among human beings. With regard to the general *nature* of these connections, i.e., to forms of "sociability," there was probably not even much novelty in so-called early social science after mid–eighteenth century.[33] Much more specifically, however, the new term "society" was meant by some important participants in the debate to denote a structure of such connections, which did not exist before a certain time, or which could not well be described by earlier languages (or both).

33. This is, for instance, what Istvan Hont claims; see "Socialist Natural Law, Commercial Society, Political Economy: A Contribution to the Idea of Social Science," paper presented at the conference on "The Rise of the Social Sciences," Uppsala, June 1993.

Key importance was given to the extension of the chains of connections due to the development of commerce, and to some extent also due to the freedom of expression and the constitution of a "public sphere,"[34] as well as to the newly relevant boundary of these extended chains created by the nation-based constitutional polity. These were, in various combinations, the decisive elements of the novel structure of connections that came to be called "society." However, though some novelty was without doubt brought about by the effects of the industrial and democratic revolutions, one may well remain very skeptical as to whether the structure of social connections was in any way decisively transformed before, at least, the middle of the nineteenth century, so as to call for an entirely new language. It was at that point that an open-minded observer such as Mohl could make an effort at a summary account of empirical phenomena of a "societal" nature.

Mohl himself provided what we could call a justification for a "weak" concept of society. To him, it sufficed that a number of durable and important phenomena existed that could be analyzed neither from the viewpoint of methodological individualism nor as derivative of the polity. In the German lands, he had difficulty getting a hearing for even this "weak" position. Elsewhere, a much "stronger" concept of society was offered. Social theorists tended to have specific views on how these phenomena formed chains of linkages and how their coherence—or, for critical theorists, their self-contradictory nature—was brought about. This is to say they had substantive theories about the solidity of society or its historical direction that made many more presuppositions than any empirical analysis of chains and webs of connections could ever confirm.

By way of conclusion, let me review the arguments in favor or against the existence of "society" from a current perspective.

The Disappearance of "Society" as an Object

It seems difficult to reject the most general proposition that there are relevant phenomena between the house (the individual, the private sphere) and the polity (the state, the public sphere). The question that needs to be raised is whether their conceptualization as being in-between these other two phenomena does not accept too much of a specific discourse, namely political theory, so as to limit the analysis of "society." This suspicion finds confirmation in some of the further issues raised.

34. Significantly, the former is more typically the Scottish-English view, the latter the French one; see Johan Heilbron, "French Moralists and the Anthropology of the Modern Era: On the Sociogenesis of the Notion of 'Interest'," in Heilbron, Magnusson, and Wittrock, *Rise of the Social Sciences,* 77–106.

In some discourses on society, this object came to take the place that "polity" held in political philosophy. One could not say that it was synonymous with polity, since the whole discursive structure was transformed, but it did serve as the term for the integration of a multitude of diverse "parts" (which could be, but did not have to be, "individuals") into a "whole." Organicist theories of society as a body are the most obvious example, even translating the metaphor of the "body politic" into a new discourse. But Durkheim's "organic solidarity" and Parsons's "system integration" also use "society" to a similar end. If one wants to do justice to the repeated postulate of the "autonomy" of society from the state, such reasoning risks a severe conflation of issues.

A weaker version of the preceding assumption is the idea that society is an effectively bounded whole, whether well-integrated with the polity or disruptive of it. Reference is then often made to some special, often explicitly *nonstate* aspect of relations *within* the polity. But there is something artificial about this conceptualization. If "societal" relations, such as "cultural" or "economic" ones, remain "within" the boundaries of a polity, then they are most often not truly of a "nonstate" character, but indeed statewise manipulated or even controlled by the state, as in the case of educational institutions or customs regulation. Otherwise they would most likely not remain confined to state boundaries. And the shift from social practices to meanings and language is, in principle, open to the same criticism as are stronger notions of society. Why should cultures and languages be more prone to boundedness and coherence than other social practices? And is the impression of their being "systems" of orderly interrelated elements given once and for all, or is it but rather a result of the historical labor of making them closed and coherent, with the nation-state being again, at least in Europe, one important source of such efforts?

The idea of a nationwide extension of such phenomena as we found it in encyclopedia definitions of society leads to a further variation of this kind of assumption. My survey found no theoretical ground for such an assumption. Instead, the historical development of sociology happened to coincide with a period of conscious attempts to "nationalize" social practices. Rather than the firm foundation of a science of society, the sociology of "national society" is itself a historical phenomenon. Such historical coincidences may then to some extent explain the longevity of the two aforementioned assumptions, since this is how the nation-state, national society, and the latter's relative cultural-linguistic homogeneity were indeed linked.

Of a different order is the use of "society" as a means to demarcate a sociological way of thinking from an individualistic one, as is prevalent in economics and psychology. Much of the sociological criticism of ontological

individualism is well justified, but its force does not really depend on holding the inverse proposition of ontological holism such as in the postulate of
"society." A sociology might well be based on a concept of sociability or social relations that allows for the open-ended nature of the chains of connections created through such relations.

If the objective of sociological analysis were defined as the investigation
of social relations, in the above sense, then some of the fundamental debates
in the history of sociology could be recast. Instead of asking whether "society" was a historically emerging phenomenon or a fundamental condition
of human life, changes in the extension of effective connections between
human beings would have to be analyzed. Furthermore, such an analysis
would have to distinguish whether such connections are material, as in a
new division of social labor, or discursive, as in a redefinition of membership
communities in terms of "collective representations" and of society as an
"object of consciousness, of volition, of thought."

Furthermore, no such identification of a particular "structure of society"
would allow us to draw immediate conclusions as to whether the position in
the structure determines the thought and action of an individual. The position may be seen as firm, stable, and socially defining, but it may also be seen
as unimportant, transient, a momentary fact. The sociological tradition
tended to regard the social and socially determined nature of human life as
one of its most important insights. Without entirely doing away with that
insight, we should rather consider it as the question sociology contributed
to social and political thought, not as the answer to all questions.

Some Inescapability of "Society" as a Concept

At the end of the twentieth century, the suspicion is widespread that "society" is a concept difficult to uphold, not least for reasons of its decreasing
empirical reference due to the internationalization, sometimes globaliza-
tion of practices, on the one hand, and alleged individualization, on the
other. Predictions of its passing away will almost certainly come true if the
concept of "society" is meant to carry several or even all of the strong assumptions it was historically endowed with. However, none of the alterna-
tives to the sociological mode of reasoning that are currently in vogue, and
certainly not rationalist individualism, will be able to deal with what could
more prudently be called the *representation of the state of social relations.*[35]

"Society" was a conceptual tool to represent the state of social relations
in a particular temporal and cognitive space. In its strong versions, it was

---

35. I owe this way of putting the issue to discussions with Luc Boltanski.

never fully convincing. However, it addressed a key issue of social thought, namely a transformation in the political problem of unity in liberal theory, in a way that tried to take certain changes in social relations into account. Tensions between these two aspects of the term, a politico-conceptual necessity and an empirical phenomenon, could never be overcome. But between the early nineteenth and the late twentieth centuries—the period of the coming into being and the possible passing away of "society"—no superior solution to this political problematic could be offered. "Society" will not entirely disappear until this has happened.

If "society" is currently out of fashion without being superseded by a more appropriate concept, this means that a political sociology that conflated issues in conceptual shortcuts has been replaced on the one hand by a return to a sociologically ill informed political philosophy, and on the other by a sociology that is blind to political issues. It is difficult to say whether such development is a liberation or a loss. It seems evident, however, that some new relation of sociology to political philosophy will have to be built that addresses the inescapable issues of politics without imposing constraints on social analysis.

ACKNOWLEDGMENTS

I would like to acknowledge the support given by the Wissenschaftszentrum Berlin für Sozialforschung and the Center for Studies in Higher Education at the University of California, Berkeley, where the first and second versions respectively of this chapter were written. This chapter has profited from comments offered by Johan Heilbron and Neil Smelser, to whom I am grateful. The quotation in the main title is by Robert von Mohl, "Gesellschafts-Wissenschaften und Staats-Wissenschaften," *Zeitschrift für die gesammte Staatswissenschaft* 2 (1851): 3–71, at 6.

"Sentimental Pessimism" and Ethnographic Experience; or, Why Culture Is Not a Disappearing "Object"

There is no way "culture" can disappear as the principal object of anthropology—or for that matter as a fundamental concern of all the human sciences. Of course it can and already has lost some of the natural-substance qualities it had acquired during anthropology's long infatuation with positivism. But "culture" cannot be abandoned, on pain of failing to comprehend the unique phenomenon it names and distinguishes: the organization of human experience and action by symbolic means. The persons, relations, and materials of human existence are enacted according to their meaningful values—meanings that cannot be determined from their biological or physical properties. As my teacher Leslie White used to say, no ape could appreciate the difference between holy water and distilled water, any more than it could remember the Sabbath and keep it holy. This ordering (and disordering) of the world in symbolic terms, this culture is the singular capacity of the human species. To demand that the study of culture be banished from the humanist disciplines, on the grounds (for example) that it is politically tainted by a dubious past, would be a kind of epistemological suicide. The anthropological sense of culture has been able to transcend the notion of intellectual refinement from which it descended (the sense of cultivation, which is still a common gloss of the term). It has been able to separate itself from the progressive ideas of "civilization" with which it was once entangled (as by E. B. Tylor). We can be sure that culture will also weather the current attempts to delegitimate it by virtue of its presumed historical associations with racism, capitalism, and imperialism.[1]

DEATH TO THE NOBLE CULTURE?

Contemporary threats to the noble culture arise in connection with its partitive manifestations as specific forms of human social life, the classic "cultures" of given communities and societies.[2] The criticisms are of two sorts. There is the short-term problem just mentioned: the moral suspicions cast on culture by a certain politics of interpretation, usually backed up by a historiography of original sin. The long-term and more serious issue concerns the continuity and systematicity of anthropology-cultures—of which the present postmodern panic about the coherence of cultural orders is, I believe, only the latest manifestation. Here the disappearing-object paradigm is surely relevant. It has always been relevant. Anthropology may be the only discipline founded on the owl of Minerva principle: it began as a professional discipline just as its subject matter was dying out. Or if the so-called primitive peoples were not actually dying, their exotic cultures were certainly disappearing ("acculturating") under the onslaught of the capitalist world order. It seemed that soon there would be nothing left to contemplate but local versions of Western "civilization." In this respect, anthropology originally took the same views on progress as the colonial masters, if with greater regrets.

But to consider first the moral-political controversies now besetting the anthropological culture concept. "Culture" is notably suspect insofar as it

1. Robert J. C. Young, *Colonial Desire: Hybridity in Theory, Culture, and Race* (London: Routledge, 1995); Christopher Herbert, *Culture and Anomie: Ethnographic Imagination in the Nineteenth Century* (Chicago: University of Chicago Press, 1991). On the general history of the term "culture" see Raymond Williams, *Keywords: A Vocabulary of Culture and Society* (new York: Oxford university Press, 1976); George W. Stocking Jr., *Race, Culture, and Evolution: Essays in the History of Anthropology* (New York: Free Press, 1968); Norbert Elias, *The Civilizing Process: History of Manners* (New York: Urizen Books, 1978).

2. As is well known in the English-speaking world, the term "culture" is now at a discount. It is used for social categories and groups of every shape and form. One speaks of the culture of almost any definable category (the "culture of drug addicts," the "culture of adolescents," etc.), of all sorts of activities (the "culture of bungee jumping," "the culture of autobiography"), and, of course, of all sorts of groups (the "culture of corporations," the "culture of the university," the "culture of the cigar factory"). The word has replaced "ethos" (we used to speak of the "ethos of the university" or the "ethos of bodybuilding") and also "psychology" (as in the "psychology of Washington, D.C.," or the "psychology of the Cold War"). At present, it is not easy to say whether all this really cheapens the anthropological concept of "culture," as it might seem, or actually strengthens and reinforces it. We will return to aspects of the modern "culturalism" later in this essay.

marks customary *differences* among peoples and groups, and especially when it thus distinguishes subordinate populations in oppressive political regimes. Culture as the *demarcation of difference* is at issue here—together with an implicit and quixotic battle against something no one really believes, that cultural forms and norms are prescriptive and admit no possibility of human agency. So for a colonized or racially-discriminated-against people, a reference to their culture—for instance, "Nuer culture" or "African-American culture"—is an operation of the hegemonic distinctions of their servitude. Hence the current critiques of the culture concept as an ideological trope of colonialism: an intellectual mode of control that has the effect of "incarcerating" hinterland peoples in their spaces of subjection, permanently separating them from the progressive Western metropole.[3] Or more generally, by its deployment to the stabilization of difference, the anthropological idea of culture authorizes the inequalities of every shape and form—including racism—inherent in the functioning of Western capitalism.[4]

The indictment of culture for its alleged complicity in some of the principal crimes of modern history rests on certain suspect forms of theoretical

3. So according to Dirks, colonialism is probably a necessary condition of the *invention* of the culture concept:

> The anthropological concept of culture might never have been invented without a colonial theater that both necessitated the knowledge of culture (for the purposes of control and domination) and provided a colonized constituency that was particularly amenable to "culture." Without colonialism culture could not have been so simultaneously, and so successfully, ordered and orderly, given in nature at the same time it was regulated by the state. Even as much of what we recognize as "culture" was produced by the colonial encounter, the concept was in part invented because of it. (Introduction to Nicholas Dirks, ed., *Colonialism and Culture* [Ann Arbor: University of Michigan Press, 1992], 3)

For other criticisms of "culture" as colonizing, see Arjun Appadurai, "Disjuncture and Difference in the Global Cultural Economy," *Public Culture* 2 (1988): 1–24, and Ronato Rosaldo, "Imperialist Nostalgia," *Representations* 26 (1989): 107–22.

4. Culture as capitalist difference, linked indissociably with racism:

> Culture never stands alone but always participates in a conflictual economy acting out the tension between sameness and difference. . . . [T]he constant construction and reconstruction of cultures and cultural differences is fueled by an unending internal dissension in the imbalances of capitalist economies that produce them. . . . Culture has always marked cultural difference by producing the other; it has always been comparative, and racism has always been an integral part of it: the two are inextricably clustered together, feeding off and generating each other. Race has always been culturally constructed.

If the very object of anthropology could be said to be cultural difference, this clearly makes it a particularly significant discipline in our contemporary culture of difference:

pleading. Culture is subjected to a double conceptual impoverishment: a reduction to a particular functional effect (marking difference), from which is construed an abbreviated narrative of misbegotten origins (in colonialism or capitalism). First, the concept is translated as an instrument of social differentiation.[5] Culture creates classes, races, colonized peoples; it is an ideological means of victimization. A move from content to supposed effects, from properties to purported purposes, this functional reduction erases just about everything anthropology seeks to understand—and in the field, struggles to know—about human cultures as forms of life. Here is one of those bargains that functionalist explanations make with the ethnographic reality, trading off most of what we know about a phenomenon in return for an understanding of it. Social institutions, modes of production, values of goods, categorizations of nature, and the rest—the ontologies, epistemologies, mythologies, theologies, eschatologies, sociologies, polities, and economies by which peoples organize themselves and the objects of their existence—all this comes down to no more than an apparatus by which societies or groups distinguish themselves from one another. And as this is what culture is really about, then anthropology's "very object" can be no more than "cultural difference."[6] Not even the explication of cultural diversity, mind you, or some such examination of what the differences (and similarities) are, but the demarcation of difference as such, as a value. On the epistemological level, contrast as a means of knowing has been turned into knowing as a means of contrast. The effect is the perverse reduction of cultural comparison to invidious distinction.[7] Second, then, the corollary to

---

but what this passes over is the way it has participated in the history of difference; which continues to repeat on us today. (Young, *Colonial Desire*, 53–54)

Perhaps so, but what this seems to pass over is a recurrent historical opposition of anthropology to racism based on constructions of culture—including an appreciation of its constructedness—that go far beyond the operation of difference.

5. For example:

Culture is the essential *tool* for making other. As a professional discourse that elaborates on the meaning of culture in order to account for, explain, and understand cultural difference, anthropology also helps *construct, produce,* and *maintain* it . . . In this regard, the concept of culture *operates* much like its predecessor—race. (L. Abu-Lughod, "Writing Against Culture," in *Recapturing Anthropology: Working the Present*, ed. Richard Fox [Santa Fe: School of American Research Press, 1991], 137–62, at 143; emphases added)

6. Young, *Colonial Desire*, 54.
7. Dumont cites Wilhelm von Humboldt:

[H]ow can one possibly know completely a nation's character without having also studied other nations with which it is in close relation? It is in contrast with them that

this resolution of the concept of culture to a politics of distinction is the temptation to derive the former from the latter, by a kind of pseudohistory of the original-sin variety. The sin of culture was indeed pride, nothing else than Western pride. An expression of capitalism's systematic creation of otherness, the conceit called culture—together with its intellectual twin, race—originated in early modern Western European relations of production. In its "genesis and semantic operation," culture bears "the stigmata of capitalism, and repeats and acts out the conflictual structure of the class system that produced it."[8] Or, as an alternative stigma, the culture idea was "produced by the colonial encounter" and "in part invented because of it"—for equally culpable purposes of discrimination and domination.[9] In sum, the functional reduction of culture to difference turns into an imagined history.

But as it actually developed in late eighteenth-century Germany, the anthropological concept of culture was connected with "yet another philosophy of history" indeed. Johann Gottfried Herder's notion of culture foresaw rather different relations between imperialism and anthropological study than are dreamed of in the current criticism:

> Our technologies are multiplying and improving: our Europeans find nothing better to do than run all over the world in a kind of philosophical frenzy. They collect materials from the four corners of the earth and will someday find what they are least looking for: clues to the history of the most important parts of man's world.[10]

Insofar as the anthropological idea of culture was originally associated with reflection on difference, it was by opposition to the colonial civilizing mission to which the concept is popularly attributed nowadays. Fact is, cultural

---

this character actually came into being and it is only through this fact that it can be fully comprehended. (in Louis Dumont, *German Ideology* [Chicago: University of Chicago Press, 1994], 120)

For a textbook demonstration of the productivity of comparison—even, in this case, semi-uncontrolled, quasi-typological comparison—see Clifford Geertz, *After the Fact: Two Countries, Four Decades, One Anthropologist* (Cambridge: Harvard University Press, 1995), esp. chap. 3, "Cultures."

8. Young, *Colonial Desire*, xx.
9. Dirks, introduction to *Colonialism and Culture*, 3.
10. Or, on the plane of *Realpolitik:*

The more we Europeans invent methods and tools with which to subjugate other continents, the more we defraud and plunder them, the greater will be their final triumph over us. We forge the chains with which they will bind us. (Johann Gottfried von Herder, *J. G. Herder on Social and Political Culture*, ed. and trans. F. M. Barnard [Cambridge: Cambridge University Press, 1969], 218)

difference in itself has no inherent value. Everything depends on who is making an issue of it, in relation to what world-historical situation. In the past two decades peoples all over the globe have been self-consciously counterposing their "culture" to the forces of Western imperialism that for so long afflicted them. Culture here figures as the antithesis of a colonial project of stabilization, since the peoples articulate it not merely to mark their identity but to seize their destiny. In an analogous way, certain German bourgeois intellectuals, bereft of power as a class or unity as a nation, answered the Enlightenment apostles of a universal "civilization"—not to forget the Anglo-French menace of industrial domination—by the celebration of indigenous national *Kulturen:*

> The princes speak French, and soon everybody will follow their example; and, then, behold, perfect bliss: the golden age, when the world will speak one tongue, one universal language, is dawning again! There will be one flock and one shepherd! National cultures, where are you?[11]

Unlike "civilization," which could be transferred to others—as in beneficent gestures of imperialism—"culture" is what uniquely distinguished a given people—as from the superficial French manners of the Prussian aristocracy. Culture comes in kinds, not degrees. As specific forms of life it is inherently pluralized, by contrast to a universal progress of reason culminating in Western European "civilization." In the late eighteenth century—as again in the late twentieth—an anthropological idea of culture emerges in a relatively underdeveloped region, in its demands of autonomy, in the face of the hegemonic ambitions of Western European imperialism:

> *Kultur* theories can be explained to a considerable extent as an ideological expression of, or reaction to, Germany's political, social and economic backwardness in comparison with France and England . . . These *Kultur* theories [Russian as well as German] are a typical ideological expression—although by no means the only one—of the rise of backward societies against the encroachments of the West on their traditional culture.[12]

11. Ibid., 209.
12. Alfred Meyer, "Historical Notes on Ideological Aspects of the Concept of Culture in Germany and Russia," in *Culture: A Critical Review of Concepts and Definitions*, ed. A. L. Kroeber and Clyde Kluckhohn, eds. (New York: Vintage Books, 1952), 403–13, at 404–5. The historic opposition of Enlightenment "civilization" and Germanic "culture" can be followed in many discussions, including *Herder on Social and Political Culture;* Philippe Bénéton, *Histoire de mots: culture et civilisation* (Paris: Presses de la Fondation Nationale des Sciences Politiques, 1975); Emile Benveniste, *Problems in General Linguistics* (Coral Gables: University of Miami Press, 1971), 289–96; Isaiah Berlin, *Vico and Herder: Two Studies in the History of Ideas* (New York: Vintage Books, 1976); Isaiah Berlin, *Against the Current: Essays in the His-*

As it was counterposed to a totalizing discourse of enlightenment, how-ever, this "culture" had to signify much more than a politics of difference. Entering the lists against the philosophes' idea of a singularly utilitarian human nature, everywhere perfectible by the exercise of right reason on clear and distinct perceptions, the Herderian "culture" entailed equally ex-pansive if suitably contrastive views of the human condition. More than that: inasmuch as the conceptions of the enlightened philosophers were thoroughly consistent with bourgeois sensibilities, the Counter-Enlight-enment anthropology developed as a *critique of capitalism*—including no-tably a critique of bourgeois individualism. Contrary to the Hobbesian origin myth, which effectively projected capitalism back to a state of nature inhabited by autonomous and self-regarding individuals competing with each other for power after power, man for Herder was always and ever a so-cial being. The human being is "actually formed in and for society, without which he could neither have received his being nor become a man."[13] So as against the large camp of philosophes who (following Locke and Hobbes) were prepared to make corporeal pleasures and pains the basis of all knowl-edge, industry, and society, Herder understood people's needs as determi-nate and limited. They were limited in the same way they were organized:

---

tory of Ideas (Harmondsworth: Penguin Books, 1982), 1–24; Isaiah Berlin, *The Crooked Timber of Humanity* (New York: Alfred A. Knopf, 1991); Isaiah Berlin, "The Magus of the North," *New York Review of Books,* October 21, 1993, 64–71; Ernst-Robert Curtius, *L'idée de civilisation dans la conscience Française* (Paris: Publications de la Conciliation Internationale, no. 1, 1929); Louis Dumont, *Essays in Individualism: Modern Ideology in Anthropological Perspective* (Chicago: University of Chicago Press, 1986); Dumont, *German Ideology,* Elias, *The Civilizing Process;* Arthur O. Lovejoy, *Essays in the History of Ideas* (Baltimore: Johns Hopkins University Press, 1948); Johann Gottfried von Herder, *Reflections on the Philosophy of the History of Mankind,* ed. Frank Manuel (Chicago: University of Chicago Press, 1968); Henri Massis, *L'honneur de servir* (Paris: Librairie Plon, 1937); Meyer, "Historical Notes"; Marshall Sahlins, *How "Natives" Think: About Captain Cook, For Example* (Chicago: Uni-versity of Chicago Press, 1995), 10–14; Jean Starobinski, *Blessings in Disguise; or The Moral-ity of Evil* (Cambridge: Harvard University Press, 1993), chap. 1); Stocking, *Race, Culture, and Evolution,* chap. 4; and George W. Stocking Jr., *Victorian Anthropology* (New York: Free Press, 1987), chap. 1. Lately, the flaw in Herderian relativity, that it promoted a sense of the incom-mensurability of cultures to an extent that could imply a lack of common humanity among different peoples, has been exposed in Berlin, *The Crooked Timber of Humanity,* 70–90 (fol-lowing an observation by Momigliano), and Anthony Padgen, "The Effacement of Difference: Colonialism and Nationalism in Diderot and Herder," in *After Colonialism,* ed. Gyan Prakash (Princeton: Princeton University Press, 1995), 129–52. This understanding was not transmit-ted in the development of the Herderian-Humboldt tradition of "culture" in anthropology (via Boas et al.). But neither has anthropology's stand to the contrary (i.e., psychic unity) pre-vented popular abuse of the significance of cultural differences—as witness the stigmatization of both anthropology and culture on that basis.

13. Herder, *Reflections,* 58.

by the various ancestral traditions that had developed in particular environments, traditions that entailed particular modes of being in, and perceiving, nature. National character included the character of the economy.
It also unified society from within, by the shared inner bonds of common
language and the distinctive worldview that each and any language conveyed. The arguments thus defied all contract theories of state and society.
For their presuppositions of the anarchy that must ensue from the pursuit
of individual self-interest, the theories of contract were unable to conceive
the formation of society except by the institution of state. But the state was
an artificial, external imposition in the Herderian anthropology—an anthropology that knew no more necessity to found society on coercion than
people needed a unified nation to know they were Germans.

As opposed to the bourgeois myths, Herder put in evidence the peoples'
own myths. Transmitted in the mother tongue and the bosom of the family,
the inherited ancestral traditions constructed the folk and their world in
relative forms of happiness and reality. In the name of the peoples' several
ideas of what there is, the Counter-Enlightenment would challenge the
universal rationality cum sensationalist epistemology of the *philosophes*.
People order their experience in the terms of their traditions: worldviews
that are, moreover, endowed with the morality and emotions of their transmission. People do not simply discover the world, they are taught it. To
speak of reasoning correctly on the objective properties of things, things as
they are known through unmediated sensory perceptions, would be out of
the question for an anthropology sensitive to the cultural organization
of knowledge. Seeing is also dependent on hearing, and in the sociology of
thought—what Herder once referred to as "the family or kinship mode of
thought"[14]—reason is entangled with feeling and bound to imagination.
So "the shepherd beholds nature with different eyes from those of the fish-
erman."[15] In Locke's view, at least one of them, the shepherd or the fisher-
man, has to be making a mistake. Yet just so, what was *error* for the
empirical philosophers was *culture* for Herder.

With good reason, one is reminded of the Boasian dictum that the seeing
eye is the organ of tradition. The good reason is that these principles of the
German Counter-Enlightenment went on to become main understandings
of the concept of culture in American anthropology. Transmitted from
Hamann and Herder through the likes of the Humboldts, Dilthey, Ritter,
Rätzel, and Bastian, they reappear, together with a soupçon of Kant and a
dash of Nietzsche, in the early twentieth-century work of Boas, Lowie,

14. *Herder on Social and Political Culture*, 163–64.
15. Ibid., 300.

Kroeber, and their American colleagues.[16] Although antagonistic to the Boas group, Leslie White added the considerations of symbolic order that made their ideas of culture complete—and, generally speaking, what "culture" remains today in American anthropology.[17]

Not all anthropologists think alike about culture, it has to be said. It needs saying because it often escapes the notice of modern (and postmodern) critics, who object to the discipline's essentialized and totalized sense of culture—on the grounds that no culture is like that. Exception to the last is apparently being taken on behalf of the anthropological tribe, which would be the only one known to science with a monologic ideology, lacking the saving political graces of contested categories, heteroglossia, or the like. But in truth, Western anthropologies have differed over "culture" throughout the twentieth century, or ever since the professional establishment of the field. Ever burdened with the sense of "high culture" as Matthew Arnold sanctified it, British anthropology, with the partial exception of Malinowski (who was Polish anyhow), could never bring itself to make culture its scientific object. Rather the social order as such was the matter of the discipline, which was accordingly designated "Social Anthropology" and situated academically as a sociology of primitive peoples. In the classic Radcliffe-Brownian perspective, "culture" or "custom" was a secondary consideration, being the ideological and historically contingent means of maintaining the social system. But the social system alone was systematic; culture was its arbitrary means of expression.

In France likewise, anthropology (the Durkheim school) was bound up with sociology. Only lately has French anthropology accepted anything resembling the American culture concept. Unlike Britain, which could at least make partial synonyms of "culture" and "civilization" (Tylor), France re-

16. Matti Bunzl, "From *Volkgeist* and *Nationalcharakter* to an Anthropological Concept of Culture," in *Volkgeist as Method and Ethic: Essays on Boasian Anthropology and the German Anthropological Tradition*, vol. 8 of *History of Anthropology*, ed. George W. Stocking Jr. (Madison: University of Wisconsin Press, 1996), 17–78.

17. Leslie A. White, *The Science of Culture* (New York: Farrar, Straus & Giroux, 1949). Because American cultural anthropology was linked to physical anthropology, a culture-nature opposition developed as a matter also of academic distinction. American anthropologists were especially sensitive to the question of what was distinctively human, thus constitutive of "culture." In this regard, they connected with a general understanding of the significance of the symbolic, as represented by Susan K. Langer, *Philosophy in a New Key*, 3d ed. (Cambridge: Harvard University Press, 1976), and Kenneth Burke, *Language as Symbolic Action: Essays on Life, Literature, and Method* (Berkeley and Los Angeles: University of California Press, 1966). American anthropology was thus preadapted to the specifically linguistic turn ushered in by French structuralism.

mained highly allergic to the culture concept until well after World War I. As is common knowledge, through the early twentieth century the national oppositions of France and Germany continued to be signified by the antithesis of *civilization* and *Kultur* formulated in the Counter-Enlightenment. It seems safe to say that French anthropology did not take culture seriously until World War II, the connection to German-American notions being mediated by Lévi-Strauss. All the same, in the last decades of the century, just when it seemed that the American sense of culture (and cultures) had become dominant, it began to unravel.

I will not rehearse the whole litany of common complaints against "culture," now heard both within anthropology and, more and more, in certain humanities circles—which nevertheless do not hesitate to call themselves "cultural studies." These complaints are sometimes coupled with the aforementioned dismissals of culture as an instrument of domination. They are distinctive, however, as epistemological reproaches, mainly having to do with the received anthropological discourses of structure and order. They fault the discipline's disposition to *overvalue order:* the perception of culture as objectified, superorganic, essentialized, stereotypic, primordial, homogeneous, logical, cohesive, bounded, or otherwise too systematic. Many of these criticisms have actually been shadowing the culture concept for a long time. Except that before they were expressed on anodyne social-science terms, such as ideal vs. actual behavior, norm vs. practice, system vs. human agency, etc., rather than the moral-political guises they assume in contemporary consciousness—that is, again, by functionalizing the original issues. Yet insofar as the modern (and postmodern) disillusionment is an expression of the loss of the object, the present anxieties about culture can be regarded as conjunctural versions of the long-term epistemological crisis about the possibility of any anthropology. Here are new translations of the anthropological nostalgia for the "vanishing primitive." The non-Western world, James Clifford acutely remarks, "is always vanishing and modernizing—as in Walter Benjamin's allegory of modernity, the tribal world is conceived as a ruin."[18]

Professional ethnography, ever since its beginnings—whether one dates these beginnings to Lewis Henry Morgan's interviews of Iroquois or the summer trips of Boas and his students to Indian reservations—has been an "archaeology of the living" (in Lévi-Strauss's phrase), a salvage effort haunted not merely by the decline of the indigenous culture but the loss even of its memories. Or again, if Malinowski's research be taken as the

---

18. James Clifford, *The Predicament of Culture* (Cambridge: Harvard University Press, 1988), 202.

origin of modern ethnography, it is sobering to reflect that his classic monograph on the Trobriands opens with these words:

> Ethnology is in the sadly ludicrous, not to say tragic, position, that at the very
> moment when it begins to put the workshop in order, to forge its proper tools,
> to start ready for work on its appointed task, the material of its study melts
> away with hopeless rapidity. Just now, when the methods and aims of scien-
> tific field ethnology have taken shape, when men [N.B.] fully trained for
> the work have begun to travel into savage countries and study their inhabi-
> tants—these die away under our very eyes.[19]

Past objects? Yes, history studies these. But how many academic disciplines
besides high-energy physics originated as the study of disappearing objects? And it now seems as if the anthropological quasi-object has crumbled
altogether, victim of the capitalist world system. Pastiches of local customs
and transnational flows, without any indigenous order or structure, the
so-called cultures are in postmodern disarray. And the anthropological nostalgia, reflecting the course of imperialism, lapses into "sentimental pessimism," as Stephen Greenblatt calls it: the collapse of other people's lives
in global visions of Western hegemony.[20]

  On the other hand, the skepticism in Greenblatt's remark suggests that
the moral hegemony of world system anthropology has been contested—
by subaltern events and voices that the sentimental pessimism had not envisioned. Until recently the positive complement of the nostalgia for the
"vanishing primitive" was a serious concern with the destruction of the
Other, probably with some hope that good would come from the documentation of global capitalism's cultural cannibalism. The problem was that in
denying any cultural autonomy or historical agency to the indigenous others, the anthropologies of the world system became too much like the colonization they justifiably detested. Our academic theories seemed to
complete in the register of superstructure the same kind of domination the
West had long imposed in economic and political practice. Supposing that
the cultural forms and purposes of modern indigenous societies had been
constructed solely by imperialism, or merely as its negation, the critics of
the global imperialism were creating an anthropology of neo-historyless
peoples. So something had to be said for the recalcitrant ethnographic facts
(if I may use such a quaint expression): the refusal of the indigenous peoples

  19. Bronislaw Malinowski, *Argonauts of the South Pacific* (1922; reprint, London: Routledge and Kegan Paul, 1961), xv.

  20. Stephen Greenblatt, *Marvelous Possessions: The Wonder of the New World* (Chicago:
University of Chicago Press, 1991), 152.

either to go away or to become just like us. It turns out these peoples were not merely disappearing a century ago, at the beginnings of anthropology, they are *still* disappearing. They are forever disappearing. The little initiation ritual we put on for first-year anthropology graduate students, exhorting them to go out and study the exotic societies while they are still there, has been repeated now every autumn for generations. For at least those peoples who physically survived the colonial onslaught are still taking cultural responsibility for what was inflicted on them. They have been struggling to incorporate the world system in an order of even greater human scope: their own system of the world.

THE INDIGENIZATION OF MODERNITY

The anthropological agenda is now the indigenization of modernity. I am not saying that ethnographic experience is solely responsible for the decline of sentimental pessimism. The issue hardly lends itself to pure induction. Probably some dialectic or pendular motion of normal social science is also involved. And the continued relevance of the moral-political context is manifest in another necessary qualification: that we shall be speaking only of the survivors. The survivors no doubt make up a minority of the social-cultural orders in existence, say, in the fifteenth century. What follows, then, should not be taken for sentimental optimism, ignoring the agonies and deaths of whole peoples from disease, violence, enslavement, removal, and the other sufferings visited all over the planet by Western "civilization."

The catastrophe has been so overwhelming that until the late 1970s or early 1980s, hardly any theoretical attention could be paid to the contrary movements that Richard Salisbury was calling "cultural enhancement." In 1981, Salisbury organized a symposium on "Affluence and Cultural Survival" for the annual meetings of the American Ethnological Society. Published in 1984[21] the symposium benefited from an introduction by Salisbury that addressed all the main themes I will again be talking about here. What especially caught Salisbury's attention, by virtue of work with the James Bay Cree as well as the Siane of New Guinea, was the apparently paradoxical enrichment of traditional culture that sometimes accompanied the integration of indigenous societies into the global economy. As Salisbury described it, this "cultural enhancement" was a selective and oriented project of all round development, reflecting customary notions of the "good

21. Richard Salisbury and Elisabeth Tooker, eds., *Affluence and Cultural Survival*, 1981 Proceedings of the American Ethnological Society (Washington: AES, 1984).

life" and associated with an explicit promotion of the indigenous "culture"—if funded materially by an articulation with the market and thus threatened ultimately by a condition of dependency.[22] Besides the Cree people, researchers in the symposium were able to make similar observations of the Huron nation, the Tsimshian, Osage, and Yemenis of the Central Highlands.[23] Robert Grumet, for example, told of a "spectacular cultural efflorescence" among Coast Tsimshian of the late eighteenth century, following upon a "massive infusion" of European trade wealth. It is interesting that in an analogous study published around the same time, Chris Gregory used the same term, "efflorescence," to describe how certain New Guinea peoples used their participation in the modern "commodity economy" to expand their traditional "gift economy."[24] But now that we come to think of it, the phenomenon is worldwide, and some places it has been going on for centuries. Certain cases of postcontact "efflorescence" or "cultural enhancement" are anthropological classics: the American Great Plains during the horse-and-gun era; the Northwest Coast potlatch; the Huron and Iroquois confederacies; the conquest kingdoms of Hawai'i, Tahiti, Tonga, and Fiji.

Because of a certain heteronomy, the variety of these local responses to the capitalist world system is too often dissolved in the sentimental pessimism of universal acculturation. "We can easily conceive of the time when there will be only one culture and one civilization on the entire surface of the earth," writes Lévi-Strauss. But for his part, he refuses to concede it, "because there are contradictory tendencies always at work—on the one hand towards homogenization and on the other towards new distinctions."[25] Even as the world becomes more integrated globally, it continues to differentiate locally—the second in some measure stimulated by the first. So within the planetary ecumene, as Hannerz and others have taught, are many new forms of life: syncretic, translocal, multicultural forms unknown to tra-

22. Richard Salisbury, "Affluence and Cultural Survival: An Introduction," in Salisbury and Tooker, *Affluence*, 1–11.

23. Colin Scott, "Between 'Original Affluence' and Consumer Affluence: Domestic Production and Guaranteed Income for James Cree Hunters," in Salisbury and Tooker, *Affluence*, 74–86; Bruce Trigger, "The Road to Affluence: A Reassessment of Early Human Responses to European Contact," in R. F. Salibury and E. Tooker, *Affluence*, 12–25; Robert S. Grumet, "Managing the Fur Trade: The Coast Tsimshian to 1862," in Salisbury and Tooker, *Affluence*, 26–34; Stephen I. Thompson, Susan Vehik, and Donald C. Swan, "Oil Wealth and the Osage Indians," in Salisbury and Tooker, *Affluence*, 40–52; Daniel Martin Varisco and Najwa Adra, "Affluence and the Concept of the Tribe in the Central Highlands of the Yemen Arab Republic," in Salisbury and Tooker, *Affluence*, 134–49.

24. C. A. Gregory, *Gifts and Commodities* (London: Academic Press, 1982).

25. Claude Lévi-Strauss, *Myth and Meaning* (New York: Shocken, 1978), 20.

ditional anthropology. Nor are the techniques for understanding the traditional anthropology-cultures always and ever relevant. In the light of the global-historical changes, the postmodernist critique of ethnography is intelligible. But the sequitur is not the end of "culture." It is that "culture" has taken on a variety of new arrangements and relationships, that it is now all kinds of things we have been too slow to recognize. Rather than celebrate (or lament) the passing of "culture," then, anthropology should seize the opportunity of renewing itself: that is, through the discovery of unprecedented forms of human culture. It is almost as if we had found life on another planet, this history of the past three or four centuries that has given other modes of life on this planet—a whole new cultural manifold.

The rest of my paper is about just such ethnographic discovery, as experienced by three excellent anthropologists: Rena Lederman, Epeli Hau'ofa, and Terry Turner. At one level, the paper is about the kind of phenomenological reduction (epoché) each went through, to emerge from his or her respective field experience with changed ideas about the nature and viability of the traditional cultures each had come to study. Each became impressed with a certain indigenization of modernity that was not imagined either in the received anthropology of cultural monads or in World System forecasts of gloom and doom. At another level then, I try to generalize about the kind of modern (and postmodern) transformations these ethnographers had encountered, the types of cultural processes they witnessed: what I shall call "developman" in Lederman's case, the "multilocal culture" (or "translocal society") in Hau'ofa's, and "culturalism" in Turner's.[26]

Rena Lederman on Mendi: Developman

Some recent enthnography from the Southern Highlands of New Guinea:

> Despite initial resistance to most things suggested by the Australians, the Anganen were soon eager for development, or at least those projects they saw as achieving this end. "Development" *(divelopman)* is a broad concept in Anganen, but one most notably measured in material goods, a process largely realized through, and symbolized by, money. Cash has many uses, of course—to set up trade stores or buy cars, cattle, consumables, pay school fees or taxes, for gambling, etc.—but its greatest significance for Anganen is its prominence in ceremonial exchange.[27]

---

26. The "multilocal culture" is perhaps more effectively described as a "multilocal sociocultural order," but I will keep it manageable by using either "multilocal culture" or "translocal society."

27. Michael Nihill, "The New Pearlshells: Aspects of Money and Meaning in Anganen Exchange," *Canberra Anthropology* 12 (1989): 144–60, at 147.

The ethnographer further observes that "exchange has actually flourished in Anganen since the arrival of the Australians, in part due to the rendering of money as a legitimate item of exchange."[28]

"Developman" *(divelopman)* is the neo-Melanesian word for it, the word that ostensibly corresponds to the Western category "development," but given the persisting differences in meaning I prefer to gloss it as it actually sounds in English: "develop *man,*" the development of people. Even when it refers to *"bisnis"* or making money, developman is characteristically realized by New Guineans as an expansion of traditional powers and values, notably through increased ceremonial and kindred exchanges. Or as a leader of the Kewa people told an anthropologist: "You know what we mean by 'development'? [in Kewa, *ada ma rekato,* lit. 'To raise' or 'to awaken the village']: building a 'house line' *(neada),* a men's house *(tapada),* killing pigs *(gawemena).* This we have done.'"[29]

"Developman" refers to a process—a passing moment of "first contact" that may endure for more than a hundred years—in which the commercial impulses excited by an encroaching capitalism are turned to the provisioning of indigenous notions of good life. In this event, European goods do not simply make the people more like us, but more like themselves. Foreign wealth is harnessed to traditional values. This is Salisbury's "cultural enhancement" or Gregory's "efflorescence," of which several macroscopic examples have already been mentioned. And as I have also published on the matter,[30] I shall try to be brief, shifting the focus to the everyday appropriation of European objects as Rena Lederman observed this for Mendi people of the Southern Highlands, New Guinea.[31] The change of scale allows us to magnify the dynamics of developman, to see in detail how the people are able to give their own meanings to foreign things.

The Mendi even made jewels of European refuse. When Lederman and her husband, Mike Merrill, first came upon this project they understandably pitied the Mendi their indigence rather than complimenting their cre-

---

28. Ibid., 144.

29. Lisette Josephides, *The Production of Inequality: Gender and Exchange among the Kewa* (London: Tavistock, 1985), 44.

30. Marshall Sahlins, "Cosmologies of Capitalism: The Trans-Pacific Sector of the World System, *Proceedings of the British Academy* 74 (1989): 1–51; Marshall Sahlins, "The Economics of Developman in the Pacific," *Res* 21 (1992): 12–25. I apologize for the overlap between the present article and these and other works. For some time I believed this piece was not destined for publication.

31. Rena Lederman, "Changing Times in Mendi: Notes towards Writing Highland New Guinea History," *Ethnohistory* 33 (1986): 1–30; Rena Lederman, *What Gifts Engender: Social Relations and Politics in Mendi, Highland Papua New Guinea* (Cambridge: Cambridge University Press, 1986).

ativity. What other conclusion could be drawn of people who fashioned armbands out of tin cans and hats out of bread wrappers?; people who had gone all their lives barefoot now walking around in galoshes several sizes too large, or perhaps in only one torn one?; people who bought expensive radios that soon broke down leaving them unable to repair them? A labor historian, Merrill supposed that if this appropriation of the rubbish of "civilization" made no functional sense, it probably did signify something—probably an invidious sense of deprivation. "One shoe," he wrote in his journal, "is of no use and in fact is probably a hindrance to walking, especially when its heel is torn out . . . But one shoe does mean something. It signifies a desire on the part of the owner to have two shoes, and not just shoes but everything else as well."[32] And for want of a shoe the culture was lost. Using an anthropology of the ancien régime, an old functionalist logic about the necessary correspondence between a type of technology and the cultural totality, the ethnographers were convinced that the desires of the Mendi for foreign material things would engage the people in the meanings and relationships of these commodities—to the extent they would jeopardize their traditional existence:

> For steel axes, textiles, cars, table service, rice and tinned fish, nails, etc. are not neutral objects. . . . They come into the area with their social origins visible and influential. . . . The meanings of the world market must, in the long run, predominate. . . . Eventually the traditional social structure will be eroded by the corrosive action of the articles which are now used in traditional ways, but which contain within them other, more powerful intentions.[33]

All the same, by the early 1980s, after a generation of experience with colonial and postcolonial government and considerable experience with the market selling both produce and labor, it hadn't happened yet. Neither the commodities nor the relations of their acquisition had transformed the Mendi structures of sociability or their conceptions of a proper human existence—*except to enlarge them.* Provisioned by increased wealth in cash, pearl shells, pigs, and foreign goods, clan ceremonials and kindred exchanges achieved unprecedented magnitudes of scale and frequency.[34] Lederman remarked that the indigenous social relations generated a far greater demand for modern currency than did the existing market outlets.[35] Reflecting on the white man's disposition to private consumption, one Mendi friend characterized the European economy as a "subsistence

32. Lederman, "Changing Times," 7.
33. Ibid.
34. Lederman, *What Gifts Engender*, 153.
35. Ibid., 232.

system," by comparison to his own people's interest in giving and receiving, a true exchange system. Now there's a howdy do.[36]

The Mendi, writes Lederman, have interacted with foreigners "while maintaining a sense of themselves." The local cultural system "is still the framework within which Mendi define, categorize, and orchestrate the new things and ways of acting to which they have been introduced during the past generation or so."[37] But notice that to invoke a cultural framework or logic in this way, as orchestrating historical change, is not to speak of the stereotypic reproduction of primordial custom. *Tradition here consists of the distinctive ways the change proceeds:* change as appropriate to the existing cultural scheme. In the New Guinea Highlands it can mean the developman of interclan ceremonial competition—pari passu with the decline of warfare. But then, the competition may be manifest in church-building projects.[38]

We are entitled to be skeptical about simplistic notions of "acculturation" as following necessarily and functionally from engagement in the market economy. Marx said in the *Grundrisse* that archaic relations of community are destroyed by money, since money *becomes* the community. Of course he did not know New Guinea people who ritually fetishize new twenty-kina notes as exchange valuables. Embodying male strength, these are the monies used by Anganen in interclan ritual exchanges, by opposition to the coins associated with women and everyday consumption. To adapt O. H. K. Spate's phrase (from Fiji to New Guinea), money here remains the servant of custom rather than its master.[39] One might well argue

---

36. Ibid., 236. No doubt the Mendi was invoking a common Highland distinction, critical to the operation of ceremonial cum social life, between giving things out in exchange and consuming them within the family. The contrasting dispositions of exchange and consumption may be further correlated with men and women respectively. Andrew Strathern, "Gender, Ideology and Money in Mount Hagen," *Man*, n.s., 14 (1979): 530–48; Nihill, "The New Pearlshells."

37. Lederman, *What Gifts Engender*, 9, 227

38. Ibid., 230. Margaret Jolly rightly complains about the Western academic inability to comprehend change as an authentic process within other traditions. Regarding the so-called natives:

If they are no longer doing "it" they are no longer themselves, whereas if colonizers are no longer doing what they were doing two decades ago, this is a comforting instance of Western progress. Diversity and change in one case connote inauthenticity, in the other the hallmark of true Western civilization. (Margaret Jolly, "Specters of Inauthenticity," *Contemporary Pacific* 4 [1992]: 49–72, at 57)

39. For excellent analyses of such processes of the integration of money in traditional relationships, see Maurice Bloch and Jonathan Parry, eds., *Money and the Morality of Exchange* (Cambridge: Cambridge University Press, 1989).

that this can only be a temporary condition, that "commoditization," "consumerism," and "dependency" will sooner or later subvert all traditional good intentions. Perhaps. But as Durkheim remarked, a science of the future has no subject matter. In the meantime, the temporal and other parameters of the phenomenon remaining unknown, developman in all its forms, viable or not, opens up a whole field of anthropological discovery.

To return then to the ethnography, Lederman and Merrill did not persist in their laments for the Mendi's economic plight, since that was not at all the significance of the people's uses of European things. Not even of their bricolage with tin cans and other Western oddments: this was no sign of humiliation or prelude to thwarted desire. Perceiving that, on the contrary, the New Guineans' relations to foreign objects entailed a kind of mastery, the ethnographers gradually abandoned their a priori dismal conclusions. The mastery was as much a matter of symbolic dexterity as it was technical: the ability of Mendi to give their own meanings to things. "People seem so easily to incorporate Western odds and ends," Lederman wrote in her field journal, "gathering them as casually as they gather bush materials." She continues:

> Here most things in the world are generally accessible. People know how to make most of the things they use. How then are Western items, so clearly different in this respect, to be dealt with? Well, as if they were "natural," of course! . . . Tolap turns the bread wrapper in his hand for a moment, considering what is to be done with it. The wrapper has no fixed purpose, but may be given one and then shaped to fit it. Is it to be burned or worn?[40]

The hunting and gathering of bread wrappers and umbrella spokes had lost its poignancy. To Lederman and Merrill it was no longer a premonition of cultural death. There was another logic, a Mendi logic, in the people's exotic improvisations. The goods were European, but not the needs or intentions. "The Mendi," Lederman reflected, "do not see these objects the same way we see them: their purposes supplied for us."[41] Their perceptions were guided by a different set of conceptions.

So Lederman sums up the Westerners' experience, the ethnographic reflex of the indigenization of modernity—about which, however, certain reservations linger, especially with the labor historian:

> On the other hand, the desire for Western products might mean something other than what Mike (and I, to a lesser extent) first thought it did. Just how powerfully *do* Western intentions assert themselves by means of their ob-

40. Lederman, "Changing Times," 8.
41. Ibid.

jects? Just how visible and influential *are* the social origins of these things? Gloomy prognosis of future trends toward severe dependency and demoralization may turn out to be accurate in any case. But an understanding of the social forces at work would be incomplete without some knowledge of what the world looks like from the perspective of rural village culture. One's own system of significance and values may not appear as overpowering and compelling to others as it does to oneself . . . Both Mike and I were to reassess our views about the meanings which Western objects possess for the Mendi after we moved out of Mendi town and lived for a time in Wepa (though, to tell the truth, we still argue about them).[42]

But for now there can be no doubting that the Mendi, like other Highland peoples—e.g., Chimbu, Hagen, Siane, and Enga, as well as Anganen— have known a developman of traditional culture since, and by means of, their articulation with the modern World System.[43] In addition to remarking on the Mendi ability to "define, categorize, and orchestrate the new things and ways of acting to which they have been introduced," Lederman draws attention to two other aspects of the continuing historicity of the traditional cultural schemata. One is that Mendi have increased the range and intensity of both small-scale reciprocity and ceremonial exchange—and thus of kinship in various modalities—in spite of pressures to the contrary from colonial and postcolonial governments, whose policies have been inspired rather by Western notions of economic "development."[44] Second, neither the fact nor the direction of this indigenous developman is new. "Tradition" was no more static in the past than it is now. The greatest developman of a Highlands sweet potato–and–pig production system, with its complementary social and ritual order, occurred during the two or three centuries before the colonial era. "Long before whites entered the Highlands," Lederman notes, "Highland children have grown up in worlds different from their grandparents." So rather than an indication of breakdown, an ability "to innovate and renovate the indigenous system" is itself a quality of that system.[45]

Epeli Hau'ofa: The Intercultural Society

Born in New Guinea of Tongan parents, educated in Papua New Guinea, Tonga, Fiji, Canada, and Australia; formerly deputy private secretary to the king of Tonga and now professor and head of the School of Social and Eco-

42. Ibid., 7–8.
43. Gregory, *Gifts and Commodities*.
44. Lederman, "Changing Times."
45. Ibid.

nomic Development, University of the South Pacific in Suva, Fiji; with a Ph.D. in anthropology from the Australian National University based on ethnographic fieldwork with Mekeo people of Papua; author of notable works of fiction as well as technical monographs on Mekeo society and Tongan economic development—Epeli Hau'ofa incarnates in his own biography a vision of an Oceanic space of life created by the free movement of island peoples that he himself articulated in 1993 in defiance of neocolonial conceptions of Pacific peoples as doomed to underdevelopment by their isolation and their multiple lacks—of land, of population, of resources, and, not least, of enterprise. As a professor in a university serving twelve Pacific island countries, Hau'ofa said he could no longer peddle this European discourse of belittlement to his students. The occasion was a public lecture, "Our Sea of Islands," delivered during celebrations of the twenty-fifth anniversary of the USP.[46] It sent shock waves through the campus. The immediate result was a small volume—*A New Oceania: Rediscovering Our Sea of Islands*[47]—which featured Hau'ofa's lecture and the responses to it by nineteen colleagues. Some were quite taken aback by Hau'ofa's "romantic idealism." Here were arguments about the cultural autonomy of ordinary people, even mytho-practical allusions that referred their current freedom of movement to the legendary travels of ancestral heroes to the heavens above and the underworlds below, while seeming to ignore the this-worldly system of neocolonial domination transmitted locally by comprador ruling classes and multinational corporations. Yet in a final reflection on the criticism, Hau'ofa drew attention to the people's own cultural consciousness—a self-reflexive use of "culture" such as we shall see breaking out the world over. He regretted that local intellectuals were ignoring their cultural traditions in favor of the apparently universal languages of political economy and political science. The indigenous scholars were speaking in an alien tongue—while ordinary islanders were adapting their ancestral discourses to their current situation:

> It is a pity that we seem to have so ignored the importance of our cultures that whenever some of us try to look at our own heritage, to the achievements of ancestors, for inspiration and guidance, we bring down on ourselves charges of romanticism, mythical consciousness, speciousness and valoriza-

---

46. The lecture was first read at the University of Hawaii, Hilo, then rewritten and delivered at the East-West Center in Honolulu, before being revised again and presented some weeks later at the University of the South Pacific.

47. Eric Waddell, Vijay Naidu, and Epeli Hau'ofa, eds., *A New Oceania: Rediscovering Our Sea of Islands* (Suva, Fiji: School of Social and Economic Development, University of the South Pacific, 1993).

tion, especially from our own people. We cringe when our culture is mentioned because we associate our traditions with backwardness and unen-
lightenment. After all, we are internationalists and progressives who think
and speak only in the universal and culture-free languages of political economy and science . . . By deliberately omitting our changing traditions from
serious discourses, especially at the School of Social and Economic Development [of the USP], we tend to overlook the fact that most people are still using and adapting them as tools for survival . . . I believe we should pay a great
deal more intellectual attention and commitment to our cultures than we
have done, otherwise we could easily become V. S. Naipaul's mimic men and
mimic women.[48]

Even Hau'ofa's retort thus drew upon traditional cultural resources. For
as many of his writings show, he knows very well the underlying skepticism that attends Polynesian systems of authority: the contradictions of
kinship and power whose traditional complement is a popular and cunning
disposition to subversion. In something of the same popular spirit, Hau'ofa
would now undermine the foreign-imperialist theories of "dependency"
according to which the island societies were too poor to achieve any sem-
blance of autonomous "development"—or then, any self-respect. "MIRAB
societies," as they were unhappily known at the USP, subsisting on migration, remittances, aid, and (overblown) bureaucracies. Through the 1980s,
Hau'ofa had been a reluctant accomplice of this ideology of despair. In 1986
for a seminar on "development" he wrote a paper on Pacific societies called
"The Implications of Being Very Small." It was a veritable catalogue of
the economic laments occasioned by this uninteresting condition. Moreover, Hau'ofa argued, the islands' geographic situation in combination with
their lilliputian proportions made their sovereignties as vulnerable to the
machinations of Pacific superpowers as their environments were to nuclear
testing "and other things large countries dare not do at home." Small might
be beautiful to some people, Hau'ofa said, "but the world at large has made
our smallness and our geographical location the roots of our predicament."[49]

Still, in this work and others there had always been a certain ambivalence in Hau'ofa's pessimism. Even the report on marketing he compiled
for the Tongan government is punctuated by gently ironic descriptions of
how the people's customary inclinations manage to undermine various de-

48. Ibid., 129.

49. Epeli Hau'ofa, "The Implications of Being Very Small," paper presented to the Tokai
University/Friedrich-Ebert-Stiftung Seminar on "Cooperation in Development: Sharing of
Experiences," Tokyo, November 1986.

velopment schemes of foreign inspiration.[50] In analogous ways, "The Implications of Being Very Small" perceived the so-called development—urbanization, expansion of the monetary sector, and the like—as a *threat of impoverishment*, inimical to the traditional "subsistence affluence" enjoyed on the islands. Indeed in certain passages, the tragic notions of smallness and insufficiency amounted to an external ideological trip laid on the island peoples by self-appointed experts in economic development:

> [I]n any publication on aid and development in the region, you will most likely read that we are tiny, scattered, resource poor, and incapable of standing on our own feet in the modern world. This idea has been so consistently inculcated into us that our own leaders and people are convinced of our insignificance and therefore, general helplessness.[51]

Yet it was in his fiction, especially the *Tales of the Tikongs*, that Hau'ofa's populist resentments came out in their strongest and most Polynesian form. I say "populist," although precisely what is Polynesian about these hilarious send-offs of "development" is that the common people do not speak from a position of class dependency; on the contrary, they are the true people of the land, by contrast and in opposition to ruling chiefs who would trace their origins to the heavens and other such foreign places. Just so the useless bureaucrats of the tiny fictional island of Tiko are continuously off to conferences in Wellington, seminars in Geneva, and training courses in London, while expatriate technical experts sent out by the Great International Organization naively fail to cope with local knowledges and subterfuges.[52] In critical respects, "development" in these pages appears as the continuation of Polynesian cosmology and polity by other means—something the people have long known how to evade. Or as Hau'ofa now sees it, the people have their own means and modes of adaptation to the modern world, independently of the projects and policies of the Development Establishment.[53]

50. Epeli Hau'ofa, *Corned Beef and Tapioca: A Report on the Food Distribution in Tonga*, Development Studies Centre Monograph no. 19 (Canberra: Australian National University, in association with the University of the South Pacific Centre for Applied Studies in Development, 1979), 4–5, 8, 119.

51. Hau'ofa, "Implications."

52. In his editor's note to the recently reprinted *Tales of the Tikongs*, Vilisoni Hereniko draws attention to the people's defiance of the potential tidal wave of development: "These are not stories of fatal impact so much as upbeat tales of indigenous responses to cultural and economic imperialism." Vilisoni Hereniko, editor's note to Epeli Hau'ofa, *Tales of the Tikongs* (Honolulu: University of Hawaii Press, 1994), vii.

53. So in "Our Sea of Islands," Hau'ofa speaks of

> ordinary people, peasants and proletarians, who, because of the poor flow of benefits from the top, scepticism about stated policies and the like, tend to plan and make deci-

We need not suppose, then, that Hau'ofa's conversion to this defiant view of the islanders' plight was as dramatic as he claims. I say "conversion" because in "Our Sea of Islands" Hau'ofa describes a trip he made in 1993 across the Big Island of Hawai'i, between Hilo and Kona, as his "Road to Damascus." Rising from the fiery depths and expanding into the sea, the flow of Kilaeua volcano under the aegis of the goddess Pele seemed to him a better metaphor of the islanders' cosmos than the political boundaries and "mental reservations" to which they had been too long confined by Western determinations of their existence. We do not now, he said, nor did we ever live imprisoned on "tiny islands in a far sea," the way it looks to Europeans. The sea is our home, as it was to our ancestors. The ancestors' world "was anything but tiny. They thought and recounted their deeds i n epic proportions."[54] They lived in great associations of islands linked by the sea—as in the kula ring, or the regional community of Tonga, Fiji, Uvea, Samoa, Rotuma, Futuna, and Tokelau—*linked* by the sea, not separated by it.

Since World War II, Hau'ofa continues, the Pacific peoples have been able to resume this traditional mastery of ocean space, if by new means, for new purposes, and to ever greater extent. They now expand their islands in novel forms:

> Everywhere they go, to Australia, New Zealand, Hawai'i, mainland U.S.A., Canada and even Europe, they strike roots in new resource areas, securing employment and overseas family property, expanding kinship networks through which they circulate themselves, their relatives, and their stories all across their ocean; and the ocean is theirs because it has always been their home.[55]

By contrast to the Western developmental notions of their minuteness, the islanders are embarked on an unparalleled "world enlargement." Rather than fixed and insufficient resources, they have gained access to the products of an international division of labor. For their

---

sions about their lives independently, sometimes with surprising and dramatic results that go unnoticed or ignored at the top. Moreover, academic and consultancy experts tend to overlook or misinterpret grassroots activities because these do not fit in with prevailing views about the nature of society and its development. Thus views of the Pacific from the level of macroeconomics and macropolitics often differ markedly from those of the level of ordinary people. (Epeli Hau'ofa, "Our Sea of Islands," in Waddell, Naidu, and Epeli, *A New Oceania*, 2–16, at 2–3)

54. Ibid., 7.
55. Ibid., 10.

"homes abroad" (Hau'ofa's term) are connected by kinship ties and an interchange of personnel—not to forget the connections of telephone, fax, and e-mail—to the island homeland that still constitutes their identity and their destiny. Nor need one speak the Western-economistic language of "remittances." The exchanges are two-sided, something like the customary reciprocity between kinsmen, including the elements of total prestation that add certain social values to the transactions. Hau'ofa tells of Tongan goods and foods flowing to Auckland and Honolulu against the reverse movement of cash, or it may be refrigerators or outboard engines. Yet the apparent "remittances" and "repayments" are the material dimension only of an ongoing circulation of persons, rights, and regards between the home islands and the homes abroad. The international boundaries and oceanic distances that, in the white man's construction of planetary space, signify difference and isolation are traversed by a specifically Tongan set of social and cultural relationships. Tongans—as also Samoans, Tuvalans, or Cook Islanders—live in multilocal communities of global dimensions. They have expanded their cultural scope and potentialities in ways that cannot be conceived by the development economics of their insignificance. Anyhow, what people could think of themselves as "remote"?

Information on the scope of the Tongan diaspora is not as accessible to me as comparable materials on the neighbor people of Samoa, so I use the latter for illustration. Figure 7.1 is a modern map of Samoa taken from a volume by F. K. Sutter pertinently entitled *The Samoans: A Global Family*. By means of photographs and texts, including brief autobiographies of many of the people depicted, Sutter presents a fascinating account of the Samoan diaspora. It should be noted that by the mid-1980s, some one-third of the Western Samoan population was living overseas, while more than 60 percent of the population of American Samoa had left for Hawaii and the U.S. mainland.[56] Western Samoans were concentrated in Auckland and Wellington, Honolulu, Los Angeles, San Diego, and the Bay Area of California; but they also lived in smaller American cities such as Oxnard, California, and as far east as New York and Chapel Hill. Altogether, Samoans could be found in some twenty states of the United States and thirty nations around the world.

In the autobiographies of this diaspora collected by Sutter, a detective in Wellington writes:

56. Paul Shankman, "The Samoan Exodus," in *Contemporary Pacific Societies*, ed. Victoria S. Lockwood, Thomas G. Harding, and Ben J. Wallace (Englewood Cliffs, N.J.: Prentice-Hall, 1993), 156–70.

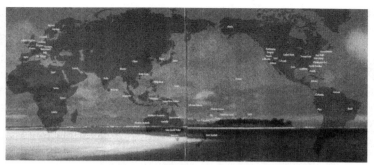

Figure 7.1.   A modern map of Samoa. (From Frederic Koehler Sutter, *The Samoans: A Global Family* [Honolulu: University of Hawaii Press, 1989], 2–3)

> I consider myself a true blue Samoan and am very proud of it . . . I've managed to get home to Samoa every 2 years. Presently I'm conducting courses at the Royal New Zealand Police College on Samoan language and culture.[57]

A blue-collar worker in Paremata, New Zealand, tells of his flight from the Tokoroa, New Zealand, Samoan community because it "had too much Samoan custom"; he was subsequently reconciled with his family.[58] The sumo wrestler Konishiki, the first foreigner to achieve champion rank in Japan, recounts the difficulties he has had with the media establishing his identity:

> For the longest time they insisted on calling me a Hawaiian. But that's finally changing. They now report I'm a Samoan born in Hawaii, and that makes me proud.[59]

Several American professional football players are represented in Sutter's text, including Mosi Tatapu of the New England Patriots, who considers himself supremely "blessed" to be a Samoan and who dedicated his 1985 Super Bowl game to his father-in-law.[60] Sutter's saga of Samoans also includes a shepherd in Invercall, New Zealand; a chief warrant officer in the U.S. Navy serving in the Philippines; a pastor in Zambia; a nun in Rome; an international civil servant working at UNESCO, Paris, to whom God granted "the gift of being born Samoan—body, mind and soul"; a brew master in Munich; an engineer in Norway; a clergyman in Jamaica; an FBI agent in Florida; a high-rise construction worker in Atlantic City; a casting

57. Misiotele, in Frederic Koehler Sutter, *The Samoans: A Global Family* (Honolulu: University of Hawaii Press, 1989), 167.
58. Alo'iai, in ibid., 168.
59. Ibid., 173.
60. Ibid., 194.

director in Hollywood; a firefighter in Colorado who believes "Samoans are able to live in and contribute to any society" and is "proud to be who I am— a Samoan"; and a doctoral student in theology in Montpelier, France, who does not forget the *aiga* (the kindred group):

> I hope that my writing in the first person will not obscure the communal support I have had from my family, my wife and her family, friends and village people. My life's tracks rest on a broad base of communal support, Samoa.[61]

There is much more here than personal nostalgia. As individuals, families, and overseas communities, the emigrants are part of a dispersed social-cultural totality centered in the homeland and united by a continuous circulation of people, ideas, goods, and money. Moving between foreign and indigenous cultural loci, adapting to the former while maintaining their commitment to the latter, Tongans and Samoans, and numerous other peoples like them, have been able to create the novel formations we are here calling multilocal cultures. "In many ways," notes an ethnographer of northern California Samoans, Craig Janes, "Samoa and San Francisco constitute a single social field in which there is a substantial circulation of members."[62] Moreover, in many regards, "Samoan migrants think they are more Samoan than the Samoans in Samoa."[63] Janes describes the San Francisco *aiga* or extended family network as taking a particular functional shape, adapted to the exigencies of the diaspora. The overseas *aiga* is marked by the solidarity of close kin of the same generation—as contrasted to the intergenerational hierarchies of the homeland—and by more frequent formal interaction with distant kin than in Samoa. The Samoan village is also adaptively reproduced: that is, as the congregation of an overseas church. Indeed Cluny Macpherson remarks of similar Samoan communities in New Zealand: "[F]or one who had been working in Samoa, it appeared that the Samoans had recreated Samoa in New Zealand, and that everything was happening very much as it did in Samoa."[64]

61. Ibid., 181.

62. Craig R. Janes, *Migration, Social Change, and Health: A Samoan Community in Urban California* (Stanford: Stanford University Press, 1990), 58.

63. Ibid., 62.

64. Cluny Macpherson, "Samoan Migration to New Zealand and Polynesian Migration and Settlement," in *New Neighbors: Islanders in Adaptation*, ed. Cluny Macpherson, Bradd Shore, and Robert Franco (Santa Cruz: Center for South Pacific Studies, University of California, Santa Cruz, 1978), 11–15. Cf. Karla Rolff, "Fa'asamoa: Tradition in Transition" (Ph.D. diss., University of California, Santa Barbara, 1978; and Lydia Kotchek, 1978, "Migrant Samoan Churches: Adaptation, Preservation, and Division," in *New Neighbors: Islanders in Adaptation*, ed. Cluny Macpherson, Bradd Shore, and Robert Franco (Santa Cruz: Center for South Pacific Studies, University of California, Santa Cruz, 1978), 286–93.

Here is a whole new field of comparative anthropology: not merely comparison between the differently situated homeland and overseas communities of the same translocal society, but among different kinds of multilocal cultural formations such as the Samoan and Tongan. George Marcus (1993) remarks on the contrasts between the Samoan overseas *collectivities*, their members moreover strongly linked to their villages of origin in Samoa, and Tongan kindred *networks*, dispersed in overseas locations and tied rather to persons than to places in Tonga—in the most successful cases, to the noble and commoner elite gathered about the royal capital of Nuku'alofa. Marcus suspects that the difference "might have something to do with the extension abroad of fundamentally different kinds of local organization at home in Tonga and Samoa."[65] Indeed, following Marcus's description, the contrastive Tongan principle is the hierarchical focus of kinship order on an elite personage—he or she in turn focused on and located about the kingship—which is what gives definition and coherence to the group as a whole. The same hierarchical principle permits the consolidation of the network's dispersed and diversified resources at its elite homeland center. When Marcus further suggests that overseas kinship networks tend to break down if they cannot convert their resources into elite status at home,[66] the continuity with the analogously dispersed, chiefly-centered lineages of ancient Tonga is plain to see. Compare Gifford's remarks on the traditional *haa* (lineage), often distributed over the archipelago but always with "a chief as a nucleus":

> Everything points to the necessity of a line of powerful chiefs for a nucleus about which the lineage groups itself. Without such chiefs it appears to wilt and die and its membership gradually aligns itself with other rising lineages.[67]

Epeli Hau'ofa was no doubt correct in asserting that Polynesians had their own structures of world enlargement well before Europeans tried to exile them to forsaken little islands set in a distant sea.

Since the late nineteenth century, moreover, world-enlarging cultures similar to the Tongan and Samoan have been evolving all over the Third World, among peoples supposedly incarcerated by imperialism and with-

---

65. George E. Marcus, "Tonga's Contemporary Globalizing Strategies: Trading in Sovereignty Amidst International Migration," in *Contemporary Pacific Societies*, ed. Victoria S. Lockwood, Thomas G. Harding, and J. Wallace (Englewood Cliffs, N.J.: Prentice Hall, 1993), 21–33, at 28.

66. Ibid., 29.

67. Edward Winslow Gifford, *Tongan Society*, Bernice P. Bishop Museum Bulletin 61 (Honolulu: Bishop Museum, 1929), 30. Gifford 1929: 30.

out hope of "development."[68] Mostly taking shape as urban ethnic out-posts of rural "tribal" homelands, these synthetic formations were long un-noticed as such by the Western social scientists studying them. Or rather, in studying urbanization, migration, labor recruitment, or ethnic formation, Western researchers presented a spectacle something like the blind men and the elephant, each content to describe the translocal cultural whole in the terms of one or another of its aspects.[69] Still, since the 1950s, space-defying communities of this kind were developing all over: in Java, Suma-tra, Kalimantan, the Philippines, Thailand, and other parts of Southeast Asia; in East, West, Central, and Southern Africa; in Egypt, Peru, Mexico, even Portugal.[70]

The stranglehold of European history on the anthropological imagina-tion was a main reason this novel cultural structure of modernity was for a

68. Perhaps similar intercultural formations have existed since antiquity in the cities of nonnational states. Another kind would be the dispersed communities of Arab and Indian traders that set up residences in China and Indonesia in the first millennium A.D.

69. The so-called circular migration—today considered by some an oxymoron and rela-beled "circulation" as distinct from (permanent) "migration"—involving a return to the homeland, or perhaps repeated cycles of this sort, especially drew attention to the network of relations between rural homelands and city homes abroad.

70. See, among others, Caroline B. Brettell, *Men Who Migrate, Women Who Wait: Popu-lation and History in a Portuguese Parish* (Princeton: Princeton University Press, 1986); Edward M. Bruner, "Urbanization and Ethnic Identity in North Sumatra," *American Anthro-pologist* 63 (1961): 508–21; R. Mansell Prothero and Murray Chapman, eds., *Circulation in Third World Countries* (London: Routledge and Kegan Paul, 1985); Walter Elkan, "Is a Prole-tariat Emerging in Nairobi?," in *Circulation in Third World Countries*, ed. R. Mansell Prothero and Murray Chapman (London: Routledge and Kegan Paul, 1985), 367–69; Victoria S. Lockwood, Thomas G. Harding, and Ben J. Wallace, eds., *Contemporary Pacific Societies: Studies in Development and Change* (Englewood Cliffs, N.J.: Prentice-Hall, 1993); J. A. Jackson, ed., *Migration*, Sociological Studies 2 (Cambridge: Cambridge University Press, 1969); Josef Gugler and William G. Flanagan, "Urban-Rural Ties in West Africa: Extent, Interpretation, Prospects, and Implications," *African Perspectives* 1 (1978): 67–78; Keith Hart, "Migration and Tribal Identity among the Frafras of Ghana," *Journal of African and Asian Studies* 6 (1971): 21–36; Graeme J. Hugo, *Population Mobility in West Java* (Yogyakarta: Gadjah Mada University Press, 1974); Michael Kearney, "From the Invisible Hand to the Visible Feet: An-thropological Studies of Migration and Development," *Annual Review of Anthropology* 15 (1986): 331–61; Cluny Macpherson, Bradd Shore, and Robert Franco, eds., *New Neighbors: Is-landers in Adaptation* (Santa Cruz: Center for South Pacific Studies, University of California, Santa Cruz, 1978); Philip Mayer, *Townsmen or Tribesmen: Conservatism and the Process of Urbanization in a South African City* (Cape Town: Oxford University Press, 1961); J. Clyde Mitchell, "Labour Migration in Africa South of the Sahara: The Causes of Labour Migra-tions," *Bulletin of the Inter-African Labour Institute* 6 (1959): 12–46; David Parkin, ed., *Town and Country in Central and Eastern Africa* (Oxford: Oxford University Press, for the Inter-national African Institute, 1975); Marc Howard Ross and Thomas Weisner, "The Rural-Urban Migrant Network in Kenya: Some General Implications," *American Ethnologist* 4 (1977): 359–75; Lillian Trager, *The City Connection: Migration and Family Interdependence in the Philippines* (Ann Arbor: University of Michigan Press, 1988).

long time conceptually underdetermined. The general presumption of Western social science was that urbanization must everywhere put an end to "the idiocy of rural life," as had happened in early modern Europe. By the very nature of the city as a complex social organism, relations between people would become impersonal, utilitarian, secular, individualized, and otherwise disenchanted and detribalized. Such was progress. Such was the course of Redfield's famous "folk-urban continuum." As the initial and final stages of a qualitative change, countryside and city were structurally distinct and opposed ways of life.[71]

True that strong empirical arguments against this rural-urban discontinuity had already been voiced by the early 1960s, based on studies of migrant communities in non-European cities. Edward Bruner explicitly criticized the Redfieldian perspective— "After the rise of cities men become something different from what they had been before"[72]—by demonstrating the continuities of identity, custom, and kinship between highland villages of Toba Batak and their urban relatives in Medan (Sumatra). Bruner offered a description of Batak unity of a kind we have already heard echoed for Samoans—indeed that was destined to be repeated the world over: "Examined from the structural point of view, the Toba Batak communities in village and city are part of one social and ceremonial system."[73] All the same, the received wisdom about the historic antithesis of village and city has made it difficult to change the gestalt, to perceive the possibility of a translocal society that could inhabit both situations and remain an interdependent social and cultural whole.

British social anthropology in Africa was long hung up on the same dualist a priori. In 1960, in an influential article that purported to sum up twenty years' research by the staff of the Rhodes-Livingstone Institute, Max Gluckman made the distinction of "townsmen" and "tribesmen" an issue of basic theoretical principle. "An African townsman is a townsman,

71. This also proves that scientific structural-functionalism is the handmaiden of unilinear evolutionism.

72. Robert Redfield, *The Primitive World and Its Transformations* (Ithaca: Cornell University Press, 1953), ix.

73. Bruner, "Urbanization and Ethnic Identity," 515; see also Edward M. Bruner, "Kinship Organization among the Urban Batak of Sumatra," *Transactions of the New York Academy of Sciences* 2 (1959): 118–25. Taking on Redfield, Bruner remarked in the introduction to the arguments from Sumatra: "Contrary to traditional theory, we find in many Asian cities that society does not become secularized, the individual does not become isolated, kinship organizations do not break down, nor do social relationships in the urban environment become impersonal, superficial, and utilitarian" (Bruner, "Urbanization and Ethnic Identity," 508). Bruner went on to show not only the similarities of urban and village Batak, but the systematic relations between them, including their interconnected economies.

an African miner is a miner: he is only secondarily a tribesman."[74] Gluck-man and his colleagues were always prepared to deny the colonial prejudice that African townsmen were necessarily "detribalized." But the tribal "classifications" one observed in the cities were distinct in function and be-havioral implication from tribalism in the country, a distinction that re-flected two different social systems. "The African in the rural area and in town," said Gluckman, "is two different men."[75]

Meanwhile many of Gluckman's students and associates were describ-ing something quite different: a synthesis of "townsmen" and "tribesmen" in a single sociocultural field that took its identity from and otherwise priv-ileged the rural homeland.[76] The conceptual status of these observations, however, is signified by their relegation to a footnote in Clyde Mitchell's 1967 review article on "Theoretical Orientations in African Urban Stud-ies." Said Mitchell: "I am excluding here those studies of migration which look upon town and country as integral parts of one social system in which townsmen and tribesmen are linked in networks of relationships in the town, in the rural areas, and between the two."[77]

But as they were more and more frequently remarked, such translocal systems soon became difficult to ignore. Explicit criticisms were made of the townsman-tribesman dualism of the Rhodes-Livingstone school, anal-ogous to the early empirical objections to the folk-urban continuum.[78] For

74. Max Gluckman, "Tribalism in Modern British Central Africa," *Cahiers d'Ètudes Africaines* 1 (1960): 55–70, at 57.

75. Ibid., 69.

76. Thus William Watson, *Tribal Cohesion in a Money Economy* (Manchester: Manches-ter University Press for Rhodes-Livingstone Institute, 1958), on Mambwe; Mayer, *Townsmen or Tribesmen*, and Philip Mayer, "Migrancy and the Study of Africans in Towns," *American Anthropologist* 64 (1962): 576–92, on Red Xhosa; J. Van Velsen, "Labour Migration as a Posi-tive Factor in the Continuity of Tonga Tribal Society," *Economic Development and Cultural Change* 8 (1960): 265–78, on Tonga of Nyasaland; and P. H. Gulliver, "Nyakusa Labour Mi-gration," *Rhodes-Livingstone Journal* 21 (1957): 32–63, on Nyakusa.

77. Clyde Mitchell, "Theoretical Orientations in African Urban Studies," in *The Social An-thropology of Complex Societies*, ed. Michael Banton (London: Tavistock, 1967), 37–68, at 61.

78. Ross and Weisner, "The Rural-Urban Migrant Network." Cf. Hart, "Migration and Tribal Identity"; Mayer, *Townsmen or Tribesmen*; and Mayer, "Migrancy and the Study of Africans." *Pace* Gluckman:

> Gluckman considered it more productive to see city and country as analytically dis-tinct. Thus two different theoretical explanations of behavior could be developed: one which was appropriate for rural life and one for urban life. We are suggesting that so-cial theory must account for behavior in both settings at the same time, partly because the migrants themselves see their behaviors in the two fields as interdependent and partly because the patterns of interaction and psychological ties between the two areas are important factors that account for attitudinal and behavioral variations through-out Africa today. (Ross and Weisner, "The Rural-Urban Migrant Network," 370–71)

one thing, the antithesis of townsmen and tribesmen was not generally known (as such) to the people themselves: not even to long-term residents of the city, members of labor unions or other urban associations—they did not forsake their tribal affiliations or their relations to rural homelands. Accordingly, study after study, and not only in Africa, spoke of the union of village people and their city relatives in "a bilocal society," "a common social field," "a single social and resource system," "a social village spread over thousands of miles," or some species of the like.[79] Indeed, many researchers perceived a tendency for the metropolitan and hinterland sectors of this unified system to become more alike, although not merely because the flow of ideas and commodities from the town was transforming the countryside. Modernization has not been the only game, even in the town. The inverse effect, the indigenization of modernity, is at least as marked—in the city and country both. In the complex dialectics of the cultural circulation between homelands and homes abroad, customary practices and relations acquire new functions and perhaps new situational forms. Van Velsen came to the interesting conclusion for Nyasaland Tonga that returning migrant workers, by competing for local political positions and taking up local land rights, while relying on kinsmen in the process, were "actively stimulating the traditional values of their rural society."[80] For the African Tonga as much as the Polynesian, kinship is often a beneficiary of modernization rather than its victim—by contrast again to the European experience and its normal social science. Wealth from the city subsidizes relationships in the village, while relatives in the city organize migration from the village. In perceptive and prescient researches undertaken in the 1960s, Keith Hart

---

79. Dawn Ryan, "Migration, Urbanization, and Rural-Urban Links: Toaripi in Port Moresby," in Contemporary Pacific Societies: Studies in Development and Change, ed. Victoria S. Lockwood, Thomas G Harding and Ben J. Wallace (Englewood Cliffs, N.J.: Prentice-Hall, 1993), 219–32, at 226; Ross and Weisner, "The Rural-Urban Migrant Network," 361; Trager, The City Connection, 194; Douglas Uzzell, "Conceptual Fallacies in the Rural-Urban Dichotomy," Urban Anthropology 8 (1979): 333–50, 343. Apart from the studies cited by Mitchell, which employ such descriptions as P. Mayer's "sets of relations" between city and country to transcend the received dichotomy, see also the works cited in note 70 above, as well as G. K. Garbett and B. Kapferer, "Theoretical Orientations in the Study of Labour Migrations," The New Atlantis 1 (1970): 179–97; Josef Gugler, "On the Theory of Rural-Urban Migration: The Case of Subsaharan Africa, in Migration, J. A. Jackson, ed., Sociological Studies 2 (Cambridge: Cambridge University Press, 1969); Frederick Errington and Deborah Gewertz, "The Historical Course of True Love in the Sepik," in Contemporary Pacific Societies: Studies in Development and Change, ed. Victoria S. Lockwood, Thomas G. Harding, and Ben J. Wallace (Englewood Cliffs, N.J.: Prentice-Hall, 1993), 233–48.

80. Van Velsen, "Labour Migration," 278. Similarly, Caroline Brettel writes of northwest Portugal that circular migration "has indeed served to perpetuate a way of life" (Brettel, Men Who Migrate, 263).

was able to show that the integration of rural and urban Frafras (Tallensi and related peoples of Ghana) was largely effected through their classic lineage system. From this Hart concluded the necessity of a new anthropological perspective, one that would transcend the correlated oppositions of the modern and the traditional, townsman and tribesman, urban and rural. He spoke rather of an "expansion of the horizons of the community":

> This expansion of the horizons of the community, in terms of the physical distribution of those who claim membership in a socially defined aggregate such as a lineage, makes it no longer easy to dichotomise, at least spatially, the traditional and the modern or even the rural and urban in Frafra life today. The world of the migrant and that of the homeland are not separable entities . . . The difficulty of separating the old and the new in the analysis of present day Frafra society, either in the national context of modern Ghana or even in the local context of the home tribal area, is illustrated by the simultaneous participation by most Frafras in both cultures, the exchange of personnel on a reciprocal basis between the home compound and southern city, the internal urbanisation of the Frafra district itself, the pervasiveness of the market economy, and especially the ease of communication between all parts of the country. When the discontinuities between town and village life have been diminished, what meaning can we legitimately give to types such as "townsmen" and "countrymen"?[81]

I hazard a few generalizations on the structure of these translocal systems as described by Hart, Hau'ofa, and many others. Culturally focused on the homeland, while strategically dependent on the peripheral homes abroad, the structure is asymmetrical in two opposed ways. Taken as a whole, the translocal society is centered in and oriented toward its indigenous communities. The migrant folk are identified with their people at home, on which basis they are transitively associated with each other abroad. These denizens of the town and the larger world remain under obligation to their homeland kinsmen, especially as they see their own future in the rights they maintain in their native place. Accordingly the flow of material goods generally favors the homeland people: they benefit from the earnings and commodities acquired by their relatives in the foreign-commercial economy. In such respects, the indigenous order encompasses the modern.

The complementary asymmetry, involving a certain superiority of the modern and the external, occurs precisely when the exchanges are referred to traditional practices of reciprocity—as Hau'ofa rightly insists they

81. Hart, "Migration and Tribal Identity," 26.

should be. Insofar as the exchanges reciprocally involve regards as well as goods, rights as well as hospitalities, the contributions of the people abroad are having a powerful effect on relationships at home. Within the reciprocity, a strategic imbalance appears: the indigenous center becomes dependent on the people abroad for its own reproduction—or it may be, for a certain develop-man. Key traditional functions, such as marital or mortuary transactions, descent and title transmissions, are subsidized by earnings in the commercial economy. Not unusually, a sojourn as a wage earner becomes a rite of manhood or qualification for leadership in the indigenous society. All this argues that certain prestigious values and powers reside in the foreign sphere. Perhaps then it is relevant to the development of translocal societies that many peoples had already accorded such virtues to external beings and things, even before colonialism introduced them to more draconic versions.[82]

The modern multilocal system also generates its own ideological forces, folklores of the internal and the external with similar capacities of moving people and goods between them. Both city and country know their contradictions: social tensions that are exacerbated by their interdependence—and then give complementary positive values to the alternative way of life. The reproduction of the homeland through emigration is accompanied by intergenerational stresses. The young break away to the larger world. In addition to the attractions of modernity, the city is perceived in the countryside as a place of freedom—freedom notably from old men and traditional constraints. Yet the centrifugal cultural and social effects are likely to be checked by the urban experience. As victims of discrimination, proletarianization, and pauperization, some important proportion of the "tribal" people in the modern sector develop a nostalgic view of their ancestral places. From the vantage of the foreign metropole, the homeland is idealized as the site of a "traditional" lifestyle: where people share with one another, where no one starves, where money is never needed. Ideological products of the intercultural system, the respective visions of each

---

82. Olivia Harris properly reminds us that

[w]hite people do not necessarily have the unique status in the constructs of the colonised that we as Euro-Americans expect and assume them to have. In many instances, white people are only one of a whole range of alien powers, and it would be a professional distortion to extract the image of white people from this wider context and fetishize them as something *sui generis*. The ways Europeans and their culture have been interpreted and incorporated depends on previously existing categories of otherness and ways of representing the alien and exotic. (Olivia Harris, "The Coming of the White People? Reflections on the Mythologisation of History in Latin America," *Bulletin of Latin American Research* 14 [1995]: 9–24, at 17)

other formulated in the modern and traditional sectors keep up the circulation between them.

Will it last? Can it last? Supposing the migrants settle permanently abroad, would not the translocal society have a sort of generational half-life, the attachments to the homeland progressively dissolving with each city-born or foreign-born generation? Will not the acculturation of the people abroad sooner or later make the diaspora irreversible, thus breaking apart the translocal society? Probably these outcomes often happen, but perhaps not as rapidly or easily as we are predisposed to believe. In Java, circular migration seems to have been in vogue since 1860; a Dutch scholar, Rannett, who studied it in 1916, thought it impeded the formation of a stable local proletariat, as the migrants who entered into the capitalist mode of production were "traditional men" with a strong stake in their villages of origin.[83] In a 1985 article entitled "Is a Proletariat Emerging in Nairobi?," Walter Elkan came to very similar conclusions about African people more than a century later.[84] Many of the African rural-urban tribal orders were already in their second or third generation by the time they were noticed by Western researchers, having begun in the 1930s or earlier. And although in recent years, "tribesmen" have been working in the cities for longer periods, perhaps for their active lifetimes, they remain as committed as ever to their native places—socially, morally, and economically.[85] If anything, studies of Luo and Kikuyu in Nairobi show that interest and investment in the rural homelands is greater in proportion to the status, stability, and remuneration of a person's urban employment. The most successful people in the city are the most engaged in the traditional order of the country—as indeed they can best afford it.[86] Or consider a New Guinea example: the Uri-

83. Graeme J. Hugo, "Circular Migration in Indonesia," *Population and Development Review* 8 (1982): 59–83, at 72.

84. Indeed, reviewing a large anthropological literature on migration and development, Kearney remarks on the general theme running through it to the effect that migrants "have not been proletarianized in any deeply ideological sense" ("From the Invisible Hand," 352).

85. Gugler, "On the Theory of Rural-Urban Migration," 146.

86. Parkin, *Town and Country;* Elkan, "Is a Proletariat Emerging?" Reflecting on the high incidence of Nairobi Luo in workers' associations, the fact that many have wives and children with them at least part of the year, that many have significant economic interests in the city, and other indices of successful adaptation to urban life, Parkin remarks:

> But there is no evidence that this intense social and economic involvement of Luo in Nairobi is resulting in a lessening of rural relationships and commitments. In fact, as Ross has noted for a different area of Nairobi, the higher-status and more securely placed townsmen among all groups are most likely to have made a number of corresponding rural commitments. They may have bought more farming land, built a house . . . and even branched out into a rural "business," such as a shop or transport service. It is clear, as we might expect, that those who most succeed in town are most likely to ex-

tai villagers in Port Moresby that Dawn Ryan has been working with since the 1960s. By the 1990s three-fourths of them were either born in a city or had long been absent from the village. Their native land rights had gone cold, and Ryan thought they realistically had no option to return to the village. Nevertheless, they were still Uritai and had intensive interaction with home villagers—some of whom, for their part, were still migrating to the town. "There has not been a progressive withering away of primary links between the village and the town."[87]

I think the secret of the seeming failure to urbanize is that there has not been a progressive withering away of the village. The translocal society may well persist so long as there is a cultural differential between the rural and the urban, or more generally between the indigenous homeland and metropolitan homes abroad. In this regard, Western history seems to afford the illusion that the social characteristics of urbanization—impersonal relations, individualism, decline of extended kinship, secularization, etc.—that these developments were sui generis effects of the city as a social-cultural order. But was not the cultural dissolution of rural life, idiocy though it may have been, a necessary condition for the city to thus work its modernizing magic? Whether through enclosures or other forms of disenfranchisement and depopulation, or through the leveling of cultural differences wrought by nationalism, the original homeland may no longer function as a distinct option and valued identity. In any event, however that may be for Europe, the peripheral sectors of translocal societies centered in the Third World are doubly alienated from the indigenous homelands: at once as culturally foreign and as sites of national or imperial hegemonic institutions. Given such cleavages, the multilocal culture, as a life-form of modernity, could outlast us all.

Perhaps the multilocal society will endure because it has become associated with the powerful movement of cultural self-consciousness now sweeping the planet. We discuss this "culturalism" in the next section. The remarks of Gupta and Ferguson on the new cultural topologies produced by postmodern diasporas make an appropriate segue to that discussion:

> The irony of these times . . . is that as actual places and localities become ever more blurred and indeterminate, *ideas* of culturally and ethnically distinct places become perhaps even more salient . . . Remembered places have often served as symbolic anchors of community for dispersed people . . . "Home-

---

ploit new economic opportunities available in their home areas. This is a familiar enough phenomenon in modern Africa which hardly needs to be elaborated. (Parkin 1975:148)

87. Ryan, "Migration, Urbanization, and Rural-Urban Links," 232.

land" in this way remains one of the most powerful symbols for mobile and displaced peoples, though the relation to homeland may be very differently constructed in different settings . . . We need to give up naive ideas of communities as literal entities . . . but remain sensitive to the profound "bifocality" that characterizes locally lived lives in a globally interconnected world.[88]

Terence Turner: The Contemporary "Culturalism"

For a long time, Terry Turner has championed an instrumental and historical view of culture—in opposition to prevailing notions of a self-determining symbolic order, dissociated from its genesis in social action and human purpose. For Turner, culture is precisely "the system of meaningful forms of social action"; so it "is to be understood essentially as the means by which a people defines and produces itself as a social entity in relation to its changing historical situation."[89] Turner has also long argued for the historical agency of indigenous peoples in the face of the capitalist world system—as opposed, that is, to the outlook that dehumanizes the peoples and ignores their struggles by conceiving them merely as the patients and objects of Western domination.[90] One of the ironies of fashionable discourses of otherness, Turner remarks, "is that it tends to exaggerate the potency of Western representations to impose themselves upon the 'others,' dissolving their subjectivities and objectifying them as so many projections of the desiring gaze of the dominant West." Moreover, such an anthropology of pessimism shares the alienation from action and the innocence of history that have too often marked the discipline's concepts of culture. Anthropology is left unable to account for its own repeated observation that

in virtually every situation of contact between tribal peoples and westernized national societies, a significant portion of the social and cultural changes in

---

88. Akhil Gupta and James Ferguson, "Beyond 'Culture': Space, Identity, and the Politics of Difference," *Cultural Anthropology* 7 (1992): 6–23, at 10–11.

89. Terence Turner, "The Politics of Culture: The Movement for Cultural Conservation as Political Cause and Anthropological Phenomenon" (unpublished MS., 1986).

90. So in 1979, Turner criticized the tendency to read the fate of all indigenous peoples from extreme cases of demoralization as a failure "to give the people who are the objects of their concern the attention and respect that their own demonstrated tenacity and capacity for struggle command." He spoke of the "terrible though seductive luxury in issues like 'genocide' and 'ethnocide' which consists precisely in the fact that in opposing the absolute dehumanization of human beings it becomes unnecessary to take positive account of the humanity (that is, the particular socio-cultural identity, with its concomitant capacity for social, political and cultural action and adaptation) of the victims" (Terence Turner, "Anthropology and the Politics of Indigenous Peoples' Struggles," *Cambridge Anthropology* 5, no. 1 [1979]: 1–43, at 4, 5).

the native society are not merely the result of deliberate and overt oppression by the national society or exploitation by representatives of international capital, but are either actively acquiesced in or spontaneously initiated by the indigenous people themselves.[91]

For some time Turner has also noted that "cultural survival" in the modern world consists of the peoples' attempts to appropriate that world in their own terms. It is not, as many have supposed, a nostalgic desire for teepees and tomahawks or some such fetished repositories of cultural meaning. A "naive attempt to hold peoples hostage to a moment of their own histories," such a supposition, Turner remarks, would thereby deprive them of history.[92]

Yet like most of us, Terry Turner came to more sophisticated conceptions of culture and history by criticizing his own past. When he first went to the Amazon in 1962—as he documents in a recent article whose subtitle is "Historical Transformations of Kayapo Culture and Anthropological Consciousness"—neither he nor the Indians understood their "culture" or their historic situation the way they both do now.[93] In 1962, the Kayapo of Gorotire village apparently lived a double life. Their own customary existence was practiced on the margins of space, time, and personhood remaining to them after they conformed to the "civilizing" demands of the larger Brazilian society. Dependent on the government, Indian agents, and Christian missionaries for medicines, guns, ammunition, and other vital commodities, they seemed to have no inclination to act otherwise. They willingly took off their penis sheaths and lip plugs and donned Brazilian clothing whenever it seemed appropriate. They adapted their ceremonies to the spatial cum cosmological constraints imposed by the Brazilian-style village in which they had been obliged to settle. But the two cultures they thus lived appeared to have no relationship: any more than, in the traditional cosmology, beings such as the Brazilians, living beyond the pale, could be equated with the "complete" and "beautiful" humanity of Kayapo. Hence the culture that came from the national and international spheres had the quality of "an alien overlay beneath which the authentic Kayapo culture still persisted. The native social and cultural forms . . . appeared to have persisted in spite of their encompassment by the situation of inter-

91. Ibid., 8.

92. Turner, "The Politics of Culture."

93. Terence Turner, "Representing, Resisting, Rethinking: Historical Transformations of Kayapo Culture and Anthropological Consciousness," in *Colonial Situations*, ed. George Stocking (Madison: University of Wisconsin Press, 1991), 285–313.

ethnic contact, rather than because of any stable or harmonious accommodation to it"[94]

So in 1962, ethnography took on the character of archaeology, excavating under the disturbed modern topsoil of acculturation for the remains of the authentic Indian. Turner points out the peculiar complicity between this sort of anthropology and its apparent subject matter. Like Kayapo culture, the anthropology of the period defined itself "in abstraction from the 'situation of contact,' as the antithesis of 'change' and the enemy of 'history[95]'" The static concepts of culture that anthropology had inherited from structural-functionalist ancestors and their like were matched only by the Kayapo's seeming inability to take reflexive consciousness of their culture—that is, as their own social product—and deploy it against the external forces and institutions that were afflicting them. Failing to objectify their culture as an instrumental value, neither could the Kayapo make their ethnic identity an assertion of autonomy.

Although by the late 1970s, Terry Turner was already contending that a conscious concept of culture would be a potent resource in indigenous peoples' struggles for "cultural survival," in the mid-1980s he was still pessimistic about the Kayapo's chances of achieving the necessary self-awareness. Indeed in 1976 he had explicitly attempted to teach an instrumental sense of culture to the Kayapo, but they didn't get it. They were a long way from grasping and applying it to their situation, he wrote in an article that appeared in 1986. "I am not saying this would be impossible," he continued, "just that with a people like the Kayapo with no critical conception of their own culture, it would be far from easy."[96]

Yet when he returned to Gorotire in 1987, some twenty-five years after his first field study, everything in this respect had changed. The word *"cultura"* (from the Portuguese) was now commonly heard. Associated with it was an entirely different relationship to the other Indian peoples, the national society, and the international system. The Kayapo were actively and creatively engaged in the interethnic field—with an eye singular to the harnessing of its powers and products to the reproduction of their own "culture." Kayapo now understood their culture, including modes of subsistence, diet, ceremonies, social institutions, the body of lore and custom— as necessary to their "life," their "strength," and their "happiness." It was common, writes Turner,

94. Ibid., 291.
95. Ibid., 292.
96. Turner, "The Politics of Culture."

to hear Kayapo leaders and ordinary men and women speaking about continuing to follow their cultural way of life and defending it against assimilative or destructive pressures from the national society as the animating purpose of their political struggle. Many, including otherwise monolingual speakers, had begun to use the Portuguese word "cultura" to subsume their mode of material subsistence, their natural environment as essential to it, and their traditional social institutions and ceremonial system. The native term for the body of lore and custom, *kukràdjà* [meaning something that takes a long time to tell] was now also commonly used in the same way, in speaking of Kayapo customary practices and lore as requiring conscious efforts on the part of the community as a whole to preserve and reproduce.[97]

Again, this does not mean a return to a state of nature (or primordial culture). The reproduction of Kayapo culture now depends on the people's ability to domesticate the means and control the forces of their historic change. Kayapo do not refuse history; they seek to take responsibility for it. They would orchestrate it by the logics of their own schemes of things. It is worth repeating: in the struggle with the modern Leviathan, *the continuity of indigenous cultures consist in the specific ways they change.*

Turner offers a remarkable example in his deft analysis of Kayapo use of video.[98] On one hand, Kayapo turn the camera against the external forces threatening them, using it to document the activities of government agents and their like—while all the time making sure that their own documentation is recorded in documentaries made for international audiences. On the other hand, they would lend a certain facticity and permanence to their own ceremonies by making a video archive of them. Here, in this internal realm, Turner shows in detail how Kayapo camera work and editing respond to the people's traditional notions of "beauty"—which is also the transformation of nature to culture. The medium is thus the message: the dependency of the Kayapo upon Brazilian society is now complemented by a sustained opposition to it—in the name of the indigenous Kayapo "culture."[99]

97. Turner, "Representing, Resisting, Rethinking," 304.
98. Terence Turner, "Defiant Images: The Kayapo Appropriation of video," *Anthropology Today* 8, no. 6 (1992): 5–16.
99. A similar double-pronged movement of cultural resistance appears in the recent activities of the Wánai, a small Carib-speaking group of the middle Orinoco (Franz Scaramelli, "Culture Change and Identity in Mapoyo Burial in the Mid-Orinoco, Venezuela," *Ethnohistory*, in press). In June 1992, over four hundred years after Spanish explorers first met them, the Wánai men painted their bodies with the warrior signs their ancestors had used and planned a surprise attack on a powerful neighboring people, the Wóthutha. In a parallel strategy, they also wrote to the official Venezuelan government agency for resolving legal disputes among Indians, denouncing the Wóthutha for taking over Wánai lands—which was also the reason they were prepared to fight. In the same year, it might be noted, an article appeared in

This kind of cultural self-awareness is a worldwide phenomenon of the late twentieth century. For ages people have been speaking culture without knowing it: they were just living it. Yet now it has become an objectified value—and the object too of a life-and-death struggle. One should not give too much credit to anthropologists and their kind for the anomalous interest and respect they have accorded native cultures. Many peoples had decades of anthropologizing without celebrating their culture, and many others have recently become conscious of it without benefit of anthropology. "Culture"—the word itself, or some local equivalent—is on everyone's lips, especially in the context of the national and global forces that menace the people's traditional existence. And if to determine is to negate, it is interesting that the negation of culture, the defining contrast, is quite often the economic values of the incoming capitalism: the way Fijians oppose "living in the way of the land" *(bula vakavanua)* to "living by money" *(bula vakailavo),* or New Guineans may contrast *kastom* to *bisnis.*

The opposition is qualified in practice, as we have already seen, since *bisnis* is characteristically harnessed to the *divelopman* of *kastom.* The means are modern, *bisnis,* but the ends are indigenous, such as the extension of kinship through customary exchange. Or in Easter Island, the contradiction is synthesized by the creation of a "Corporation for Cultural Protection," which has successfully resisted several unwelcome projects of the Chilean colonial government. "We are different from Chileans in language, culture, and way of thinking," says the founder of the so-called corporation, Rodrigo Paoa. He continues:

> As a Rapa Nui [Easter Islander], if I want to sleep, I sleep. If I want to eat, I eat. I can spend a whole week without spending any money. If we are not careful, people will turn this island into another Hawaii or Tahiti, where the only important thing is money.[100]

The erstwhile victims of colonialism and imperialism have discovered their "culture." Hawaiians too make claims in its name, as do Australian Aboriginals, Inuit, Ojibway, Iroquois, Swazi, Ibo, Iban, Sami, Malays, Yakuts—peoples from all over the Third and Fourth Worlds. (A Burmese example appears in figure 7.2.) The cultural humiliation inflicted in the colonial period no longer grips these peoples as it too often had. Colonially inspired defamations of the "pagan," pre-European past are rapidly going

---

the *Revista Española de Antropologia Americana* entitled "The Last Wánai." It announced the "cultural extinction" of this people.

100. Nathaniel C. Nash, "As World Crowds In, an Island Shields Its Culture," *New York Times,* international ed., 6 Feb 1993, p. 2.

Figure 7.2.   Preservation of culture in Burma. (Photo by Peter Sahlins)

out of style—especially among the younger people. In a curious role reversal, the younger generations are often champions of "tradition" and spon-
sors of its revival. More than likely they are opposed by elders who had
accommodated to the white man and internalized the latter's reproaches
against the ways of the ancestors. But now, as Lamont Lindstrom observed
of Tanna islanders (Vanuatu), the received moral contradiction between
kastom and modernity is collapsing.[101]

And it is only a seeming contradiction too, that some of the most visible defenders of traditional culture are sophisticated students of the
Western world order. Often they are the most "acculturated" persons:
such as the artist and self-styled "cultural practitioner" of Zuni who dec-
orates the walls of the local Catholic church with paintings of kachinas. Or

101. Lamont Lindstrom, "Leftamap Kastom: The Political History of Tradition in Tauna,
Vanuatu," *Mankind* 13 (1982): 316–29, at 325. See also Kay B. Warren, "Transforming Memories and Histories: The Meanings of Ethnic Resurgence for Mayan Indians," in *Americas:
New Interpretive Essays,* ed. Alfred Stepan (New York: Oxford University Press, 1992), 189–
219.

consider Dr. Andrew Lakau of Enga Province in the New Guinea Highlands, shown in figure 7.3 in the dress he donned to receive his doctorate of philosophy from the University of Queensland in December 1994. The newspaper report indicated he preferred this garb to the usual academic gown "because of the great significance of the occasion for himself, his family and his culture."[102] But are not such masters of the global and the local in the best position to mediate between them? Besides, the contradictions at issue are as much those of class and social exclusion originating in the culture of dominance: the structures of discrimination from which the indigenous culture movement often develops its leaders and its passions.

The "return to the source," as Amilcar Cabral noted, is generated in the injustices of the colonial outposts of world capitalism.[103] Martyr of the Guinea Bissau liberation movement, Cabral was one of the first to speak of the role of culture in the anticolonial struggle. It enters in a twofold way. First, in the "social and cultural drama" of indigenous people who have risen into the urban middle classes or the colonial elite. Unable, however, to cross "the barriers imposed by the system," to be truly integrated or participate in the ruling Western order, some important number of the indigenous bourgeoisie are marginalized by their own successes. "So they turn towards the other pole of the social and cultural conflict in which they are living—the mass of the people."[104] Second, then, the people's struggle for liberation is a culture war. For precisely what has been attacked by the system of foreign domination—by the capitalist economy and the Western development ideology, by the disciplines of taxation, census, sanitation, missionization, and the other means of colonial control—*what has been attacked is the people's own form of life*. Colonialism is a massive process of *cultural hegemony*. Hence culture, Cabral wrote, "has proved to be the very foundation of the liberation movement."[105] And by reclaiming the people's control of their own existence, the liberation will restore the historicity of their culture. Because a society "that really succeeds in throwing off the foreign yoke reverts to the upward paths of its own culture, the struggle for liberation is above all an act of culture."[106]

The same political sense of culturalism continues to be echoed today, in the

102. *The Australian*, 21 December 1994.

103. Amilcar Cabral, "The Role of Culture in the Battle for Independence, *UNESCO Courier*, November 1973, 12–20; Amilcar Cabral, *Return to the Source: Selected Speeches* (New York: Monthly Review Press, 1973).

104. Cabral, "The Role of Culture," 15.

105. Ibid., 16.

106. Ibid.

Figure 7.3.   Andrew Lakau of Enga Province in the New Guinea Highlands in the dress he donned to receive his doctorate of philosophy from the University of Queensland. (From the *Australian*, 21 December 1994)

so-called postcolonial era, by African intellectuals. So is the appreciation of the future of tradition. "Culture is not only a heritage"; as Paulin Hountondji says, "it is also a project."[107] It is, as Elikia M'Bokola says, a demand for specifically African forms of modernity:

Throughout the twentieth century, culture has been the terrain on which Africans have chosen to carry out the struggle for the recognition of their dignity, which involves much more than the mere recognition of their civic and political rights. Since the independence, the continent's most legitimate spokespersons have never ceased drawing attention to culture as a constitutive particularity of contemporary Africa, either to underscore that in a "meeting of give and take," it is precisely culture that would constitute the continent's specific contribution, or to demand that development mold itself to the demands of African culture.[108]

107. Paulin Hountondji, "Culture and Development in Africa: Lifestyles, Modes of Thought, and Forms of Social Organization," UNESCO World Commission on Culture and Development CCD-IV/94/REG/INF.9, 1994, p. 2.
108. Elikia M'Bokola, "Culture and Development in Africa," UNESCO World Commission on Culture and Development, CCD-IV/941REG/INF.3, 1994, p. 1.

Reports of Maya peoples of Guatemala in the late 1980s through early 1990s make exactly the same point.[109] In the rural highland community studied by Kay Warren, the famous civil-religious hierarchy, with its set of traditional authorities and ritual practices, which appeared to collapse in the early 1970s, was revived in a translated form in 1989—precisely by the Catholic Action group that had originally done it in. The *costumbre* so recently demeaned and dismissed by many Mayas was now revitalized as something uniquely valuable. (Here again, just as in the Pacific islands, the contradiction between custom and modernity dissolves.) Meanwhile, the development and defense of a pan-Maya culture by a rising class of university-trained intellectuals and professionals occurred in the cities, as John Watanabe has documented. Regardless of their success in the national context, these Maya "have chosen not only to retain their ethnic identity but also struggle to have their languages, cultures and histories recognised as valid subjects within the academy—and worthy of respect within the larger society."[110] Moreover, Maya culturalism continues to evolve despite the criticisms of national intellectuals of the left and the right both. For Marxists, "Maya culture" is either a false consciousness of the Indians' plight or merely the ideological trappings of a popular resistance to colonialism and class oppression. In either event, by its ethnic particularization it detracts from the general unity of the proletarian revolutionary struggle. On the right, for conservatives, the emphasis on Indianness is an embarrassment to the nation and an impediment to its integrity. All the same, the indigenous intellectuals have gone ahead with the establishment of centers for Mayan studies, for the promulgation of Mayan languages and literatures, for the decolonization of Mayan history, and for anthropological research into their own communities and traditions.[111]

CONCLUSION

The historical significance of the late twentieth-century culturalism remains to be determined. Certain dimensions of anthropological interest, however, are already apparent. One is that the cultural self-consciousness

109. Warren, "Transforming Memories and Histories"; John M. Watanabe, "Unimagining Maya: Anthropologists, Others, and the Inescapable Hubris of Authorship," *Bulletin of Latin American Research* 14 (1995): 25–45.

110. Watanabe, "Unimagining Maya," 31.

111. Ibid., 32–33. On the analogous development of Mixtec culturalism centering in the Mexican-U.S. borderland, but elaborating on the heritage of the southern Mexican homeland, see Kearney, "From the Invisible Hand," 355.

of the indigenous peoples is an aspect of the global expansion of the Western-capitalist order—especially of its newer modes of colonization, commodification, and communication. But again, it is not a totally conservative reaction, a return to some sort of pre-European, primordial condition. On the contrary, the return to the source is coupled to desires for the technical, medical, and other material "benefits" of the world system. The Inuit do not want to give up their snowmobiles or the Fijians their outboard engines; they only want to use these for their own purposes—visiting their kinsmen, for example. Again: they would encompass the global order within their own cosmos.

As local peoples thus create differentiated spaces within the world ecumene, the planetary structure of culture changes.[112] We have a new global organization of human culture. At the world level, humanity is beginning to realize itself culturally as a species being—at least in the mode of similarity or mechanical solidarity. Yet at the same time, as the global cultural flows are locally diverted, they are diversified according to particular cultural schemes. The new world organization is a Culture of cultures.

For anthropology the novelty has been accompanied by a certain irony. Faced with a world Culture of cultures, a development that swept away the old anthropology-cultures—the apparently bounded, coherent, and sui generis "systems" of yore—the discipline was seized by a postmodern panic about the possibility of the culture concept itself. Just when the peoples they study were discovering their "culture" and proclaiming their right to exist, anthropologists were disputing the reality and intelligibility of the phenomenon. Everyone had a culture; only the anthropologists could doubt it. Still, the discipline's epistemological hypochondria seems to have been produced by this global reorganization of culture, not by some sort of inherent disorder in the phenomenon—of which nothing in the way of human knowledge could be said. Happily then, the nonexistentialist philosophy is passing. Now there are all kinds of interesting cultural processes and relationships to explore.

112. Ulf Hannerz, *Cultural Complexity: Studies in the Social Organization of Meaning* (New York: Columbia University Press, 1992).

8   Jed Z. Buchwald

How the Ether Spawned the Microworld

Consider figures 8.1 through 8.3. All three were produced by a scanning tunneling microscope, or STM, in IBM laboratories in 1988 and 1989. According to IBM's public release, figure 8.1 constitutes "the first direct observation of the buckling of the surface layer of gold atoms."[1] Figure 8.2, writes the research scientist whose team produced it, is the first-ever image of "the internal structure of an isolated molecule."[2] Figure 8.3 immediately captures the eye of the historian of chemistry. According to IBM publicity, it is one of "the first pictures that show how atoms are arranged in individual benzene molecules. . . . Clearly visible is the benzene molecule's distinctive ring shape, reconfirming what nineteenth century German chemist August Kekulé envisioned in a dream."[3]

These images require a great deal of specimen preparation, though perhaps not more than what must be done to produce a good electron micrograph of, say, a fly's eye. Many like them have been made in the last seven years. Two among them are particularly interesting: the atomic logo (figure 8.4) and molecular man (figure 8.5), by Peter Zippenfeld. Figure 8.4 was produced in 1990. IBM publicity describes it as follows: "Scientists at IBM's Almaden Research Center, San Jose, Calif., have demonstrated that individual atoms can be moved across a surface and positioned where desired with great accuracy. The first structure produced was the letters 'IBM' in xenon,

1. IBM release dated Atlanta, Georgia, 7 March 1989. From the Research Division, Almaden Research Center, San Jose, California.

2. Conference handout by Dr. R. J. Wilson, IBM Almaden Research Center, 17 January 1989.

3. IBM release dated San Jose, California, 18 July 1988. From the Research Division, Almaden Research Center, San Jose.

Figure 8.1.   Gold. (Courtesy of IBM Corporation, Research Division, Almaden Research Center)

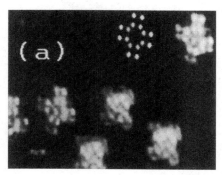

Figure 8.2.   Copper phthalocyanine. (Courtesy of IBM Corporation, Research Division, Almaden Research Center)

Figure 8.3.   Benzene. (Courtesy of IBM Corporation, Research Division, Almaden Research Center)

shown here as it was being created in a series of four images."[4] Figure 8.5, which IBM refers to as the "molecular man," "was crafted of 28 individual molecules one molecule at a time by Peter Zippenfeld."

The IBM logo in xenon, which was widely reproduced in newspapers,

4. IBM release entitled "IBM Scientists Position Individual Atoms—First to Build Structures One Atom at a Time."

shows that the atom has today become something much more influential than a tool of abstraction and arcane laboratory procedure. It has become an advertising medium. Like the electromagnetic spectrum before it, the atom roots firmly in human reality only as it becomes a prosaic acquaintance. Who after all would wonder about atomic reality when anyone can go to the local superstore to buy a "Zippenfeld Atomic Etcher" for touching up the family's greeting cards?

Zippenfeld's *Amazing* ATOMIC ETCHER!!

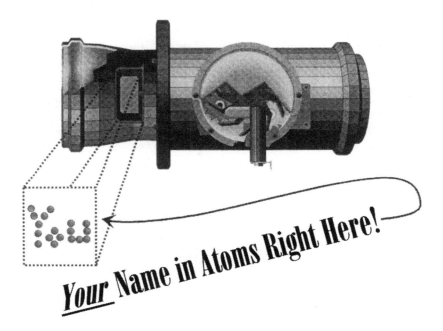

Despite the atom's mundane presence today, less than a century ago it was the subject of intense discussion among many philosophically minded scientists, who debated its very existence. The present viewpoint began to emerge during the early 1900s, when measurements of Avogadro's number, as well as of other microphysically significant quantities, cohered in-

(1)   (2)

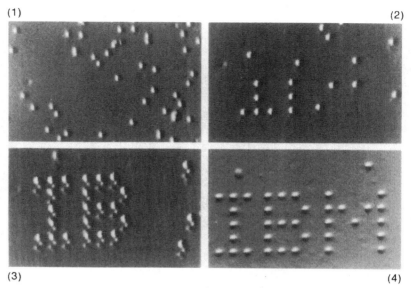

(3)   (4)

Figure 8.4.   The atomic logo. (Courtesy of IBM Corporation, Research Division, Almaden Research Center)

creasingly among one another, thereby solidifying scientists' confidence in the microworld's reality so that, by 1917, little doubt remained among physicists concerning atomic existence. A practical concern with the characteristics of atomic entities had by then replaced doubts about their reality, and physicists were beginning increasingly to use these entities in efforts to deduce macroscopic properties from their behavior in bulk. By the end of the Second World War a substantial part of physics concerned the microcosm.

Since the beginning of the twentieth century physicists have used microphysical entities to build quantitative theories that are linked to instruments and devices such as the remarkable STM. Although a considerable amount of discussion has been devoted to the many interesting philosophical problems that microphysical hypotheses raise, until recently not so much attention has been paid to the historical origins of this practical microphysics. By "practical" I mean a microphysics that is used as an appliance for the development of theories, as well as for the analysis, construction, and deployment of devices and instruments. A microphysics, in other words, that forms part of a scientist's normal toolbox.

The central image of twentieth-century physics represents the universe as a set of discrete, interacting entities, vastly smaller than the human scale,

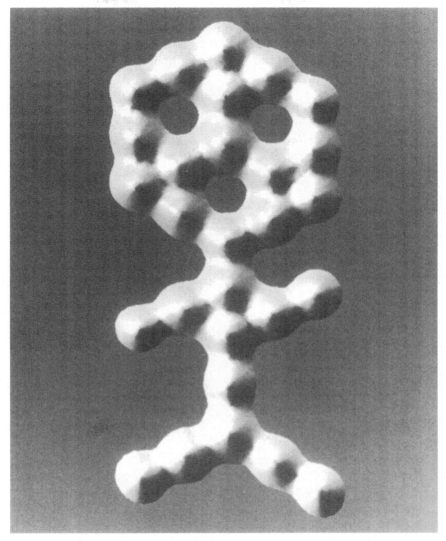

Figure 8.5.   Molecular man, by Peter Zippenfeld. (Courtesy of IBM Corporation, Research Division, Almaden Research Center)

that, when gathered in appropriately constituted groups, form material objects as we know them. During this period physicists have developed a set of conceptual, mathematical, and instrumental tools for handling microphysical entities that have enabled them to build a working science founded directly on the properties ascribed to these objects. The STM is one such device, and we can use it to open a perspective on the present character of microphysical practice, which will enable us to see that microphysics evolved as a practice, and not solely as a set of specific concepts, from considerably different antecedents.

The word "practice" appears frequently in contemporary history of science and merits some discussion, particularly when, as here, it will be used to develop a distinction that can too easily be unclear. In examining a physicist's production we can usually, though not necessarily always (or easily), delineate the specific theoretical concepts, mathematical techniques, exemplary (or canonical) experiments, instruments, effect-producing devices, and so on with which the physicist is concerned. These elements might be said to constitute en masse the physicist's particular "practice," in the sense of daily routine or methods. That is not, however, the sense in which I wish here to use the word, because it includes at once too much and too little.

By grouping together a possibly disparate set of techniques, "practice" in this sense fails to distinguish among elements that may in fact have little substantive connection with one another. Physicist A, for example, may share with physicist B a common experimental practice, but may differ with his colleague over the appropriate mathematical tools to be used in analyzing experimental data. Or B may hold considerably different views about the intestine working of a piece of common apparatus than A does, even though both work with the equipment in much the same way. Perhaps they do not differ with one another in these respects, but they are for a panoply of reasons preoccupied with different effects.

One might solve this inadequacy of distinction by developing a typology of practice (e.g., theoretical, experimental, mathematical, instrumental, etc.), which is certainly a meaningful and informative thing to do, but in that case the term would include too little to encompass our needs. By dividing a physicist's workaday world into heterogeneous elements in this way, we would almost certainly miss the possible existence of a powerful motivating, and structuring, force that unites them, or certain aspects of their work, in a common endeavor. Mathematical technique may be particularly adapted to the character of the physicist's theoretical scheme, which may in turn reflect, and be reflected in, either the form of apparatus in which the physicist is most interested, or else in the manner in which the apparatus is deployed to investigate problems that are considered interesting. Or all of these several elements may be in the service of a particular kind of physics, one that potentially encompasses an indefinitely large set of distinguishable elements (theoretical, instrumental, mathematical, etc.) that have in common certain structural characteristics that reveal themselves in the working physicist's activity on paper and in the laboratory. Any particular set of such elements might reasonably and productively be said to constitute their user's "normal toolbox," but here we will reserve the phrase, like its correlate in action "practice," for a more general group of characteristics that unites these sets among one another.

In the specific case of microphysics, one aspect of the practice that I have in mind can be illustrated by comparing the STM with another device of twentieth-century physics, indeed a canonical device (which is itself far from unitary), namely the high energy collider, in respect to their rather different microworlds. The STM shows and makes objects by what one might call *constructive manipulation:* Zippenfeld's molecular man is for example built out of things moved into place by the STM, and it is also made visible by the very same device that is used to construct it. The structures dealt with by the STM are in fact new kinds of things altogether, designer molecules as it were, whose existence and character are manifested by the STM itself. The atoms of the STM's world are in themselves unitary objects; they are unreduced building blocks that the STM uses to make molecules. This marks a signal difference between the STM and the particle accelerator, in respect both to conception and to instrumental behavior, that is worth briefly exploring for our purposes.

To high-energy people the STM's startling images are in fact pictures of a higher-order world that can be analyzed in itself only by means of laws that, from the high-energy perspective, and however obtained, must necessarily be approximations to an underlying reality. That reality can be reached only through huge machines—colliders—that fabricate a universe of immense energies. IBM's scanning tunneling microscope does not do anything like this, and there is a corresponding difference between the two devices. Collider objects, unlike STM objects, come into being through a process that might be called *destructive manipulation:* entities produced by the device are destroyed within it in order to produce other entities that a distinct component of the apparatus detects. STM science examines and manipulates undivided existing things; collider science seeks information by ripping things apart.

Although the one builds where the other dismembers, both the STM and the collider are nevertheless designed with specific microproperties in mind, and they are intended to produce knowledge by working with and on objects in their respective microworlds—and not on the things that these objects may themselves constitute when collected in heavily populated groups that erase the identity of their constituents. The goal of the kinds of physics represented by these sorts of machines is accordingly not to explain relations known independently to hold true outside their worlds. The aim is rather to engage directly with their proper microcosms, which is prototypical of twentieth-century physics. This is a fundamental characteristic of what I mean here by the phrase "microphysical practice." My purpose in what follows will be to examine whether anything at all analogous to that practice existed in the nineteenth century.

We might begin our quest for a nineteenth-century microphysical practice by looking directly for images that are similar to those produced today by the STM. If we did, then we would naturally think first of nineteenth-century chemists, many of whom did draw pictures of a microcosm that were intended to be structural representations of molecules. However, the comparison with twentieth-century practice fails. In addition to the obvious and centrally important difference between a drawing and an automatic image, there is also a difference between twentieth- and nineteenth-century attitudes toward the microcosm. Mary Jo Nye, for example, has remarked that, although "most chemists were convinced by the existence of isomers and optical rotatory power that 'molecules' exist as real, three-dimensional structures of some sort," nevertheless "their schemata of representation were heuristic, often short-lived, and intended to be so,"[5] whereas neither STM nor collider images are thought of in that way. We can use Nye's remark to introduce a series of distinctions that will help us better to understand just what was done during the nineteenth century.

The kind of activity engaged in by nineteenth-century chemists, though quite varied, might best be called *reductive explanation.* Here microstructural configurations are sought that elucidate chemical behavior or that give meaning to formulaic representations, with the connection being somewhat loose and tentative. This kind of work did not involve either *destructive* (in the high-energy physics sense) or *constructive* (in the STM sense) *manipulation* because entities were neither being ripped apart nor

5. Mary Jo Nye, "Explanation and Convention in Nineteenth-Century Chemistry," in *New Trends in the History of Science* (Amsterdam: Rodopi, 1989), 182. Mary Jo Nye's books and articles have cast a sharp light on these developments and must be carefully studied by any historian or philosopher who wants to understand the tremendous complexity and variability of time of microworld argumentation and analysis. See especially her *From Chemical Philosophy to Theoretical Chemistry* (Berkeley and Los Angeles: University of California Press, 1993). See also Alan Rocke, *Chemical Atomism in the 19th Century: From Dalton to Cannizzaro* (Columbus: Ohio State University Press, 1984), who argues for a subtle understanding of the chemical molecule during the nineteenth century.

produced as microobjects. Chemical analysis and synthesis might loosely be said to correspond to destruction and construction, but my point here is precisely that many chemists of the day did not deploy working laboratory or paper tools that were tightly bound to a microworld, even when they believed in it. For them such things might be the stuff of (tentative) explanation but they were generally not the core of either laboratory practice or, for the most part, of paper work.

Chemistry was not, however, the only discipline for which pictures of a microcosm were produced during the nineteenth century. Consider for example the drawings in figure 8.6. Unlike the images produced by the STM, these four figures were drawn by hand; no machine intervened between them and the conception held by their author, who created them in 1842. They were nevertheless intended to represent something entirely similar in kind to the STM's molecular images, for they are seriously intended pictures of a microcosm. Although physicists in the 1840s were no more able

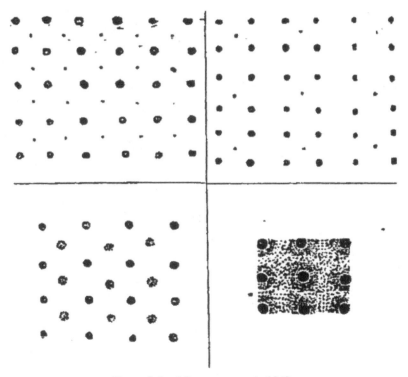

Figure 8.6.   Microstructure in 1842.

than chemists of the day to use microobjects for purposes of material fabrication, nor did they have machines to make pictures of the microcosm, their embrace of mathematical reasoning did nevertheless produce attempts at what might be termed reproduction, rectification, and production.

Suitably inclined practitioners of microphysics might undertake three distinct kinds of activities in respect to the microworld of the day. Attempts might be made to *reproduce* known phenomena from microphysical structures, and to specify in so doing the limitations inherent in the reproduction. Reproduction can be either quantitative or qualitative. If qualitative, this may be the same as *reductive explanation,* in which structural configurations are sought that elucidate known behavior or that give meaning to formulaic representations, with the connection being somewhat loose and tentative. Attempts to repair defective knowledge by *rectifying* it on the basis of such things might also be undertaken; here the limitations imposed by reproduction might turn into corrections to known laws. Rectification can be either qualitative or quantitative. Finally, there may be attempts to generate entirely novel effects or connections among effects by *productions* based on microprocesses. Usually quantitative, production can be qualitative as well.[6] These kinds of work may frequently involve the use of *mathematical tools* for the production and manipulation of what might best be termed a *paper microcosm.* Indeed, it seems to me that there is a reasonable if difficult sense in which one can speak about paper experimentation. Like laboratory work, paper work deploys certain kinds of tools in flexible ways, but, again like laboratory work, it is also subject to constraints imposed by those tools. And, not infrequently, unexpected results appear. These three forms of microactivity are highly characteristic of twentieth-century physics, and they took place as well during the nineteenth century. There are, however, signal differences in reproductive and productive practice between then and now that can best be explicated through examples.

To provide a first sense of these differences consider a case drawn from Lord Kelvin's widely influential 1884 *Baltimore Lectures on Molecular Dynamics and the Wave Theory.* The title at least sounds promising. Here, perhaps, we might find something like the modern attitude toward the hid-

---

6. The distinction between rectification and production cannot be made rigorous, because many rectifications might be considered novel constructions. Take, e.g., the corrections to Cauchy's dispersion formula (discussed below) implied by micro-based analyses after the mid-1870s. The laws governing anomalous dispersion were not well known, and so here reproduction could only be qualitative. One might consider the laws that were deduced to be rectifications of Cauchy's formula, and this does have logic to it, because Cauchy's formula can be deduced from them as a special case. On the other hand, the laws also permitted the existence of bodies with altogether novel dispersing properties, for productive implications.

den world. And indeed we do seem to. At one point, late in the series, Kelvin told his audience a story about a discussion he had with himself and with Rayleigh:

> I have brought a book which I intended to make the subject of our lecture. I am afraid it will be passed over. The book is Stokes' paper "On the Metallic Reflexion exhibited by certain Non-metallic Substances." I only wanted to tell you that his molecular theory [referring to previous days' lectures] explains the colors of aniline and this wonderful thing that Stokes experimented on— this safflower-red. I wanted to read about the bright lines in the light reflected from safflower-red discovered by Stokes. I was thinking about this three days ago, and said to myself, there must be bright lines of reflexion from bodies in which we have these molecules that can produce intense absorption. Speaking about it to Lord Rayleigh at breakfast, he informed me of this paper of Stokes and I looked and saw that what I had thought of was there. *It was known perfectly well, but the molecule first discovered it to me.*[7]

This is a lovely instance of serendipitous *reproduction;* we might even want to call it *production,* because Kelvin had apparently not previously read Stokes's paper. This greatly excited him. He continued:

> I am exceedingly interested about these things, since I am only beginning to find out what everybody else knew, such as anomalous dispersion and those quasi colors and so on. There is no difficulty about explaining these things; *we can predict them from the consideration of the molecule without experimental knowledge.*[8]

That is, we do not have to discover this thing from experiment; we can predict it well before we enter the laboratory. This sounds familiar, and in retrospect it is—because we today expect microprocesses to provide just this kind of potentially novel information.

Kelvin's enthusiasm is deceptive. Rather than signifying an attitude familiar to us, it betrays just how uncommon this sort of expectation was at the time; he was so enthusiastic precisely because he had found an utterly uncommon situation, one in which the power of a microstructure proved itself in novel application. It is, however, less important to emphasize Kelvin's surprise than to remark something that makes his little homily unfamiliar even to us, who are no longer surprised by microwork. In the final sentence

7. Kelvin, *Baltimore Lectures on Molecular Dynamics and the Wave Theory of Light* (stenographically reported in 1884 by A. S. Hathaway; published under the same title, but with numerous changes, London: C. J. Clay and Sons, 1904), 282; emphasis added. Page citations are to the 1884 version.
8. Ibid.; emphasis added.

of the paragraph he remarked, "And here again is a thing that suggested it-self to me, that most probably there are bodies in which light is propagated faster than in the luminiferous ether." This too was suggested by Kelvin's "molecular" constructions. Nor was Kelvin unusual in making the connec-tion; amalgams of molecules with ether constituted the ground for a great deal of speculative microwork in the late nineteenth century.

By "speculative" I mean that Kelvin's amalgams of ether and molecules had very little purchase on the vast array of experimental and mathemati-cal work that was also being done during the 1880s. They were, in that sense, rather suggestions offered in partial explanation than they were the foundations for a working microphysics. By the end of the 1890s Kelvin-like amalgams had begun to disappear from the public literature, and from private investigations as well. At the same time, the beginnings of a practi-cal, if highly heterogeneous, microphysics were forged, a microphysics that provided conceptual, mathematical, and instrumental tools for further in-vestigations. In that sense—in the sense of practice—the microworld first became strikingly real among physicists during the 1890s. If we are to un-derstand how this transformation occurred then we must also understand how it was (frequently) bound on paper and (occasionally) in the laboratory to a world that we no longer believe to exist at all, the world of the ether. In what follows I argue that the modern era of practical microphysics, as it be-gan to emerge in the 1890s, was built out of, and in agonistic engagement with, the tools and concepts provided by ether dynamics.

A practical microphysics, in the sense meant here, has at least *reproduc-tive* power, and possibly *productive* power as well. That is, it can be used to reproduce relations that are already thought to hold true, and it may also be used to produce entirely novel ones. Now according to the lore of contem-porary high-energy physics, all events are due to the behavior of elemen-tary particles; any laws (all necessarily approximate) that hold in our instrumentally accessible world must accordingly be the result of processes that occur at this microlevel. A similar belief held sway during the first half of the nineteenth century. According to a belief that took firm shape in France during the 1820s, and that was subsequently pursued in Britain and Germany until the late 1840s (and even afterward in Germany), all optical events are also due to the calculable behavior of elementary particles; thus, any (necessarily approximate) laws that apply in the world of optical instrumentation must result from processes occurring among these parti-cles. There is, however, one small difference between the two convictions. High-energy particles constitute matter; the earlier particles constituted ether. Figure 8.6, which was drawn by a contemporary practitioner of this sort of physics, does not, for example, represent just a material crystal; it represents a lattice of ether atoms interspersed among material particles, with which they interact.

Today's material atom or molecule plays a central role in a myriad of disciplines, from high-energy physics through chemistry to molecular biology. The shape and texture of much contemporary science depends directly upon the vivid presence of this microworld. At least one of the central novelties of twentieth-century physics—quantum mechanics—was designed explicitly for the microworld, while the other—relativity—was tied from its beginnings to the invisible realm. It is of course historiographically dangerous to simplify the complexity of an era by choosing to identify this rather than that set of developments. Nevertheless, and however complex the practice of an era may be, the historian who immerses himself or herself in the debates, calculations, laboratories, and rhetoric of an era can identify coherent schemata that serve well to illuminate scientific work. Such systems—varied in origin and nature though they undoubtedly are—as quantum mechanics and relativity work well for organizing historical understanding. The historian of nineteenth-century physics can similarly identify and describe a set of useful schemes, including (at least) wave optics, thermodynamics, field theory, and continuum mechanics. Each of these areas achieved a highly developed quantitative character during this period, and all four were seated strongly in the laboratory as well.

Of these four schemes, only wave optics made either productive or reproductive use of microphysical reasoning. Thermodynamics, though born in reasoning that did invoke the microworld, made little subsequent use of it until the invention of statistical mechanics, which itself achieved widespread use only at the beginning of the twentieth century. Field theory, particularly as practiced in Britain and (later) in Germany, avoided the microworld as a matter of principle for many decades. And continuum mechanics, to which I shall return in a moment, was developed separately from a molecular mechanics of bodies, though that very separateness did give rise to a research tradition (particularly in France) that attempted to explore their relationship.

Wave optics was early connected, both productively and reproductively, to a microworld. But that world, which we have already glimpsed, was hardly like the modern one, for it was profoundly linked to an entity that vanished from science during the first decade of the twentieth century— the ether. Nevertheless, many of the reproductive and constructive practices that did come into common use by century's end for dealing with the material microworld were first produced for handling ether problems. Stories of ether mechanics have long pedigrees; most historians, including myself, have told such tales. We can step back a bit and examine what we have learned from these stories that casts light on our aim here.

In extremely broad outline we can divide the role that ether played in wave optics into two distinct categories, with temporal and spatial overlap between them. The earliest was developed by the French founder of wave optics, Augustin-Jean Fresnel, and was avidly pursued in France and Germany for about thirty years. The second appeared first in Britain a decade after the principles and practice of wave optics had stabilized and was pursued as well in Germany. To understand the role that a microphysics played in Fresnel's work, and in that of the many who, like the mathematician Augustin-Louis Cauchy, followed a similar path, we must be careful to understand what it did not do. Specifically, there is no evidence that microreasoning acted *constructively* in Fresnel's production of the seminal principles and practices of wave optics. He did not develop his understanding of the signal concepts of front and phase, or of the astonishingly fruitful notion that the optical disturbance lies in the front and not orthogonal to it, on the basis of any microreasoning at all. That is, although Fresnel undoubtedly did think from the very beginning of his research that the ether, which carries optical waves, consists of particles that interact with material molecules, and although that image was immensely important to him in suggesting the possibility and plausibility of a wave optics, he did not use it to produce the underlying principles and practices of his novel physics for light.

But Fresnel did use the image of a vibrating, mechanical ether when he produced a mathematics for the optics of certain crystals, for this image suggested to him the possibility of constructing a particular series of surfaces that succeeded in capturing known, and in predicting novel, crystal behavior. Although suggestion was all that the image provided here,[9] the ether, conceived as a lattice of interacting points, was becoming constructively important to Fresnel in the years just before his death. He had by then begun looking to it to provide explanations of, and even formulas for, phenomena that otherwise escaped the wave theory's principles, in particular the dispersion of light. In 1823 he advanced a qualitative explanation of dispersion, based on the spacing and on the forces between the mutually repelling particles of the ether, that became immensely influential in the 1830s when it was taken up by Cauchy.[10] During the 1830s optical theory became for a time nearly synonymous with Cauchy's ether dynamics. It is

9. For details see Jed Z. Buchwald, *The Rise of the Wave Theory of Light: Optical Theory and Experiment in the Early Nineteenth Century* (Chicago: University of Chicago Press, 1989), chap. 12.4.

10. For a brief discussion of Cauchy's structure see ibid., app. 6. Full details, as well as discussion of the structure's historical consequences, are in Jed Z. Buchwald, "Optics and the Theory of the Punctiform Ether," *Archive for History of Exact Sciences* 21 (1980): 311–73. See also Amy Dahan Dalmedico, *Mathématisations: Augustin-Louis Cauchy et l'École Française* (Paris: Librairie Albert Blanchard, 1992).

precisely here that we find a twentieth century–like belief in the power of microcosmic analysis, for Cauchy and many others (primarily in England and France) attempted to *produce* new physics out of the ether and to *reproduce* known laws from it as well.

We need not here explore in any detail the work that Cauchy and the others pursued in order to capture its character. In essence, these micro-modelers began with Cauchy's lattice structure of point particles and imposed three sorts of restrictions on the general system: the particles were arranged according to certain symmetries (for application to crystal optics), the distances between the particles were adjusted (for application to dispersion), or the particle-particle force was manipulated (for several purposes). In the case of crystal optics, the procedure was able to recover a close approximation to the surfaces that Fresnel had earlier conjectured, and that worked well empirically. As for dispersion, this procedure was the first one actually to produce a successful formula. That formula is still referred to in many physics texts, where it is correctly attributed to Cauchy. And it emerged directly out of micromodeling for the ether.

Despite these successes, which were not uncontested at the time, the ethereal point lattice did not become embedded in an ongoing, stable research tradition, and its form of micromodeling had effectively vanished by midcentury. There were many reasons for this, not the least being that the structure was extraordinarily difficult to manipulate mathematically, so that the empirical consequences that could be derived from it were meager indeed. Furthermore, many approximations had to be made, and conditions applied, to yield any results at all, although neither approximation nor condition could easily be justified by reference to conditions or to phenomena that went past the particular problem at issue. Beyond these issues of limited application and mathematical difficulty lay others that concerned the increasing discomfort that many British mathematicians felt in working with a system grounded on points and forces. Although such a system entailed ordinary differential or difference equations, at this very time mathematical attention in Ireland, Scotland, and England—driven by many factors—was grappling instead with the consequences of building structures directly on the basis of partial differential equations. These kinds of equations, when thought to be (as they increasingly were) fundamental, represent continuous effects, such as surfaces moving through space, whereas Cauchy's lattice was essentially and necessarily discrete—it yielded a moving surface only through approximation, and on the basis of conjectured relations.

A major new form of optical practice accordingly evolved in Britain. It appeared first in the work of the English mathematician George Green and, shortly thereafter, in that of the Irish mathematician James MacCullagh in the 1830s; their work was followed in the 1850s by that of the great master of continuum mechanics, George Stokes, and of others who followed the path they delineated. In these works the ethereal microcosm in the form of lattice equations played no role at all; indeed, the lattice was conspicuously avoided. Instead, equations were obtained by manipulating an unreduced "potential" function from which the medium's partial differential equation, as well as appropriate boundary conditions, could be derived by one or another form of the principle of virtual work, which had become important in late eighteenth-century French engineering mechanics. That function took on energetic significance after the 1850s and mutated accordingly.

These two approaches to producing and to reproducing effects by paper computation may, from our perspective, seem scarcely different because both, after all, evidently deploy a structure—the ether—that is no longer thought to exist. There is much warrant for this conflation. Like lattice mechanics, potential theory and its lineal descendants could reduce known effects to special ether configurations and could even (as when, for example, adapted in 1879 by George FitzGerald to electromagnetics)[11] construct novel ones as well. Those who practiced it worked with a now-discarded entity, just as much as those who practiced lattice mechanics. However, unlike Cauchy's lattice structure, potential theory and its conceptual descendants were not at all based on the microcosm—any more than continuum equations for a liquid need be based on a micromodel.

The major effect for our purposes here of treating the ether as a continuous structure was to make the material microcosm *itself* a less-than-fundamental entity. This had not been the case in Cauchy's lattice dynamics, for that system had placed the ethereal and material microworlds on precisely the same footing. The displacement of matter from its position at the foundation of physical speculation and even practical work had many effects during the years between 1850 and 1890 or so. For example, systems were envisioned in which material molecules were treated as structures formed in and out of the continuous ether—Kelvin's famed vortex atoms of the 1860s are only the best-known instance of such speculations. In work like this, the structure of the ether proper was scarcely touched at a fundamental level; it had ceased to be what it was for Cauchy, a highly malleable tool for productive purposes, with material microstructures existing as independent entities of essentially the same kind as the ether particles with which they interact. In work like Kelvin's fifty years later, and even more so

---

11. George FitzGerald, "On the Electromagnetic Theory of the Reflection and Refraction of Light," *Philosophical Transactions of the Royal Society of London* 171 (1880): 691–711.

in work by many of his Cambridge-trained Maxwellian contemporaries, material structures are not at all considered to be independent entities that interact either with one another or with the ether; they are both in and, more significantly, *of* ether, which meant in practical terms that matter itself was thought to be the exhibition by ether of local structures. There is no independent but equal material system in this vision, as there had been for Cauchy. There is only ether—subject, certainly, to regional changes in energy properties—but not to be thought of as being acted *on* by matter. As a result, there was comparatively restricted freedom to proffer seriously intended material microstructures—for any such structure had to be consonant with the ether's fundamental composition. This kind of work was quite common until the 1890s, and not only in Britain, and as long as it remained common the microworld could have little seat in physical practice of the era.

When and why did the modern attitude and practice toward the material microcosm come to working life? I want to suggest that the kind of microphysical model making that emerged during the 1890s, and that forms the bedrock of modern physics,[12] was created neither in France nor in Britain, but in the Netherlands and Germany, through the creative merger of these two very different nineteenth-century traditions, both of which had been pursued there, and both of which had been transformed by local factors. Experiment, generated independently of these abstract considerations, stimulated the transformation.

In 1870, the Danish physicist Christian Christiansen dissolved the aniline dye fuchsin in alcohol, shined light on the solution and rediscovered the peculiar optical phenomenon of anomalous dispersion. His discovery fed directly into a long-standing interest in Germany in metallic reflection, an interest that had there been connected to Cauchy's type of ethereal manipulation by (among others) Franz Neumann. The link between the two traditions we have been discussing was first forged in Berlin by Hermann Helmholtz in 1875 on the basis of considerations developed by Wolfgang von Sellmeier in 1872 in an effort to account for Christiansen's discovery. It was then elaborated by many others. Kelvin's 1884 discussions were for ex-

---

12. By "modern physics" I have in mind the admittedly loose assertion that dominant elements in the physics community since, at the latest, the mid-1930s have believed that *fundamental physics* is identical with the physics of particles at high energies, that all other physical processes are the results solely of fundamental interactions, and that all other physical laws should ideally be obtained by approximation from high-energy interactions. If, as is of course pragmatically the case, laws cannot be obtained in that way, then these laws are suspect. The image is of an austerely hierarchical world in which influence goes in one direction only—from high-energy physics upward.

ample based on what he himself had picked up of this by then decade-old trend.[13]

The essence of Sellmeier and Helmholtz's invention was deceptively simple. Forget, they suggested, about playing around with the ether itself, or at least don't do it very much. Take a different approach, along the following lines. First write down a pair of coupled equations of motion, one for the ether, and the other for matter. Of the two equations, the one for the ether is considered to be inherently fundamental, subject perhaps to some local alterations but in practical terms not much to be monkeyed with. The ether's equation acts, that is, as a practically fixed tool that constrains reproductive and productive behavior. In this way of working the ether is treated as effectively continuous, governed by an undeduced partial differential equation that contains a driving term which represents an interaction between it and the material system that is embedded in it. That system is not thought to be continuous, and the equation that governs its motions is to be altered as circumstances demand. The material system is in fact to be played with in very much the same way that lattice mechanicians in the 1830s and 1840s played with both ether and matter equations, a form of activity that was reasonably well known in German literature of the period. Only, instead of putting the two systems on the same footing, they are separated both mathematically and physically. Lattice considerations, though not techniques, accordingly passed over to material structures, whereas potential theory was relegated entirely to the now-fixed ether, where it became essentially sterile.[14]

Throughout the 1880s, particularly in Germany, a great deal of mathematics, and some experimental work, orbited around finding appropriate material equations for optical processes on the basis of Helmholtz's linked equations or variants of them. During the 1890s the material equations became explicitly microphysical, eventually evolving into molecular models. As this occurred, the link between the two equations changed from representing a mechanical force between a duo of substancelike systems (the one malleable, the other essentially fixed in nature) into an asymmetric inter-

13. For discussion of these developments, and the ones involving Drude mentioned below, see Jed Z. Buchwald, *From Maxwell to Microphysics: Aspects of Electromagnetic Theory in the Last Quarter of the Nineteenth Century* (Chicago: University of Chicago Press, 1985), pt. 5.

14. The connection to ether lattice considerations is quite explicit in Sellmeier, who must argue his way to the conclusion that punctiform properties for the ether can be ignored, whereas similar properties for matter must be taken directly into account. See Sellmeier, "Ueber did durch aetherschwingungen erregten Körpertheilchen und deren Ruckwirkung auf die ersten, besonders zur Erklärung der Dispersion und ihrer Anomalien," *Annalen der Physik und Chemie* 145 (1872): 399–421, 520–49.

action in which one of the two systems (the ether) *never* changes its character, and is therefore not, properly speaking, to be thought of as acted on. By limiting play to material structures these methods strongly affected practice in ways that lattice mechanics, with its vast array of arrangements available for both ether and matter, had not. Microphysical modeling for matter in the modern sense emerged, I want to suggest, precisely within the space opened by the foreclosure of reproduction and production for ether.

Nevertheless, the transition took place only with difficulty, uncertainty, conceptual obscurity, and instrumental instability. The microworld first became pragmatically real to many German physicists circa 1893. Its practical birth was tightly bound to a particular experimental question, namely, what happens to light when it is reflected from the surface of a magnet. This may sound like a rather limited question, but it is in fact the most general possible one to ask because every relevant variable is implicated in the problem. We need not go into details beyond considering a number of salient points. German physicists were by the early 1890s quite familiar with the twin equations in optics but had not for the most part deployed them in electromagnetics, which reflects the very recent insertion into their world of the electromagnetic field by Heinrich Hertz. Among them, and particularly by the young Paul Drude, magneto-optic processes were thought to involve just one constant beyond those that are normally required to analyze the reflection of light from metals, and this constant (like the others) was considered to have only *macroscopic* significance. In a series of occasionally heated exchanges with other physicists Drude insisted on the empirical adequacy of this approach, until in 1893 he was presented with results from the Dutch physicist Pieter Zeeman, who had not been involved in the controversy (and therefore had no particular stake in its outcome), that seemed even to Drude to demand two (and not just one) new constants. Given the contemporary context provided by Helmholtz's structure for anomalous dispersion, which had hitherto been kept apart from magneto-optics, Drude decided that this requirement meant it was not possible to avoid introducing a separate material equation of motion, for which Drude provided an incipient *microphysics,* one that he developed in more elaborate detail in his 1894 *Physik des Aethers* (Physics of the ether).[15]

Nevertheless, we do not have as yet a full-fledged microphysics even here, because before the early 1900s neither Drude nor anyone else, with the sole exception of the Dutch physicist Hendrik Antoon Lorentz, at-

15. Paul Drude, *Physik des Aethers auf elektromagnetischer Grundlage* (Stuttgart: F. Enke, 1894).

tempted to justify the combining of material and field equations in the ways that this emerging practice demanded.[16] It would be misleading to think of German attitudes in the 1890s as representing a halfway point between the macro- and the microworld, because there is no connected road that goes from the one to the other. For several years, German physicists, who had after all only recently learned about fields from Hertz, worked with a set of techniques that provided good results but that, they explicitly acknowledged, lacked mutual coherence. This began to change in the early 1900s, particularly after Lorentz's comprehensive restructuring of electrodynamics on the basis of electron theory in 1903, and after Drude's own further work in the electron theory of metals. Or, better put, what changed was the sense of incoherence, which gave way instead to a belief that microtechnique was beginning to constitute a world that could be used both productively and reproductively. This occurred as the physics of the electron became central to novel research, as electromagnetics displaced mechanics as the major unifying image for many physicists, and as microconsiderations were bound to an increasingly broad array of results that derived from considerably different phenomena.

Our discussion to this point has not taken into account the many factors that play into this startling efflorescence of microtechniques at century's end. We have not discussed nor even mentioned until now cathode, X, and Becquerel rays, though there can be little doubt that the rapid binding of these novel effects to the microworld caused it to bloom on paper and in the laboratory. These experimental works took place, however, *after*, not before, the early changes that we have examined. The practical effects of that change were rather strikingly in place by about 1895, and it may very well be that the binding just mentioned occurred so rapidly in part *because of* this preexisting practice. More importantly, we have not examined the local conditions in Northern Europe that led to the creative merger that produced practical microphysics. Among other things, one should discuss the intense and evolving production of ionic physics and how that work linked to the optical investigations I have mentioned. Further, it seems highly probable that the specific characteristics of the Berlin professional world—its conflicts and its demands—influenced Helmholtz's creation of his twin equations but at the same time distanced them from the microworld it-

---

16. Lorentz's work early on (from 1878) diverged from German and British practice, introducing both ether and matter in nonstandard ways. For a brief discussion see Jed Z. Buchwald, "The Michelson Experiment in the Light of Electromagnetic Theory before 1900," in *The Michelson Era in American Science, 1870–1930*, ed. S. Goldberg and R. H. Stuewer, American Institute of Physics Conference Proceedings, vol. 179 (New York: American Institute of Physics, 1988), 55–70.

self.[17] In the Netherlands, on the other hand, where Lorentz knew and used Helmholtz's work, and where the physicist Johannes Diderik van der Waals explicitly addressed microstructure, different factors—intellectual, professional, and cultural—were at work. Neither Lorentz nor van der Waals, for example, worried much about influences on Dutch physics that connected to odious cultural beliefs, whereas Helmholtz worried extensively about such things, which, as he saw it, connected to a kind of physics that was steeped in the microworld.[18]

Let us close by developing this Dutch-German difference, which is quite apparent in the 1890s, in a bit more detail. In Helmholtz's work, and in much work that derived from it during that decade, the precise manner in which microprocesses give rise to novel effects through their links to the ether remained profoundly, and probably deliberately, obscure. More to the point, for Helmholtz, for those who were trained by him, and for those he influenced it was not considered good form to ground fundamental physics on fixed qualities of invisible particles. The Göttingen physicist Wilhelm Weber, and those he had trained and influenced, believed otherwise, and one of Helmholtz's long-standing concerns had been to undercut Weberian physics. To Helmholtz, the microworld could be discussed and even fruitfully deployed, but only insofar as its properties were *not* taken to be fundamental: for him microproperties were not to be taken as prior to all macroproperties, for his microcosm was not intended for use in producing and reproducing the phenomenal world's qualities. In his 1888–89 winter lectures on electromagnetism Helmholtz remarked, for example, that the old indestructible electric and magnetic fluids had now become merely "constants of integration"—this despite the fact that he himself had spoken in his 1881 Faraday lecture of an "atom" of electricity. But what one did with an "atom" of electricity depended on what properties it was endowed with, and Helmholtz endowed it with little more than that of being a quantized exchange function for specifying the electric character of an ion. His

17. For an example of how the Berlin professional ethos and the kind of physics promulgated there by von Helmholtz affected one young physicist, see Jed Z. Buchwald, *The Creation of Scientific Effects: Heinrich Hertz and Electric Waves* (Chicago: University of Chicago Press, 1994).

18. See Jed Z. Buchwald, "Electrodynamics in Context: Object States, Laboratory Practice, and Anti-Romanticism," in *Hermann von Helmholtz and the Foundations of Nineteenth-Century Science*, ed. D. Cahan (Berkeley and Los Angeles: University of California Press, 1993), 334–73, where I argue that Helmholtz's special and persistent aversion to Weber's microphysics is connected to Helmholtz's intense distaste for the epistemological, social, and cultural positions of one of Weber's most ardent supporters, the notorious xenophobe and anti-Semite, K. F. Zöllner, with whom Helmholtz battled. No comparable cultural antagonisms existed in the Netherlands.

understanding of the microcosm was evidently quite complicated and in-imical to the notion that the macroworld should be constructed out of it. He shared this position with many of his British contemporaries, and he sought to imbue German physics with it.

The Dutchman Lorentz, and for that matter his fellow countryman van der Waals, never had this antagonism to the fundamental character of microstructures that animated Helmholtz. Weberian electrodynamics was never for Lorentz intrinsically objectionable, as it was for Helmholtz, and he was consequently entirely prepared to draw upon its physical im-agery—though Lorentz's career in fact began in 1875 with a thorough pre-sentation and application of Helmholtz's own electrodynamics to optics. The many intellectual, cultural, and professional factors that were impli-cated in Berlin physicists' attitudes to the microworld were simply not present in Lorentz and van der Waals's Holland. It is perhaps not surprising at all that the microworld first became intensely real—that is, first became a working paper and laboratory tool for physicists—in early 1890s Netherlands, for there, and only there, was it possible to assimilate ele-ments from several different physics cultures and to brew out of them a novel amalgam.

Nevertheless, the microworld hardly became real all at once. Tools never do. It takes time to forge them, time to learn how to use them, and time to learn their strengths and limitations. Lorentz himself reached fast for a thoroughgoing microphysics in electromagnetism, but even he did not so rapidly develop an altogether consistent and appropriate set of mathemati-cal and conceptual equipage. In Germany, where the cultural ground was considerably less fertile, microtools began to appear shortly before 1895 in a deeply limited way, tied strongly to Helmholtz's microphysically ambiva-lent twin equations. But, one might argue, it was rather these limited and conceptually fuzzy methods that first brought the microworld to life out-side the Dutch environment, and it was these methods that first permeated German practice in the 1890s in such influential texts as Paul Drude's *Physik des Aethers*. Yet even here it was a Dutch physicist—Pieter Zee-man—who provided the experimental data on magneto-optic reflection that forced Drude, among others, to deploy the twin equations in electro-magnetism.

The dramatic productions and investigations of rays—X, Becquerel, and cathode—that occurred after mid-decade were swiftly bound to this grow-ing micropractice throughout Europe. However, we can also place at this time the origins of a powerful divide within physics that grew more intense during the next two decades and that reached extraordinary proportions during and after the Second World War. Most microwork, including

Lorentz's before the end of the 1890s, was essentially reproductive and productive. Its purpose was to elucidate or to create the behavior of macroscopic objects. This began to change around 1900, when interest on paper and in the laboratory turned strikingly to the microobjects themselves. Over the next decade and a half, younger physicists, as well as many older ones, became unconcerned in their practice with reproduction and production except as evidence for the power of microtechnology. Texts, for example, tend by 1910 or so to treat reproductive and productive argument as a sort of escalator to move the physicist from field equations to microprocesses. These arguments were by then not the aim, the essential goal of cutting-edge practice.

It is not altogether clear how and why this tremendous shift took place so rapidly and so thoroughly, though it is possible to construct the many strands that constituted the working world of microphysics by the late 1920s. Novel paper and laboratory tools that enabled physicists to pose and to solve consensually gripping problems were certainly produced. It is perhaps also the case that by the mid-1890s new paper problems were increasingly hard to come by, and new laboratory devices, excepting Hertz's dipole radiator and detector, were equally rare. This had all begun to change by the end of the decade. Let me close, then, by suggesting (weakly) that the modern microworld became dominant not solely, or even primarily, because late nineteenth-century physics had failed to account for something that everyone felt had to be accounted for. It came to dominance at least in part because it provided interesting work for physicists to do and to compete over.

Theodore M. Porter

Life Insurance, Medical Testing, and the Management of Mortality

Scientific objects are not made only by scientists. Especially when we look to research involving potential use in a military, industrial, commercial, medical, agricultural, educational, regulatory, political, or bureaucratic setting, we see that they are shaped by the interests and expectations of diverse actors. The impersonal forms of knowledge that are often identified with basic science may be more closely associated with the distance and distrust characteristic of these less detached forms of research. The desire to gain access to timeless truths about nature, independent of merely human hopes and wishes, is one incentive to objectivity, but opposition of a broadly political kind to the subjective and the personal is perhaps by now a more powerful one.[1] A shift away from informal expert judgment and toward reliance on quantifiable objects is a way of privileging public standards over private skills.

This paper is about the creation and maintenance of stable mortality rates by American life insurance companies. In the nineteenth century, the companies defended themselves against the increase of mortality among policyholders that would result from admission of "bad lives" by accepting only candidates of robust constitution and good moral character. In the twentieth century, as this form of selection became administratively unworkable and politically suspect, they turned increasingly to instrumental measures and statistical tables, involving quantities such as blood pressure or weight in re-

This paper is similar to one written for a project on the human genome initiative and life insurance, to be published in Alexander Capron, ed., *Fate and Fairness: Genetic Knowledge and the Future of Insurance.*

1. I argue this way in my book *Trust in Numbers: The Pursuit of Objectivity in Science and Public Life* (Princeton: Princeton University Press, 1995).

lation to height. These, in turn, depended on new medical practices and new professional roles. The companies enlisted as allies a teeming assortment of scientific entities in order to defend standard mortality rates against the corrosive tendencies of bureaucratic intrusion and distrust.

FORMS OF INFORMATION

Insurance, with banking and accounting, is the classic information industry. The really outstanding exemplar of the alliance of life insurance and formal knowledge is the life table. A life table defines, for some category of people, the number out of a birth cohort of one hundred thousand that can be expected to remain living at each birthday up to age one hundred. Mathematicians began preparing such tables to help establish rates for annuities and insurance contracts in the seventeenth century. In the nineteenth, a whole profession grew up whose main business was to prepare life tables and to calculate premiums based on mortality experience and investment returns. The origins of the mortality table can be connected with the founding of mathematical probability in the seventeenth century. Stable mortality rates were widely advertised in the nineteenth as evidence that even human life is subject to statistical laws, which do not depend on our will.

The calculation of life tables epitomizes the role of formal methods and explicit knowledge in life insurance. In other respects, life insurance long remained dependent on highly personal ways of knowing. English actuaries in the mid–nineteenth century insisted that even their calculations required expert judgment, based on reasonable expectations of investment returns and anticipations of future mortality experience. But this was not just a matter of forecasting. The validity of the mortality tables depended on skillful selection of what were called "quality lives." If only healthy, virtuous, prosperous people were admitted to life insurance, it was generally possible to maintain death rates below those for the general population. It was universally understood among actuaries that without careful selection, life insurance would be bought preferentially by people who feared they were in poor health and who on this account would die at higher rates than the tables predicted. This came to be called "adverse selection."[2]

Adverse selection, we may note, is itself a problematic entity, whose existence is often challenged. Insurers invoke it to justify policies that exclude from coverage or demand much higher premiums from those who are sub-

2. On life tables, calculation, and judgment in life insurance, see Lorraine Daston, *Classical Probability in the Enlightenment* (Princeton: Princeton University Press, 1988), 174–82; Timothy Alborn, "The Other Economists: Science and Commercial Culture in Victorian England" (Ph.D. diss., Harvard University, 1991).

ject to increased risk, and who thus have the greatest need for insurance. What good is insurance, people ask, if it is available only to those who are unlikely to collect? Why not calculate charges based on mortality in the general population, and issue policies to anyone willing to pay the premium? The companies respond that it isn't so simple because policies will be sought preferentially by those who know themselves to be at greater risk.

If adverse selection cannot be held at bay, the actuaries tell us, life tables will lose their validity. But the selection of lives has always been a difficult business. In the nineteenth and early twentieth centuries, even more than now, it was very difficult for any medical expert to know as much about applicants' general health as they knew about themselves. Beyond that, insurance companies faced severe problems of trust. The applicant's personal physician felt far more loyalty to his patient than to a company in another city, and would rarely refuse to attest to his good health. If the company kept its own medical expert, he would probably see the applicant only once. On this account he had to rely heavily on the candidate's word. This, it seemed, was scarcely adequate. To be sure, an outright lie could in some cases void an insurance contract. The real problem was not demonstrable falsehoods, but rather an almost universal reluctance to speak openly of stubborn aches and pains or undiagnosed medical conditions.

Partly on this account, selecting among applicants for an insurance contract was rather like admitting someone to a gentleman's club. The company had to be satisfied that the applicant was honorable and trustworthy. In England, according to a description from 1843, it was not uncommon for the directors to inquire into intimate details of his life. He was asked if he had suffered certain named diseases. He was required to supply references to a physician and to a private friend, who were then asked not only about his general state of health, but also about his character and his manner of living. He was generally expected to make an appearance before a medical officer appointed by the company, and often also before the company's assembled directors, who might well overrule a favorable medical opinion. The actuary Charles Ansell explained to a select committee of the House of Commons in 1843 the "advantage which is sometimes derived from men of the world seeing the lives which are proposed for assurance; and that is, that men's habits are frequently indicated by their appearance; and it often leads to inquiries as to the parties' habits of life."[3] In American companies too, it

3. *First Report of the Select Committee on Joint Stock Companies, Parliamentary Papers,* House of Commons, 1844, vol. 7, 1, 147–48. I discuss these issues in "Precision and Trust: Early Victorian Insurance and the Politics of Calculation," in *The Values of Precision,* ed. M. Norton Wise (Princeton: Princeton University Press, 1995), 173–97. On problems of trust in science, especially where it intersects with public life, see my *Trust in Numbers.*

seems, medical knowledge was subordinate to a general assessment of dignity and morality. For example, even after the State Mutual Life Insurance Company appointed a medical director, in 1865, the medical condition of the candidate remained "quite subordinate to the consideration of his general character, reputation, and financial responsibility. So that the selection of lives was conducted very much as the dissection of character in the sewing circle of the village church."[4] Moral character meant worthiness to associate with a select society, but it was also taken as a good predictor of longevity. "The man who is known as a club-man, a free liver, is generally an early dier."[5]

This aspect of underwriting lasted far into the present century, and has never entirely disappeared. Insurers assume that applicants for policies know more about their health than the companies can readily find out, and that on this account they must be eternally vigilant against adverse selection. So they look carefully for any sign that the applicant is trying to trick them, or regards the insurance contract as a speculative investment rather than protection against loss of income. They generally refuse to sell policies that exceed some multiple of present income, or that name as beneficiary a person with no financial interest in the life of the insured. Often they examine minutely the financial affairs of the proposed purchaser of a large policy, partly to establish insurable interest but partly also in an attempt to evaluate character.

In the years following World War II, before the Fair Credit Reporting Act made it much more difficult to rely on hearsay, this form of surveillance approached perfection. Inspection agencies like the Retail Credit Company in Atlanta employed thousands of investigators to look into the private and financial lives of applicants for life insurance. The applicants were subjected to something like a security clearance. The inspectors inquired about their religious, social, and political affiliations. They identified close friends and business associates, then asked those people about the applicant's occupational duties, financial circumstances, drinking habits, and general reputation. They were keenly interested in any evidence of irregular sexual habits. Naturally the informants were often reluctant to divulge too much about their friends and neighbors. So the inspectors had to cultivate the art of lie detection. If more than one of the acquaintances hesitated momentar-

4. Homer Gage, "Address by the President," *Abstract of the Proceedings of the Association of Life Insurance Medical Directors of America* [hereafter *PALIMDA*] 5, twenty-eighth annual meeting, 1917, 5–17, at 6. Homer Gage, who by this time was medical director of the same company, is here reporting what his father had told him.

5. Z. Taylor Emery, discussion of T. F. McMahon, "The Use of Alcohol and the Life Insurance Risk," *PALIMDA* 2, twenty-second annual meeting, 1911, 466–77, at 476.

ily or smiled suspiciously when asked, say, about use of alcohol, the applicant might be charged a higher premium or denied coverage.[6]

These rather personal and often highly intrusive investigations were one way of dealing with the problem of trust faced by underwriters. Another was to move toward objectivity, which in the medical domain generally means reliance on laboratory tests and instrument readings. Medical tests have been actively promoted by insurers. As late as 1950, practitioners of insurance medicine commonly wrote that theirs was the only medical specialty that took seriously the goal of long-term prognosis. They had a hand in the development and standardization of most of the important prognostic instruments up to that time. Those instruments, in turn, created new measures, or rather measures of new objects.

THE MEDICAL DIRECTOR

It seems natural that life insurers would concern themselves with medical prognosis, but it was not self-evident that this would be expressed as a preoccupation with instrument readings. Nineteenth- and even early twentieth-century practice put more emphasis on a highly sensitized touch ("medical tact") and an ability to elicit from the patient an informative history than on anything an instrument could show.[7] For purposes of prognosis, the touch was given less emphasis, but the medical history of the patient was generally regarded as the best source of relevant information. By late in the nineteenth century, an incipient profession, or at least a job classification, had been created to assess the health of candidates for life insurance. This was the medical director. Initially, medical directors examined most applicants personally before making a recommendation to the company directors. This informal style became impossible in the later nineteenth century, as the number of people insured by the more successful companies grew into the hundreds of thousands and then to the millions, and sometimes stretched over a whole continent.[8] In the United

6. On inspection agencies see P. G. Sanford, "The Relationship between the Inspection Agencies and the Underwriter," *Journal of Insurance Medicine* 1, no. 2 (1946): 38–39. For examples of the range of considerations deemed relevant to a decision to underwrite see "Case Clinic Discussion," *Journal of Insurance Medicine* 1, no. 2 (1946): 45–53, and the richly detailed discussion in Harry Dingman, *Risk Appraisal* (Cincinnati: National Underwriter Company, 1946), 158–82, 240–45.

7. Christopher Lawrence, "Incommunicable Knowledge: Science, Technology, and the Clinical Art in Britain, 1850–1914," *Journal of Contemporary History* 20 (1985): 503–20.

8. In 1910, according to industry statistics, there were 209 companies in the United States, which had written 7.0 million policies for a total face amount of $13.2 billion. By 1925 this had

States, at least, the directors were required increasingly to rely on widely scattered examiners. At first these were picked by agents, whose commissions depended on the actual issuance of a policy. We can assume, as did the main offices of the companies, that these examiners were chosen on the basis of their willingness to issue favorable recommendations and to overlook all but the most flagrant health defects. Early in the twentieth century most medical directors moved toward appointing one or a few authorized examiners in each city or region. Even so, agents developed various tricks to influence their decisions, or to get the more rigorous examiners dismissed by claiming that they refused to make convenient house calls or were widely disliked.

The Association of Life Insurance Medical Directors of America held its first meeting in 1890. Its members were physicians. Although they took themselves seriously as professionals, they did not consult independently, but were employed by insurance companies. They were generally quite separate from the upper reaches of management. By this time, the office of medical director was clearly distinguished also from that of the actuary. Actuaries were trained in mathematics, not medicine. They did not concern themselves with individual cases, but rather tried to collect statistical data on suitably chosen categories of people to permit a more nuanced determination of mortalities. This contrasted sharply with the traditional individualism of medical doctors. Already by 1890, though, medical directors had begun to find the methods of the actuary appealing. Individualism created a variety of problems. If one company rejected an applicant while another, known to be equally strict, accepted him, the medical director could anticipate complaints from the applicant and from the insurance agent. One of the main reasons for establishing an organization of medical directors was evidently to avoid such inconsistencies, to work toward greater uniformity by sharing information and by negotiating standard methods.

The greatest obstacle to reliance on personal judgment, though, was the overwhelming problem of distrust. In the first-ever presidential address before the association, in 1895, Edgar Holden of the Mutual Benefit Company in Newark emphasized precisely this point. "If we did not seek the well, the ill would seek us, and these are rarely blind to the benefits of our posthumous philanthropy . . . In the old days of the London Equitable, the

---

grown to 247 companies, 23.9 million policies, and $58.8 billion. James S. Elston, "The Development of Life Insurance in the United States in the Last Ten Years," *Transactions of the Actuarial Society of America* 27 (1926): 330–80, at 332.

applicants came personally before the directors, and after a period of proba-
tion were accepted or declined." In the more anonymous world of the fin de
siècle, he and his colleagues faced a

> bias of judgment where self-interest in involved. The agent who is to receive
> seventy or eighty or five hundred dollars in case of acceptance, or the doctor
> who has at stake the good-will and patronage of a friend, or who fears to of-
> fend or disappoint the agent who favors him, can hardly give full weight to a
> scrofulous joint, a fistula, a necrosed bone, or even a damaged lung; and, as a
> consequence of distorted vision, a man is presented as five points above the
> line who is infinitely below it.

The companies, he concluded, should push for uniformity of selection criteria.[9]

KNOWLEDGE AND AGENCY

The proceedings of the association of medical directors were marked as
"printed for private circulation." Perhaps for this reason they were remark-
ably candid, sometimes almost conversational, until at last in the 1940s the
contributions were reconfigured to conform to the genre of the scientific
paper, so that the most sensitive business of the medical directors no longer
got into print. Their greatest problem was the management of human con-
cealment, dissimulation, and fraud. Applicants did not go out of their way to
be helpful, and occasionally were duplicitous. Medical examiners in the
field were sometimes incompetent, often misinformed, and always subject
to pressures that discouraged them from being perfectly candid with the
head office. The highest place in the demonology of medical directors,
though, was occupied by the insurance agent.[10]

The agent was absolutely indispensable to the success of insurance companies.[11] He had to be a master of persuasion. He had to return to reluctant
customers again and again, ever armed with new arguments to convince
them of their need for life insurance. His attention was focused by need, for
he almost certainly worked on commission. Irresponsible selling by agents
helped to trigger a wave of public investigations in 1905, leading to greatly

9. [Edgar Holden], "President's Address," PALIMDA 1, sixth annual meeting, 1895,
pp. 51–57.

10. On the difficulties of gaining reliable life insurance information, see W. E. Thornton,
"Some Medical Relationships of Life Insurance," PALIMDA 22, 1935, 244–65.

11. On the importance of agents for insurance see Viviana A. Rotman Zelizer, Markets and
Morals: The Development of Life Insurance in the United States (New York: Columbia University Press, 1979), chap. 7.

expanded state regulation. Even afterward, for several decades, most agents had little or no special education in insurance and worked largely outside the control of the companies.[12] It was received wisdom among the medical directors, and very likely was true, that life insurance was almost always sold, not bought. When, rarely, a candidate for life insurance appeared on his own initiative, rather than being solicited by an agent, this was automatic grounds for suspicion: "If an applicant voluntarily applies to have his life assured, his case is one which needs watching," explained Arthur B. Wood, actuary for the Sun Life Insurance Company. The best strategy for managing adverse selection was to peddle this product mainly to reluctant customers, men who hadn't previously thought they needed life insurance.

One category of humanity that was rarely approached by agents before the mid–twentieth century was women. On this account, women were generally not sold insurance, but decided on their own to buy it. Wood mentioned this circumstance to account for unfavorable results of insuring their lives, despite longevity in general equaling or exceeding that of men.[13] To be sure, one had also to take into account the hazards of childbirth. The medical directors considered this as another form of adverse selection. Women, after all, tended to buy insurance when birth hazards were greatest, and Wood attributed to them an almost occult sensitivity to "weakness or hidden disease tending to the shortening of life."[14]

The persistence and salesmanship of the agents was not directed only to potential customers. One medical director, O. M. Eakins of the Reliance Life Insurance Company, called them "the Temperamental Hallucinaries who constitute the Nerve Complex of his Company . . . An Agent who has once tasted the blood of a Medical Director is left with an insatiable thirst."[15] Agents quickly sniffed out physicians who conducted loose examinations, and directed all their business to them. The companies tried to combat this by appointing one or a few medical examiners for every area, and requiring that all business go through them. If that examiner seemed too harsh, the agents would try to get him dismissed by complaining that he was not popular in the

---

12. On the Armstrong Committee, in New York, which carried out the original and most important of these investigations, see Morton Keller, *The Life Insurance Enterprise, 1885–1910* (Cambridge: Harvard University Press, 1963), chap. 15. On agents, J. Owen Stalson, *Marketing Life Insurance: Its History in America* (Cambridge: Harvard University Press, 1942), esp. 593–604.

13. Arthur B. Wood, "Some Suggestions Regarding the Means of Detecting Adverse Selection," *PALIMDA* 2, twenty-second annual meeting, 1911, 435–57, at 454, 449.

14. Wendell M. Strong and Faneuil S. Weisse, "Women as Life Insurance Risks," *PALIMDA* 16, 1929, 307–13, at 308–9. My study of these proceedings up to 1945 suggests that all or virtually all life insurance medical directors were men.

15. O. M. Eakins, "Medical Directors," *PALIMDA* 10, 1923, 128–37, at 135.

community, or refused to travel to conduct examinations, or kept examinees waiting so long that they began to feel second thoughts about their need for life insurance. In short, the agents had many tricks to expand their influence over these medical officers, making it difficult or impossible for the company to rely on anything so loose as the professional judgment of examining physicians as a basis for accepting and rejecting insurance applications.

The best alternative, it seemed, was to reconfigure the medical examination as an exercise in following rules. This should lead to greater consistency, thus closing off an opening that agents loved to exploit. And in general, rules grounded in actuarial data enhanced the credibility of decisions. R. L. Lounsberry of Security Mutual explained: "If you are able to say to your agent that the mortality on this class will be approximately so and so, that it is not a guess but known that it is true, because you have studied thousands of insured lives and know just what you are talking about, you have convinced him that his business is not being treated arbitrarily or empirically."[16]

NUMERICAL METHOD

The numerical method for rating life insurance applicants was worked out in the first decades of the present century by Oscar Rogers, a physician and medical director, and Arthur Hunter, an actuary. It was controversial at first, largely because it was seen as a mechanical alternative to judgment, and hence a threat to the expertise of the medical directors. Rogers and Hunter always denied this, insisting that it did not replace judgment but provided a baseline for its exercise. Still, the misperception was excusable. When their company, New York Life, hosted the annual meeting of the medical directors, their boss delivered an opening address that dismissed the whole enterprise as subjective and outdated:

> In a sense I am sorry for most of you boys, because I think you are being pushed out of your jobs. I can see what is coming about. You know the statistician and the chemist and the actuary are going to put you out of business . . . The old day of selecting risks by the whim of the Medical Director and accepting an applicant or not accepting him according to whether the Doctor's breakfast was agreeing with him . . . is passing . . . The risk is selected on the facts as they are. It is not a matter of opinion or guess. It is action based on statistics . . . [Before long the medical director] will be as dead as the dodo bird. That is the modern way of doing things.[17]

16. Dr. Lounsberry, comment on Oscar H. Rogers, "Medical Selection," *PALIMDA* 2, nineteenth annual meeting, 1908, 214–46, at 240.
17. Darwin P. Kingsley, welcoming address, *PALIMDA* 9, 1922, 6–8.

The medical directors did not entirely disagree. They uttered remarks like the following: "Uniformity" is the "groundwork of all science." "The selection of lives for insurance should come pretty near to being an exact science." "Standardization is the order of the day. It has come in manufacture, in business, in commerce, in education, everywhere." "[T]he standardization of examiners in medical insurance is desirable."[18] They were, however, little inclined to consider that their profession might be rendered unnecessary. Medical directors generally regarded numbers as aids to judgment. Another medical employee of New York Life, Rufus W. Weeks, wrote: "Every one who has dealt with statistics in a critical way knows that figures in a state of nature are great liars; that they need as much taming and training to make them tell the truth as a heathen needs at the hands of a missionary to make him into a working Christian."[19] The medical examiners worked to make the results of medical examinations suitable for actuarial analysis.

According to Oscar Rogers, New York Life first set up uniform standards for evaluating risk in 1903. Everyone associated with the effort insisted that it was entirely successful, but this was not the kind of competitive advantage that his company guarded carefully. Rogers was a determined advocate of scientific openness. Like many medical directors, he wanted to reduce the frequency with which candidates were rejected by one company and then accepted as standard risks by another. Rogers explained his method as the form of evaluation naturally employed by any mind.

> If, as we are passing judgment upon a risk, we analyze our mental processes, we shall find that each favorable or unfavorable factor, as it makes an impression on our mind, causes our judgment to lean in favor of or against the risk, and as we proceed our judgment oscillates first in one direction and then in another until, having completed our review, the conclusion which we arrive at is, as it were, the resultant of these oscillations. It is as if each factor had a certain positive or negative value and that our final judgment were the algebraic sum of these forces.[20]

18. Dr. Stebbins, "Practical Suggestions Concerning Life Insurance," *PALIMDA* 1, 6th annual meeting, 1895, 69–80, at 70; Robert L. Burrage, "Recent Standards in Medical Selection," *PALIMDA* 1, thirteenth annual meeting, 1902, 305–22, at 315; H. K. Dillard and J. P. Chapman, "Standardization of Medical Examination Blanks," PALIMDA, 8, 1921, 169–187, at 179; Robert M. Daley, organizer, symposium on "Medical Directors," PALIMDA, 15, 1928, 12–57, remark by Daley, at 16.

19. Rufus W. Weeks, "Remarks on Dr. Eugene Fisk's Paper," *PALIMDA* 2, eighteenth annual meeting, 1907, 164–67, at 164–65.

20. Oscar H. Rogers, "Medical Selection: As Influenced by the Specialized Mortality Investigation," *PALIMDA* 2, seventeenth annual meeting, 1906, 6–26, at 7.

Unfortunately, these spontaneous mental processes were poorly standard-ized. We need, Rogers continued, "a selection from which . . . the personal equation shall be eliminated,—a selection in which the fluctuating judg-ment of the individual shall as far as possible be replaced by the result of sta-tistical investigation."[21]

The association of medical directors actively supported this effort. Some-times alone, and sometimes in collaboration with the Actuarial Society of America,[22] it sponsored a series of studies of various risk factors, and at-tempted to measure the mortality associated with each. The work of Rogers and Hunter was part of the same endeavor. They argued that any underwrit-ing decision should consider nine factors: build, family history, physical condition, personal history, habits, occupation, habitat, moral hazard or in-surable interest, and plan of insurance applied for. All of these could be stud-ied statistically except perhaps moral hazard (referring here to the dangerous or immoral habits of the insured, and especially to their fraudulent claims). Rogers and Hunter published extensive tables based on the experience of their own and other companies, assigning percentages of increased risk to various occupations, physical impairments, disease histories, and the like. Their idea was that a medical director or clerk in his office could consider each factor in turn, adding 20 percent for a risky occupation and 25 percent for a marginally elevated blood pressure, then subtracting 10 percent for an excel-lent family history, until, at the conclusion of the exercise, one had a single number expressing the relative risk associated with this particular life.

The numerical method of Rogers and Hunter had, by 1920, become the standard one for life insurance underwriting. It was, of course, not simply a new protocol to be followed by the medical director in evaluating applicants for policies. It required the collection by agents and medical examiners of information in a form that could be fed into these routines of calculation. Standard policies were perhaps not so much affected, since they were issued on lives that displayed no serious impairments. The most significant effect of the numerical method, and of the extensive mortality investigations that permitted it to be implemented, was to facilitate a vast expansion of "sub-standard" underwriting. The life insurance tradition, which treated the in-sured as members of a club, tended not to discriminate among the members except according to age.[23] An applicant who had moved to the tropics, or was associated with the liquor trade, or engaged in shady financial dealings, or

21. Oscar H. Rogers, "Medical Selection," PALIMDA 2, nineteenth annual meeting, 1908, 214–31, at 227–28.

22. E. J. Moorhead, Our Yesterdays: The History of the Actuarial Profession in North America (Schaumburg, Ill.: Society of Actuaries, 1989), 111–12.

23. This goes back to the very beginnings of life insurance; see Geoffrey Wilson Clark,

had been treated for syphilis, was generally excluded from this club for some combination of moral and medical reasons. But if bartenders were simply assigned a rating of 80 percent, and former syphilitics another number based on the nature of the symptoms and the time elapsed since treatment, policies could be issued on their lives after all. Rogers considered it a worthy ideal to offer every candidate for life insurance a policy, even if it had to be an expensive one. This was not quite realizable in practice because of exceptionally severe adverse selection among the most substandard candidates. That is, someone offered a policy with a premium several times as great as standard ones would probably decline the policy unless he recognized that his health was extremely precarious. People with such knowledge would die at an even higher rate than the tables predicted. For the less severely afflicted, though, this new regime of standard tests and calculated premiums permitted life insurance to be offered at elevated rates to many who previously would simply have been declined.

Two other consequences of the numerical method, and the regime of objectivity with which it was associated, deserve mention. One is that it reduced the level of expertise associated with life insurance underwriting. Rogers and Hunter bragged that in their office, most cases could now be handled by clerks, and did not require the attention of the medical director or of medically trained assistants at all. This, they emphasized, included a large fraction of substandard cases, which could now be priced by consulting a table. Skeptics countered that clerks were employed also in offices that did not rely on mechanical methods, but Rogers effectively countered that at New York Life a singularly high percentage of cases could be handled routinely.[24] Even with the numerical method, though, clerks had to be capable of recognizing problematical circumstances requiring expert judgment. As the companies grew, even clerks without medical training nevertheless acquired such expertise. These clerks were the original life insurance "underwriters." Still, the specification of routines and rules circumscribing the exercise of judgment involved an element of deskilling.

That deskilling reached also to the office of the medical examiner. The doctor in the field, whose reliability and integrity might be doubted at the head office, was increasingly asked to report only the results of tests and measurements, and not to give an opinion as to the worthiness of a candi-

"Betting on Lives: Life Insurance in English Society and Culture, 1695–1775" (Ph.D. diss., Princeton University, 1993).

24. Oscar H. Rogers and Arthur Hunter, "The Numerical Method of Determining the Value of Risks for Insurance," *PALIMDA* 6, thirtieth annual meeting, 1919, 99–129, at 102. See also discussion, 129–73.

date for life insurance. We don't trust our medical examiners to decide what is important, or to make our decisions for us, explained Edwin W. Dwight of the New England Mutual in 1920. "We require that our men shall give us the facts and let us decide as to their importance." T. H. Rockwell of the Equitable added that the policies of the head office change, and that the medical director has access to data that the examiner does not. All he wanted from them was the "exact facts," to be recorded on a standardized form.[25] Rogers urged that the method should depend only on tests that could be carried out following a "very simple method, a method which our examiners can use, which will enable us to differentiate between the serious and insignificant types of abnormality."[26]

Inevitably, another important consideration was to cover up weaknesses that the agents might exploit whenever a potential commission was denied them because someone was rejected for life insurance. Eakins emphasized this feature in a defense of the numerical method: our agents understand, he wrote, "that opinions or pleadings or threats or even that lever of last resort, competition, do not change the answer of an arithmetical sum." It would be especially effective, urged Dr. Hamilton of the Sun Life, if all companies adopted the same numbers. Diversity in substandard ratings, he explained, is always perplexing to agents, and is "utterly unintelligible" to most applicants, who "hold the haphazard decisions in contempt."[27] A rigorous numerical method would help eliminate inconsistencies among the various companies. J. Allen Patton of the Prudential, writing in 1929, identified the numerical method as a major cause of progress toward the "millennium" of "uniformity in action" on life insurance applications. Without it, there would be no end to the complaints of agents when candidates were rejected by their own company and then accepted by some other with a reputation for strict standards.[28] Finally, reliance on a numerical method supported by statistics provided a defense against regulatory authorities who

25. See comments by E. W. Dwight, p. 65, and T. H. Rockwell, p. 68, following paper by Brandreth Symonds, "The Value of the Medical Examiner's Opinion," *PALIMDA* 7, thirty-first annual meeting, 29–57. Keller, *Life Insurance Enterprise*, 37, remarks that the massive growth and consequent bureaucratization of life insurance companies around the turn of the century led quite generally to an effort to concentrate power in the head office. On standardized forms see JoAnne Yates, *Control through Communication: The Rise of System in American Management* (Baltimore: Johns Hopkins University Press, 1989).

26. Oscar Rogers, comment in *PALIMDA* 15, thirty-ninth annual meeting, 1928, p. 223.

27. O. M. Eakins, "Numerical Ratings in the Selection of Risks," *PALIMDA* 7, thirty-first annual meeting, 1920, 194–98, at 197; H. C. Scadding, "An Attempt on the Part of Canadian Companies to Secure Greater Uniformity in the Treatment of Impaired Risks," *PALIMDA* 9, 1922, 229–53, comment by Dr. Hamilton, p. 254.

28. J. Allen Patton, "Life Insurance Selection," *PALIMDA* 16, 1929, 11–16, at 14.

suspected that rated policies were merely a ploy of the companies to take advantage of the vulnerable and to generate excess profits.[29]

INSTRUMENTS AND STANDARDIZATION

Charles F. Martin, medical director of the State Life Insurance Company, thought the numerical system rested on a fundamental mistake. He argued that "subjective symptoms and complaints" are more informative than stethoscope readings for evaluating the condition of the heart. A few measures taken hastily can never provide what medical directors really need, a sense of the general health of the candidate for insurance. Rogers was not wholly unsympathetic. "With regard to Dr. Martin's paper, I agree with every bit of it from the medical point of view and disagree with a great deal of it from the point of view of Life Insurance . . . We cannot subject applicants for insurance to the same kind of scrutiny that a physician in practice is able to employ when he is studying his patients." The medical examiner only gets to see the applicant once. And the patient who is open with his physician conceals things from the insurance company.[30] On this account, nothing was more necessary for insurance examinations than simple, reliable instruments, that did not depend on refined skills or subtle judgment on the part of medical examiners and that got around the habitual reticence of applicants. The companies needed objects to measure and tools to measure them.

The value of objective measures for insurance evaluation was already understood in the mid–nineteenth century. The prototype of such information required no special instruments at all. A contributor to the publication of the newly founded (English) Institute of Actuaries, in 1850, urged that build could be an important datum for actuaries to consider in evaluating candidates for life insurance.

> If, for example, a proposal be sent from the country, backed merely with the opinion of a referee whom we do not know, that "no signs of disease are discoverable," and that "the proposer has a robust appearance," our knowledge of the tendencies of his constitution is small indeed. But if to this it be added that he is five feet eight inches high and weighs eleven stone [154 pounds], we

29. See the testimony of Rufus Weeks, chief actuary of New York Life, in some very important New York hearings on the life insurance industry: *Testimony Taken before the Joint Committee of the Senate and Assembly of the State of New York . . . [on] Life Insurance Companies* (4 vols., Albany, N.Y., 1906), vol. 2, pp. 1109–12.

30. Charles F. Martin, "A Suggested Re-adjustment of Our Views on Cardio-vascular Examinations for Life Insurance," *PALIMDA* 3, twenty-second annual meeting, 1911, 376–93; Oscar Rogers discussion of above, 406–11, at 406–7.

feel a certain degree of safety in accepting him . . . I may add, too, it facilitates much the explanations of the reasons for a refusal or acceptance, which the directors will sometimes require from their medical advisors; for it depends on reasoning comprehensible to all, and capable of reduction to figures . . . and can avoid the vagueness of a mere negative opinion.[31]

Rogers and his collaborators agreed, and the early meetings of the medical directors in the first decades of the twentieth century included frequent discussion of the relation between build and mortality. Their proceedings contain extensive tables, and recommendations for dealing with especially risky features of build like protruding bellies. Rogers's first substantial paper was a study in 1901 of "Build as a Factor Influencing Longevity."[32] As late as the 1940s, the relation of weight to height remained the prototype of medical information in insurance, for it was easily ascertained and did not vary greatly from moment to moment or day to day. Build, argued Herbert Dingman in his 1946 textbook, ought to be

the starting basis in appraising risk. It is the one phase of insurability where reliable and positive information is obtainable in 100 per cent of cases. Physical condition may be subject to error or opinion. Habitat and occupation are subject to change. Habits and moral hazard are subject to misinterpretation. Personal history is subject to misstatement. Family history and racial history are subject to misadvertence.[33]

Initially, the companies regarded the mean as the normal, and imposed unfavorable ratings on both sides. Low weight was taken as implying danger of consumption (tuberculosis), while a high weight in relation to height implied a greater risk of heart trouble. Gradually, unfavorable ratings were associated more and more with overweight. Agents, of course, complained that in every particular case under consideration, the candidate was not fat but merely big-boned and muscular. In reply the medical directors collected evidence that even men of this sort were subject to heightened risk. None of this was self-evident before the investigations by insurance companies began. While it was widely understood that obesity might contribute to early death, stoutness was taken to indicate robust good health. In estimating risk from height and weight, insurance companies were not drawing on established medical

31. Dr. Chambers, "Corpulence in Connection with Life Assurance," *Assurance Magazine* 1 (1850–51), 87–89, at 88.
32. Oscar H. Rogers, "Build as a Factor Influencing Longevity," *PALIMDA* 1, twelfth annual meeting, 1901, 280–88.
33. Dingman, *Risk Appraisal*, 10–11. "Racial history" probably referred to the race or ethnicity of ancestors.

knowledge. Information on build became a reliable basis for projecting differential mortality only as the result of efforts by the companies themselves.

Blood pressure, even more than weight, was a favorite topic of investigation by medical directors looking for objective means to evaluate the health of applicants. Historian Audrey Davis has argued that the sphygmomanometer became a standard medical instrument largely in consequence of its use in life insurance examinations.[34] The work presented to meetings of American medical directors supports her thesis. The use of this instrument, championed most effectively by John W. Fisher of the Northwestern, was part of a campaign by insurance companies to transform medical practice. A revealing exchange took place among the medical directors in 1901. Fisher was asked if the companies pay extra to physicians to get a blood pressure test. He replied that they generally did not. Should companies furnish the instrument? No, he answered, "anymore than they would furnish a stethoscope. We expect that our examiners are up-to-date men and have got these instruments." Potential revenues from insurance examinations were among the major inducements for a physician to acquire and use such instruments, hence to enlist himself among the up-to-date. The Northwestern, Fisher explained, began in 1907 to require a measurement of blood pressure for all applicants older than forty, and sometimes for younger ones as well. He considered the sphygmomanometer indispensable.[35]

The mere decision to use an instrument did not suffice to make it workable. Even though insurance examinations could be an appreciable source of revenue for individual doctors, and indeed for the whole profession, there was some problem getting the instruments into the hands of examining physicians. At first there were no reliable sphygmomanometers that could be transported to an applicant's home and used there. Rogers exploited the occasion of a paper by Fisher to display to the assembled medical directors a new, portable instrument of his own invention, which he intended to license to a manufacturer.[36] He and others continued to work at new im-

34. Audrey B. Davis, "Life Insurance and the Physical Examination: A Chapter in the Rise of American Medical Technology," *Bulletin of the History of Medicine* 55 (1981): 392–406. On the adoption of medical tests within a company see Shepard B. Clough, *A Century of American Life Insurance: A History of the Mutual Life Insurance Company of New York* (New York: Columbia University Press, 1946), 288–90.

35. See discussion of sphygmomanometer in *PALIMDA* 2, twentieth annual meeting, 1909, p. 256a; J. W. Fisher, "The Diagnostic Value of the Use of the Sphygmomanometer in Examination for Life Insurance," *PALIMDA* 2, twenty-second annual meeting, 1911, 393–406.

36. Rogers comment on Fisher, "Diagnostic Value," 406–11. A year later he confessed that he had discovered certain defects, and he presented what he hoped was a corrected instrument: Rogers comment on J. W. Fisher, "Fuller Report upon the Use of the Sphygmomanometer in Examinations for Life Insurance," *PALIMDA* 3, twenty-third annual meeting, 90–94, at 94.

provements. There was also the problem of standardizing, of working out a method that would produce relatively uniform results from examiner to examiner. This was particularly difficult because, it seemed, any particular individual might show highly variable readings depending on the state of his nerves and other factors. Medical examiners were encouraged in case of high or borderline blood pressure readings to repeat the measurement one or more times, perhaps even on another day. The agents were quick to discover this variability also, and soon were offering advice to their customers on how to get the blood pressure down. One medical director reported in 1928 the case of an agent who was paying his medical examiners a second examination fee, two dollars, to do a new blood pressure test when the first was too high.[37]

Finally, it was necessary to collect data establishing a relationship between blood pressure readings and mortality. The only source of such data was the companies themselves. Fisher presented results from the Northwestern. Hunter and Rogers, as always, worked to assemble results from various companies, and also to learn the mortality experience of rejected applicants, especially those with high blood pressure. At first, only systolic pressure was routinely measured, and until about 1930 the directors were uncertain of the significance of diastolic pressure readings. Soon the companies began compiling risk tables for both measures. By the late 1930s, "hypertension" was emerging as a national problem calling for systematic testing and public health measures.[38] The dangers of hypertension were discovered by insurance companies twenty years before they came to the attention of clinicians. The measurement of blood pressure thus came into medicine not as a consequence of disinterested medical research or of the concern of physicians for their individual patients. Rather, it arose as part of the effort by life insurance companies to develop better and more objective means of mortality prognosis.[39]

The corresponding point cannot perhaps be made so strongly about electrocardiography, but here again the insurance examination contributed

37. Davis, "Life Insurance," 400–401; Dr. Alton, comment in symposium on "Medical Directors," *PALIMDA* 15, 1928, p. 34.

38. J. W. Fisher, "Blood-Pressure Mortality Statistics—August, 1907 to August, 1913," *PALIMDA* 2, twenty-fourth annual meeting, 1913, 246–49; Arthur Hunter and Oscar H. Rogers, "Blood Pressure as Affected by Sex, Weight, Climate, Altitude, Latitude, or by Abstinence from Alcoholic Beverages,'" *PALIMDA* 6, thirtieth annual meeting, 1919, 92–97; Oscar H. Rogers and Arthur Hunter, "Mortality Statistics of Impaired Lives (No. 2)," *PALIMDA* 10, 1923, 43–51; Oscar H. Rogers and Arthur Hunter, "Systolic and Diastolic Blood Pressures Higher than the Average for Age," *PALIMDA* 13, 1926, 170–86.

39. S. B. Scholz, "Preliminary Notes and Report of the Chairman, Blood Pressure Committee," *PALIMDA* 26, 1940, 256–68, at 260.

greatly to the development and diffusion of the method. Robert Frank points out in a fine historical essay that instruments for measuring the action of the heart were, from the beginning, animated by a desire for precise records that "were not dependent on the acuity of the cultivated sense of the physician."[40] H. F. Taylor, associate medical director of Aetna, argued in 1931 that the companies were losing money because of cardiovascular irregularities, and might have to stop underwriting cases involving cardiac pathology. The best aid to diagnosis, a detailed history taken by a trained cardiologist, is rarely available to insurers. But, he added, they can take advantage of "the less important *instrumental* aids to diagnosis," and in particular of electrocardiography. Harry Ungerleider of the Equitable added in a comment that the "information derived from electrocardiography may be limited, but it is reliable." Taylor recommended that an electrocardiograph be required for all applicants over fifty years of age or who sought more than twenty-five thousand dollars of insurance.[41] Haynes Harold Fellows, also of the Metropolitan, agreed.

> We all know that it is often a most difficult problem to obtain enough medical data to arrive at a fair conclusion in the case of certain applicants for insurance. Medical histories are withheld, sometimes unwittingly and sometimes deliberately. I doubt if the applicant ever draws the attention of the examining physician to any known physical defects, and it is not entirely impossible that occasionally there is an attempt to mask or cover physical impairments ... Necessarily, then, we must have an economical, practical, fact-finding method of examination, taking advantage of the useful developments in medicine as they appear.[42]

At the same time, the early 1930s, insurance companies were also exploring the use of X rays to assess the condition of the heart and lungs. Both X rays and electrocardiograms had the advantage that they produced a physical trace. The specialists in the instruments did not have to be relied on for an expert opinion. They should, argued W. E. Thornton of the Lincoln Na-

40. Robert Frank, "The Telltale Heart: Physiological Instruments, Graphic Methods, and Clinical Hopes, 1865–1914," in *The Investigative Enterprise: Experimental Physiology in Nineteenth-Century Medicine,* ed. William Coleman and Frederic L. Holmes (Berkeley and Los Angeles: University of California Press, 1988), 211–90, at 212. On medical instruments and quantitative measures see also Joel D. Howell, *Technology in the Hospital: Transforming Patient Care in the Early Twentieth Century* (Baltimore: Johns Hopkins University Press, 1995).
41. H. F. Taylor, "The Value of Electrocardiography in Medical Underwriting," *PALIMDA* 18, 1931, 165–79, at 165; discussion by Harry Ungerleider, 180.
42. Haynes Harold Fellows, "Effect of Electrocardiographic and X-ray Examination upon Ordinary Tissue," *PALIMDA* 18, 1931, 202–8, at 202.

tional in 1935, be required to send along their X-ray films and graphical traces, and in this way to "concede the superiority of our own judgments within our own field." They should also "be required to conduct and report their inquiries along standardized lines where such have been determined by insurance medicine." Once again, the medical directors were working to shape the uses of the instruments, and were by no means passive beneficiaries of medical progress.[43] Earl C. Bonnett of the Metropolitan explained in 1940 that a decade earlier his company had not known how to distinguish a normal from an abnormal electrocardiogram. Frank Wilson, professor of medicine at the University of Michigan, argued a few years earlier that it was up to the insurance companies to define the bounds of normality, since they alone possessed large numbers of normal electrocardiograms.[44]

These, it should be added, were not easily obtained. The instrument was still not widely available in 1930. Demand by insurance companies was important for its spread. Insurers displayed so much enthusiasm for these tests, wrote Ungerleider, that there was some inclination to let the tail wag the dog.[45] The demand by insurers for neutral information was by this time extraordinarily wide-ranging. At the same time as the companies were helping to transform medical practice in the direction of systematic reliance on tests and measurements, their demand for rapid information processing was helping to create and to shape modern computing technology.[46]

The intense search by insurance professionals for reliable and objective ways of predicting age at death reflected above all a desire not to be at a disadvantage relative to the knowledge possessed by applicants about their own health. They left few stones unturned. In 1949, when the *Journal of Insurance Medicine* sponsored a "Symposium on Prognosis," one of the invited participants was the famous Duke parapsychologist J. B. Rhine. He spoke optimistically about the potential value of extrasensory perception (ESP) as a precision instrument to aid in selecting among candidates for life insurance.[47]

43. Thornton, "Medical Relationships," 238, 239.

44. Earl C. Bonnett, "Notes on One Method of Underwriting Electrocardiograms," *PAL-IMDA* 26, 1940, 120–22; Frank N. Wilson, "Recent Progress in Electrocardiography and the Interpretation of Borderline Electrocardiograms," *PALIMDA* 24, 1937, 96–153, at 96–97.

45. Harry Ungerleider, comment on Albert O. Jimenis and Edmund W. Wilson, "Influence of Electrocardiogram or X-ray on Underwriting Decisions," *PALIMDA* 23, 1936, p. 40. On the role of insurance companies in learning how to interpret electrocardiograms see Ungerleider, "The Prognostic Implications of the Electrocardiogram," *Annals of Life Insurance Medicine* 1 (1962): 131–44.

46. JoAnne Yates, "Co-evolution of Information-Processing Technology and Use: Interaction between the Life Insurance and Tabulating Industries," *Business History Review* 67 (1993): 1–51.

47. J. B. Rhine, "ESP and Prognosis," *Journal of Insurance Medicine* 4, no. 2 (1949): 16–17.

ENTITIES

Statisticians have sometimes exaggerated the orderliness of mortality numbers. They have also tried to lower death rates, and hence to improve the statistics, through public-health interventions, or watched them deteriorate as a result of epidemics or uncontrolled violence. The deconstructive gaze of science studies is not required to demonstrate that mortality is somewhat irregular and highly dependent on economic and social policies and on medical and police systems. Still, death is as certain as taxes. Medical officers and actuaries working for life insurance companies did not have the power to alter the mortality rates of a whole population, except perhaps very slowly.

But the mortality of the larger population was not their main concern, and its stability did not suffice to guarantee predictable death rates within their institutions. In the absence of something approaching universal coverage, a company could achieve stable, predictable mortality only by actively shaping the covered population. In the United States, at least, the information required to achieve this could not easily be obtained, at least not on a large scale, because the networks of local knowledge were unreliable. As an alternative, insurance professionals worked to create new, less personal forms of knowledge appropriate for conditions of dissimulation and distrust.

That distrust cannot be regarded as intrinsic to the logic of life insurance, nor even as a necessary response to competitive pressures. Mortality indicators were not developed in some eternal quest for a competitive edge, but rather through a collaboration among companies that might almost be regarded as anticompetitive collusion. The medical directors and their bosses were troubled that their customers possessed secret knowledge, and were using it to take advantage of them. Instrument readings, they thought, could help to restore the balance. To this end, they helped to make blood pressure readings and electrocardiograms into routine medical tools. In the process, they contributed to the triumph of instruments and measurements over the informal judgment of individuals and their doctors. The push for objectivity in medicine and even in science should be regarded in part as a set of partial solutions to some pervasive problems of trust.

Objectivity, in this sense, meant first of all a denial of the personal and the subjective, but it depended on the creation of a range of scientific objects. These objects should be credited not to metaphysics, but to management. Among them were mortality rates that could be held stable for select, managed populations. The insurance doctors also codified and defined meanings for a variety of bodily measures, including healthy and dangerous ratios of weight to height, good and bad blood pressure readings, and normal and abnormal electrocardiograms. These entities in turn depended on new instruments, such as sphygmomanometers and EKG machines. They required also the coming into being of new kinds of people: technicians to operate the instruments, specialists in computation to manipulate the numbers, and even a subtly redefined physician, who has learned to rely less on the feel of the pulse and the complaints of the patient and more on laboratory reports. New scientific objects come into being only in alliance with the right kinds of instruments and the right kinds of people. The linkage is not so tight as to exclude the detection of new objects by old researchers or of old objects by new instruments, but we can identify conditions that favor the multiplication of entities. Here, in insurance medicine, a leading role was played by large, resolutely impersonal institutions. Similar considerations apply, though often more subtly, to basic scientific research as well.

It would be about as helpful to argue that the blood had no pressure before insurance companies began pushing the use of sphygmomanometers as it would be to claim there was no mortality before there were actuaries. The "coming into being" of quantitative entities like these should rather be understood in terms of a selection among alternative ways of knowing. Often, as in the present case, that choice is shaped by a variety of interests and constraints. The shift toward standardization and objectivity in this story of insurance medicine cannot be regarded as the inevitable product of science or modernization or bureaucratic rationality. Rather, it was an adaptation to a very particular context of use, in which agents, physicians, business executives, government regulators, and even the unforthcoming applicants played as decisive a role as did medical directors and actuaries.

10 Bruno Latour

On the Partial Existence of Existing and Nonexisting Objects

PROLOGUE: DID RAMSES II DIE OF TUBERCULOSIS?

In 1976, the mummy of Ramses II was welcomed at a Paris air base with the honors due to a head of state, greeted by a minister, trumpets, and the Republican Guards in full attire. As hinted at in the fiery title of *Paris-Match*—"Nos savants au secours de Ramsès II tombé malade 3000 ans *après sa mort*" (Our scientists to the rescue of Ramses II, who fell ill three thousand years *after his death*)—something is at stake here that defies the normal flow of time.[1] Sickness erupts after death and the full benefit of modern technology arrives a tiny bit too late for the great king. In this stunning picture (figure 10.1), the mummy is being operated upon on the surgical table, violently lit by floodlights, surrounded by "our scientists" in white coats wearing masks against contagion (either to protect Ramses against their modern-made germs or to protect themselves from Pharaoh's curse). After careful examination, the verdict of the postmortem ("post" indeed!) is offered: Ramses II had very bad teeth, a terrible deformation of

---

This chapter remains close to the paper written for the conference that is at the origin of this book. A much modified version, more technical and more philosophical, has been published as chapter 5 of *Pandora's Hope: Essays in the Reality of Science Studies* (Cambridge: Harvard University Press, 1999).

1. In spite of the flippant titles usual for *Paris-Match*, a reading of the text shows that it is not actually the king who has become sick after his death, but rather the mummy, from an infection by a fungus. I nonetheless have kept the first interpretation, associated with the image, because of its ontological interest. All the details on the mummy transportation and cure can be found in Christiane Desroches-Noblecourt, *Ramsès II, la véritable histoire* (Paris: Pygmalion, 1996).

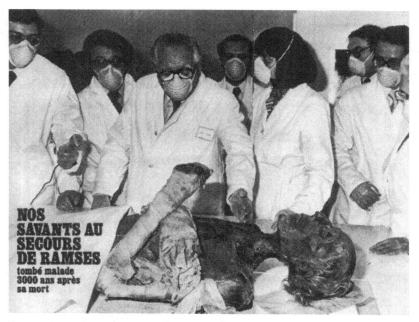

NOS
SAVANTS AU
SECOURS
DE RAMSES
tombé malade
3000 ans après
sa mort

Figure 10.1.   Our scientists to the rescue of Ramses II, who fell ill three thousand years *after his death.* (From *Paris Match,* September 1956)

the spinal cord that caused extreme pain. Too late for an intervention. But not too late to claim still another triumph for French physicians and surgeons, whose reach has now expanded in remote time as well as in remote space.

The great advantage of this picture is that it renders visible, tangible, and material the expense at which it is possible for us to think of the extension in space of Koch's bacillus, discovered (or invented, or made up, or socially constructed) in 1882. Let us accept the diagnosis of "our brave scientists" at face value and take it as a proved fact that Ramses died of tuberculosis. How could he have died of a bacillus discovered in 1882 and of a disease whose etiology, in its modern form, dates only from 1819 in Laënnec's ward? Is it not anachronistic? The attribution of tuberculosis and Koch's bacillus to Ramses II should strike us as an anachronism of the same caliber as if we had diagnosed his death as having been caused by a Marxist upheaval, or a machine gun, or a Wall Street crash. Is it not an extreme case of "whiggish" history, transplanting into the past the hidden or potential existence of the future? Surely, if we want to respect actors' categories, there must be in the Egyptian language a term and a set of hieroglyphs, for instance

"Saodowaoth," that define the cause of Ramses' death. But if it exists it is so incommensurable with our own interpretations that no translation could possibly replace it by "an infection of Koch's bacillus." Koch bacilli have a local history that limits them to Berlin at the turn of the century. They may be allowed to spread to all the years that come *after* 1882 provided Koch's claim is accepted as a fact and incorporated later into routine practices, but certainly they cannot jump back to the years *before*.

And yet, if we immediately detect the anachronism of bringing a machine gun, a Marxist guerilla movement, or a Wall Street capitalist back to the Egypt of 1000 B.C., we seem to swallow with not so much as a gulp the extension of tuberculosis to the past. More exactly, for *this* type of object at least, we seem to be torn between two opposite positions. The first one, which would be a radically anti-whiggish history, forbids us from ever using the expression "Ramses II died of tuberculosis" as a meaningful sentence. We are allowed only to say things like "our scientists have started in 1976 to interpret Ramses II's death as having been caused by tuberculosis but, at the time, it was interpreted as being caused by 'Saodowaoth' or some such word. Saodowaoth is not a translation of tuberculosis. There is no word to translate it. The cause of Ramses' death is thus unknown and should remain irretrievable in a past from which we are infinitely distant." The second solution is a sort of self-confident, laid-back whiggism that accepts tuberculosis and Koch's bacillus as the long-expected and provisionally final revelation of what has been at work all along in the course of history. Saodowaoth and all such gibberish disappear as so many mistakes; what really happened is eventually exposed by "our brave scientists."

Fortunately, there is another solution that is revealed by this picture and by the work that has been carried out, for a generation now, on the practice of science. Koch bacillus can be extended into the past to be sure—contrary to the radical anti-whiggish position—, but this cannot be done *at no cost*. To allow for such an extension, some work has to be done, especially some laboratory work. The mummy has to be *brought into contact* with a hospital, examined by white-coat specialists under floodlights, the lungs X-rayed, bones sterilized with cobalt 60, and so on. All this labor-intensive practice is quietly ignored by the whiggish position, which speaks of the extension in time as if it were a simple matter, requiring no laboratory, no instrument, no specially trained surgeon, no X rays. What is made clear by the *Paris-Match* picture is that Ramses II's body can be endowed with a new feature: tuberculosis. But none of the elements necessary to prove it can themselves be expanded or transported back to three thousand years ago. In other words, Koch's bacillus may travel in time, not the hospital surgeons, not the X-ray machine, not the sterilization outfit. When we impute retroactively a modern shaped event to the past we have *to sort out* the fact—Koch bacillus's devastating effect on the lung—with that of the material and practical setup necessary to render the fact visible. It is only if we believe that facts *escape* their network of production that we are faced with the question whether or not Ramses II died of tuberculosis.

The problem appears difficult only for *some type* of objects and only for the *time* dimension. Obviously, no one could have the same worry for a machine gun, unless we invent a time capsule. It is impossible for us to imagine that a machine gun could be transported into the past. Thus, technological objects do not have the same popular ontology and cannot travel *back* into the past under any circumstances, which might be one way of saying that the philosophy of technology is a better guide for ontology than the philosophy of science. For technology, objects *never* escape the conditions of their productions. An isolated machine gun in the remote past is a pragmatic absurdity—and so, by the way, is an isolated machine gun in the present without the know-how, bullets, oil, repairmen, and logistics necessary to activate it. Another advantage of a technological artifact is that we have no difficulty in imagining that it rusts away and disappears. Thus it always remains tied to a circumscribed and well-defined spatiotemporal envelope.[2] An isolated Koch bacillus is also a pragmatic absurdity since those types of facts cannot escape their networks of production either. Yet we seem to believe they can, because for science, *and for science only,* we forget the local, material, and practical networks that accompany artifacts through the whole duration of their lives.

Of course, we have learned, after reading science studies of all sorts, that facts cannot, even by the wildest imagination, escape their local conditions of production. We now know that even to verify such a universal fact as gravitation we need somehow to connect the local scene with a laboratory through the crucial medium of metrology and standardization. And yet, we rarely believe this to be the case in the *remote future*—there seems to be a time when the Koch bacillus proliferates everywhere without bacteriological laboratories—and in the *remote past*—there seems to be no need for a network to attach Ramses II to a diagnosis. Unlike technological artifacts, scientific facts seem, once we wander away from the local conditions of production in the past as well as in the future, to free themselves from their spatiotemporal envelope. Inertia seems to take over at no cost. The great lesson

2. Except in the Frankensteinian nightmares. See my *Aramis or the Love of Technology,* trans. Catherine Porter (Cambridge: Harvard University Press, 1996). On the layering aspect of technologies see the marvelous novel by Richard Powers, *Galatea 2.2* (New York: Farrar, Strauss and Giroux, 1995).

of the picture shown above is that extension in the past, extension in the future, and extension in space may require the *same* type of labor. In the three cases, the local scene should be hooked up to laboratory practice through some sort of extended or standardized or metrologized network. It is impossible to pronounce the sentence "Ramses II died of tuberculosis" without bringing back all the pragmatic conditions that give truth to this sentence.

In other words, provided that (1) we treat all scientific objects like technological projects, (2) we treat all expansion in time as being as difficult, costly, and fragile as extensions in space, and (3) we consider science studies to be the model that renders *impossible* the escape of a fact from its network of production, then we are faced with a new ontological puzzle: the thorough historicization not only of the *discovery* of objects, but of those objects *themselves*. By learning the lesson of this picture, we might provide a network account of reality that would escape both whiggish and radical anti-whiggish metaphysics.

PURGING OUR ACCOUNTS OF FOUR ADVERBS: NEVER, ALWAYS, NOWHERE, EVERYWHERE

To formulate the question of this essay, let me generalize the two questions of the prologue (What happened after 1976 to "Saodowaoth," the name wrongly given to the cause of Ramses' death? Where were the Koch bacilli *before* 1882 and 1976?):

• Where were the objects that no longer exist when they existed in their limited and historically crooked ways?
• Where were the objects that now exist before they acquired this decisive and no longer historical mode of existence?

I will not try to answer these questions at the philosophical and ontological level,[3] which I could call "historical realism"—*not* historical materialism!—in which the notions of events, relations, and propositions play the dominant role. My goal in this essay, although theoretical, is not philosophical. I simply want to dig out the theory of "relative existence" embedded in what could be called the "best practice" of historians of science and science studies. Not that I want to give them a lesson. I am simply interested in mapping a common ground, a common vocabulary, that would be *intermediary* between the practice of historical narrative in the social his-

---

3. For this see my *Pandora's Hope: Essays in the Reality of Science Studies* (Cambridge: Harvard University Press, 1999).

tory of science on the one hand and the ontological questions that are raised by this practice on the other. My idea is simply that in the last twenty years historians of science have raised enough problems, monsters, and puzzles, such as that of Ramses II's cause of death, to keep philosophers, metaphysicians, and social theorists busy for decades. The *middle ground* I want to explore here could at least prevent us from asking the wrong questions of the historical narratives at hand, and should help focus our attention on new questions hitherto hidden by the fierce debates between realism and relativism.

To give some flesh to the theoretical questions raised here, I will use, not the case of Ramses II (about which I do not know enough), but the debates between Pasteur and Pouchet over spontaneous generation. I do not wish here to add anything to its historiography, but to use it precisely because it is so well known that it can be used as a convenient topos for all readers.[4]

What is relative existence? It is an existence that is no longer framed by the choice between never and nowhere on the one hand, and always and everywhere on the other. If we start by having to choose between these positions imposed upon us by the traditional formulations of the philosophy of science, we cannot hope to fulfil the goals of this book. Pouchet's spontaneous generation will have *never* been there *anywhere* in the world; it was an illusion all along; it is not allowed to have been part of the population of entities making up space and time. Pasteur's ferments carried by the air, however, have *always* been there, all along, *everywhere,* and have been bona fide members of the population of entities making up space and time long before Pasteur. To be sure, historians can tell us a few amusing things on why Pouchet and his supporters wrongly believed in the existence of spontaneous generation, and why Pasteur fumbled a few years before finding the right answer, but the tracing of those zigzags gives us no new essential information on the entities in question. Although they provide information on the subjectivity and history of *human* agents, history of science, in such a rendering, does not provide any other information on what makes up *nonhuman* nature. By asking a nonhuman entity to exist—

4. John Farley, "The Spontaneous Generation Controversy—1700–1860: The Origin of Parasitic Worms," *Journal of the History of Biology* 5 (1972): 95–125; John Farley, *The Spontaneous Generation Controversy from Descartes to Oparin* (Baltimore: Johns Hopkins University Press, 1974); Gerald Geison, *The Private Science of Louis Pasteur* (Princeton: Princeton University Press, 1995); Richard Moreau, "Les expériences de Pasteur sur les générations spontanées: Le point de vue d'un microbiologiste," parts 1 ("La fin d'un mythe") and 2 ("Les conséquences"), *La vie des sciences* 9, no. 3 (1992): 231–60; no. 4 (1992): 287–321; Bruno Latour, "Pasteur and Pouchet: The Heterogenesis of the History of Science," in *History of Scientific Thought*, ed. Michel Serres (London: Blackwell, 1995), 526–55.

or more exactly to have existed—either never-nowhere or always-every-where, the epistemological question limits historicity to humans and arti-facts and bans it for nonhumans.

Contrary to this popular version of the role of history in science, it could be said that the new social or cultural history of science is defined by the *generalization of historicity*, usually granted only to social, technological, and psychological agency, to natural agencies. No one, even his French worshipers, will ask the question, "Where was Pasteur *before* 1822?" Or will require Pouchet to have been nonexistent in 1864—when he disputes Pasteur's findings—under the pretext that he was defeated by Pasteur. Rel-ative existence is exactly what we are used to dealing with in human his-tory; it is also what we take for granted for technological artifacts. None of the social and technical events making up a historical narrative have to be put into the Procrustean bed of never-nowhere or always-everywhere. Ex-isting somewhat, having a little reality, occupying a definitive place and time, having predecessors and successors: those are the normal ways of de-lineating the spatiotemporal envelope of history. These are exactly the kind of terms and expressions that should be used, from now on, for spontaneous generation itself and for the germs carried by the air.

Let me try a very sketchy history, the narrative of which relies on this symmetrical historicization. Spontaneous generation was a very impor-tant phenomenon in a Europe devoid of refrigerators and preserves, a phe-nomenon everyone could easily reproduce in one's kitchen, an undisputed phenomenon made more credible through the dissemination of the micro-scope. Pasteur's denial of its existence, on the contrary, existed only in the narrow confines of the rue d'Ulm laboratory, and only insofar as he was able to prevent what he called "germs" carried by the air to enter the culture flasks. When reproduced in Rouen, by Pouchet, the new material culture and the new bodily skills were so fragile that they could not migrate from Paris to Normandy and spontaneous generation proliferated in the boiled flasks as readily as before. Pasteur's successes in *withdrawing* Pouchet's common phenomenon from space-time required a gradual and punctilious *extension* of laboratory practice to each site and each claim of his adver-sary. "Finally," the whole of emerging bacteriology, agribusiness, and med-icine, by relying on this new set of practices, eradicated spontaneous generation, which, using the past perfect, they had transformed into some-thing that, although it had been a common occurrence for centuries, was now a belief in a phenomenon that "had never" existed "anywhere" in the world. This expulsion and eradication, however, required the writing of textbooks, the making of historical narratives, the setup of many institu-tions from universities to the Pasteur Museum. Much work had to be done—has still to be done, as we will see below—to maintain Pouchet's claim as a belief in a nonexistent phenomenon.

I put "finally" above in quotation marks, because if, to this day, you reproduce Pouchet's experiment in a defective manner, by being, for instance like me, a poor experimenter, not linking your bodily skills and material culture to the strict discipline of asepsis and germ culture learned in microbiology laboratories, the phenomena supporting Pouchet's claims will still appear.[5] Pasteurians of course will call it "contamination," and if I wanted to publish a paper vindicating Pouchet's claims and reviving his tradition based on my observations no one would publish it. But if the collective body of precautions, the standardization, the disciplining learned in Pasteurian laboratories were to be interrupted, not only by me, the bad experimenter, but by a whole generation of skilled technicians, then the decision about who won and who lost would be made uncertain again. A society that would no longer know how to cultivate microbes and control contamination would have difficulty in judging the claims of the two adversaries of 1864. There is no point in history where a sort of inertial force can be counted on to take over the hard work of scientists and relay it for eternity.[6] For scientists there is no Seventh Day!

What interests me here is not the accuracy of this account, but rather the *homogeneity* of the narrative with one that would have described, for instance, the rise of the radical party, from obscurity under Napoleon III to prominence in the Third Republic, or the expansion of Diesel engines into submarines. The demise of Napoleon III does not mean that the Second Empire never existed; nor does the slow expulsion of Pouchet's spontaneous generation by Pasteur mean that it was *never* part of nature. In the same way that we could still, to this day, meet Bonapartists, although their chance of becoming president is nil, I sometimes meet spontaneous generation buffs who defend Pouchet's claim by linking it, for instance, to prebiotics and who want to rewrite history again, although they never manage to get their "revisionist" papers published. Both groups have now been pushed to the fringe, but their mere presence is an interesting indication that the "finally" that allowed philosophers of science, in the first model, definitively to clean the world of entities that have been proved wrong was too brutal. Not only is it brutal; it also ignores the mass of work that still

5. I had the chance in 1992 for the twenty-fifth anniversary of my center to redo those experiments in the company of Simon Schaffer. See the essay in this volume by Hans-Jörg Rheinberger.

6. See the interesting notion of "grey boxes" in Kathleen Jordan and Michael Lynch, "The Mainstreaming of a Molecular Biological Tool," *Technology in Working Order: Studies of Work, Interaction, and Technology,* ed. G. Button (London: Routledge, 1993).

has to be done, daily, to activate the "definitive" version of history. After all, the Radical party disappeared, as did the Third Republic, for lack of massive investments in democratic culture, which, like microbiology, has to be taught, practiced, kept up, sunk in. It is always dangerous to imagine that, at some point in history, *inertia* is enough to keep up the reality of phenomena that have been so difficult to produce. When a phenomenon "definitely" exists this does not mean that it exists forever, or independently of all practice and discipline, but that it has been entrenched in a costly and massive *institution* that has to be monitored and protected with great care (see below). This is a lesson that was learned the hard way both by democrats who saw the Third Republic flounder in the hands of Vichy, and by the historians who saw, to their dismay, the negationists gain credit in France. "Inertia," obviously, was no protection against reopening of controversies.

DEMARCATION IS THE ENEMY OF DIFFERENTIATION

How can we now map the two destinies of Pasteur's and Pouchet's claims without appealing to the two dragons, the Faffner of never-nowhere and the Fasolt of always-everywhere? Do we have to embrace a simpleminded relativism and claim that both arguments are historical, contingent, localized, and temporal, and thus cannot be differentiated, each of them being able, given enough time, to revise the other into nonexistence? This is what the two dragons claim, or more exactly roar threateningly. Without them, they boast, only an undifferentiated sea of equal claims will appear, engulfing at once democracy, common sense, decency, morality, and nature . . . The only way, according to them, to escape relativism is to withdraw from history and locality every fact that has been proven right, and to stock it safely in a nonhistorical nature where it has always been and can no longer be reached by any sort of revision. *Demarcation*, for them, is the key to virtue and, for this reason, historicity is then maintained only for humans, radical parties, and emperors, while nature is periodically purged of all the nonexistent phenomena that clutter Her. In this demarcationist view, history is simply a way for humans to access nonhistorical nature, a convenient intermediary, a necessary evil, but it should not be, according to the two dragon keepers, a durable mode of existence for facts.

These claims, although they are often made, are both inaccurate and dangerous. Dangerous, because, as I have said, they forget to *pay the price* of keeping up the institutions that are necessary for maintaining facts in durable existence, relying instead on the free inertia of ahistoricity. But, more importantly for this book, they are *inaccurate*. Nothing is easier than to differentiate in great detail the claims of Pasteur and Pouchet. This differentiation, contrary to the claims of our fiery keepers, is made even more telling once we abandon the boasting and empty privilege they want for nonhumans over human events. Demarcation is here the enemy of differentiation. The two dragons behave like eighteenth-century aristocrats who claimed that civil society would crash if it was not solidly held up by their noble spines and was delegated instead to the humble shoulders of many commoners. It happens that civil society is actually rather better maintained by the many shoulders of citizens than by the Atlas-like contortions of those pillars of cosmological and social order. It seems that the same demonstration is to be made for differentiating the spatiotemporal envelopes deployed by historians of science. The common historians seem to do a much better job at maintaining differences than the towering epistemologists.

Let us compare the two accounts by looking at figure 10.2. In those diagrams existence is not an all-or-nothing property but a relative property

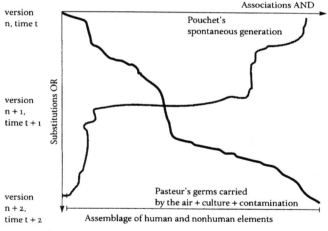

Figure 10.2.   Relative existence may be mapped according to two dimensions: association (AND), that is, how many elements cohere at a given time, and substitution (OR), that is, how many elements in a given association have to be modified to allow other new elements to cohere with the project. The result is a curve in which every modification in the associations is "paid for" by a move in the other dimension. Pouchet's spontaneous generation becomes less and less real, and Pasteur's culture method becomes more and more real after undergoing many transformations. (From Bruno Latour, *Pandora's Hope* [Cambridge: Harvard University Press, 1999], 159; copyright © 1999 by the President and Fellows of Harvard College. Reprinted by permission of Harvard University Press)

that is conceived of as the *exploration* of a two-dimensional space made by association and substitution, AND and OR. An entity gains in reality if it is associated with many others that are viewed as collaborating with it. It loses in reality if, on the contrary, it has to shed associates or collaborators (humans and nonhumans). Thus, these diagrams do not consider any final stage in which historicity will be abandoned to be relayed by inertia, ahistoricity, and naturalness—although very well known phenomena like black-boxing, socialization, institutionalization, standardization, and training will be able to account for the smooth and ordinary ways in which they would be treated. Matters of fact become matters of course. At the bottom of the diagram, the reality of Pasteur's germ carried by the air is obtained through an ever greater number of elements with which it is associated—machines, gestures, textbooks, institutions, taxonomies, theories, and so on. The same definition can be applied to Pouchet's claims, which at version *n*, time *t*, are weak because they have lost almost all of their reality. The difference, so important to our dragon keepers, between Pasteur's expanding reality and Pouchet's shrinking reality is then pictured adequately. But this difference is only *as big as* the relation between the tiny segment on the left and the long segment at the right. It is *not an absolute* demarcation between what has never been there and what was always there. Both are relatively real and relatively existent, that is extant. We never say "it exists" or "it does not exist," but "this is the collective history that is enveloped by the expression 'spontaneous generation' or 'germs carried by the air.'"

The second dimension is the one that captures historicity. History of science does not document travel *through* time of an already existing *substance*. Such a move would accept too much from the dragons' requirements. History of science documents the modifications of the ingredients composing an association of entities. Pouchet's spontaneous generation, for instance, is made, at the beginning, of many elements: commonsense experience, anti-Darwinism, republicanism, Protestant theology, natural history skills in observing egg development, geological theory of multiple creations, Rouen natural museum equipment, etc.[7] In encountering Pasteur's opposition, Pouchet alters many of those elements. Each alteration, substitution, or translation means a move onto the vertical dimension of the diagram. To associate elements in a durable whole, and thus gain existence, he has to modify the list that makes up his phenomenon. But the new elements will not necessarily hold with the former ones, hence a move through the diagram space that dips—because of the substitution—and

---

7. Maryline Cantor, *Pouchet, savant et vulgarisateur: Musée et fécondité* (Nice: Z'éditions, 1994).

may move toward the left because of lack of associations between the newly "recruited" elements.

For instance, Pouchet has to learn a great deal of the laboratory practice of his adversary in order to answer the Academy of Sciences commissions, but, by doing this, he loses the support of the academy in Paris and has to rely more and more on republican scientists in the provinces. His associations might extend—for instance he gains large support in the anti-Bonapartist popular press—but the support he expected from the academy vanishes. The compromise between associations and substitutions is what I call exploring the socionatural phase space. Any entity is such an exploration, such an experience in what holds with whom, in who holds with whom, in what holds with what, in who holds with what. If Pouchet accepts the experiments of his adversary but loses the academy and gains the popular antiestablishment press, his entity, spontaneous generation, will be a *different* entity. It is not a substance crossing the nineteenth century. It is a set of associations, a syntagm, made of shifting compromise, a paradigm,[8] exploring what the nineteenth-century socionature may withhold. To Pouchet's dismay, there seems to be no way from Rouen to keep the following united in one single coherent network: Protestantism, republicanism, the academy, boiling flasks, eggs emerging de novo, his ability as natural historian, his theory of catastrophic creation. More precisely, if he wants to maintain this assemblage, he has to shift audiences and give his network a completely different space and time. It is now a fiery battle against official science, Catholicism, bigotry, and the hegemony of chemistry over sound natural history.[9]

Pasteur also explores the socionature of the nineteenth century, but his association is made of elements that, at the beginning, are largely distinct from those of Pouchet. He has just started to fight Liebig's chemical theory of fermentation and replaced it by a living entity, the ferment, the organic matter of the medium being there not to cause fermentation, as for Liebig, but to feed the little bug that no longer appears as a useless by-product of fermentation but as its sole cause.[10] This new emerging syntagm includes many elements: a modification of vitalism made acceptable against chemistry, a reemployment of crystallographic skills at sowing and cultivating entities, a position in Lille with many connections to agribusiness relying

8. In the linguist's usage of the word, not the Kuhnian one.

9. We should not forget here that Pouchet is not doing fringe science, but is being pushed to the fringe. At the time, it is Pouchet who seems to be able to control what is scientific by insisting that the "great problems" of spontaneous generation should be tackled only by geology and world history, not by going through Pasteur's flasks and narrow concerns.

10. See Latour, *Pandora's Hope,* chap. 4.

on fermentation, a brand-new laboratory, experiments in making life out of inert material, a circuitous move to reach Paris and the academy, etc. If the ferments that Pasteur is learning to cultivate, each having its own specific product—one for alcoholic fermentation, the other for lactic fermentation, a third for butyric fermentation—are also allowed to appear spontaneously, as Pouchet claims, then this is the end of the association of the entities already assembled by Pasteur. Liebig would be right in saying that vitalism is back; cultures in pure medium will become impossible because of uncontrollable contamination; contamination itself will have to be reformatted in order to become the genesis of new life forms observable under the microscope; agribusiness fermentation would no longer be interested in a laboratory practice as haphazard as its own century-old practice; etc.

In this very sketchy description, I am not treating Pasteur differently from Pouchet, as if the former were struggling with real uncontaminated phenomena and the second with myths and fancies. Both try their best to hold together as many elements as they can in order to gain reality. But those are not the *same* elements. An anti-Liebig, anti-Pouchet microorganism will authorize Pasteur to maintain the living cause of fermentation and the specificity of ferments, allowing him to control and to cultivate them inside the highly disciplined and artificial limits of the laboratory, thus connecting at once with the Academy of Science and agribusiness. Pasteur too is exploring, negotiating, trying out what holds with whom, who holds with whom, what holds with what, who holds with what. There is no other way to gain reality. But the associations he chooses and the substitutions he explores make a different socionatural assemblage, and each of his moves modifies the definition of the associated entities: the air, as well as the emperor, the laboratory equipment as well as the interpretation of Appert's preserves, the taxonomy of microbes as well as the projects of agribusiness.

SPATIOTEMPORAL ENVELOPES, NOT SUBSTANCES

I showed that we can sketch Pasteur's and Pouchet's moves in a symmetrical fashion, recovering as many differences as we wish between them without using the demarcation between fact and fiction. I also offered a very rudimentary map to replace judgments about existence or nonexistence by the spatiotemporal envelopes drawn when registering associations and substitutions, syntagms, and paradigms. What is being gained by this move? Why would science studies and history of science offer a better narrative to account for the relative existence of all entities than the one offered by the notion of a substance remaining there forever? Why should adding the strange assumption of historicity of things to the historicity of humans simplify the narratives of both?

The first advantage is that we do not have to consider physical entities such as ferments, germs, or eggs sprouting into existence as being radically different from a *context* made of colleagues, emperors, money, instruments, body practices, etc. Each of the networks that makes up a version in the diagram above is a list of heterogeneous associations that includes humans and nonhuman elements. There are many philosophical difficulties with this way of arguing, but it has the great advantage of requiring us to stabilize neither the list of what makes up nature nor the list of what makes up context. Pouchet and Pasteur do not define the same physical elements—the first one seeing generation where the other sees contamination of cultures—nor do they live in the *same* social and historical context. Each chain of associations defines not only different links with the same elements, but different elements as well.

So, historians are no more forced to imagine one single nature of which Pasteur and Pouchet would provide different "interpretations" than they are to imagine one single nineteenth century imposing its imprint on historical actors. What is at stake in each of the two constructions is what God, the emperor, matter, eggs, vats, colleagues, etc. are able to do. To use a semiotic vocabulary, *performances* are what is needed in those heterogeneous associations, and not *competences* implying an hidden substrate or substance. Each element is to be defined by its associations and is an event created at the occasion of each of those associations. This will work for lactic fermentation, as well as for the city of Rouen, the emperor, the rue d'Ulm laboratory, God, or Pasteur's and Pouchet's own standing, psychology, and presuppositions. The ferments of the air are deeply modified by the laboratory at rue d'Ulm, but so is Pasteur, who becomes Pouchet's victor, and *so is the air* that is now separated, thanks to the swan neck experiment, into what transports oxygen on the one hand and what carries dust and germs on the other. In the narratives of historians of science, historicity is allocated to *all* the entities.

Second, as I said above, we do not have to treat the two envelopes asymmetrically by considering that Pouchet is fumbling in the dark with nonexisting entities while Pasteur is slowly targeting an entity playing hide-and-seek, while the historians punctuate the search by warnings like "cold!," "you are hot!," "you are warm!" Both Pasteur and Pouchet are associating and substituting elements, very few of which are similar, and experimenting with the contradictory requirements of each entity. The envelopes drawn by both protagonists are similar in that they are a spatiotemporal envelope that remains locally and temporally situated and empirically observable.

Third, this similarity does not mean that Pasteur and Pouchet are building the *same* networks and share the *same* history. The elements in the two associations have almost no intersection—apart from the experimental setting designed by Pasteur and taken over by Pouchet (none of the experimental designs of Pouchet was replicated by Pasteur, revealing a clear asymmetry here). Following the two networks in detail will lead us to visit completely different definitions of nineteenth-century socionature (as I have shown elsewhere, even the definition of Napoleon III is different).[11] This means that the incommensurability itself between the two positions—an incommensurability that seems so important for moral judgment—is itself the product of the slow differentiation of the two networks. In the end—a local and provisional end—Pasteur's and Pouchet's positions are incommensurable.

Thus, there is no difficulty in recognizing the differences in two networks once their basic similarity has been accepted. The spatiotemporal envelope of spontaneous generation has limits as sharp and as precise as those of germs carried by the air and contaminating microbe cultures in medium. The abyss between the claims that our two dragons challenged us to admit under threat of punishment is indeed there, but with an added bonus: the definitive demarcation where history stopped and naturalized ontology took over has disappeared. The advantage is important in rendering networks comparable at last because it allows us to go on qualifying, situating, and historicizing even the *extension* of "final" reality. When we say that Pasteur has won over Pouchet, and that now germs carried in the air are "everywhere," this everywhere can be documented empirically. Viewed from the Academy of Sciences, spontaneous generation disappeared in 1864 through Pasteur's work. But partisans of spontaneous generation lasted a long time and had the sentiment that they had conquered, Pasteur's chemical dictatorship receding into the fragile fortress of "official science." So they had the field to themselves, even though Pasteur and his colleagues felt the same way. Well, the comparison of the two "extended fields" is fea-

---

11. Bruno Latour, *Pasteur: une science, un style, un siècle* (Paris: Librairie académique Perrin, 1994). Pouchet, for instance, writes a letter to the emperor asking him for support in favor of spontaneous generation. Pasteur, the same year, also writes to ask for the emperor's support but this time to ask for his money, not for his opinion about the controversy. Do they write to the same emperor? No, since one is supposed to have an opinion and the other one money, one—Pouchet's emperor—is supposed to invade science and rectify the bad judgments of scientists, while the other is supposed to strictly respect the demarcation between science and politics but fully to support the former, keeping his opinions to himself.

sible without recurring to some incompatible and untranslatable "paradigms" that would forever estrange Pasteur from Pouchet. Republican, provincial natural historians, having access to the popular anti-Bonapartist press, maintain the extension of spontaneous generation. A dozen microbiology laboratories withdraw the existence of this phenomenon of spontaneous generation from nature and reformat the phenomena it was made of by the twin practices of pure medium culture and protection against contamination. The two are not incompatible paradigms (in the Kuhnian sense this time) by nature. They have been *made* incompatible by the series of associations and substitutions constructed by each of the two protagonists. They simply had fewer and fewer elements in common.

The reason why we find this reasoning difficult is that we imagine for microbes a substance that would be a little bit *more* than the series of its historical manifestations. We might be ready to grant that the set of performances remains always inside of the networks and that they are delineated by a precise spatiotemporal envelope, but we cannot suppress the feeling that the substance travels with fewer constraints than the performances. It seems to live a life of its own, having been, like the Virgin Mary in the dogma of Immaculate Conception, always already there, even before Eve's fall, waiting in Heaven to be translocated into Anna's womb at the right time. There is indeed a *supplement* in the notion of substance, but we should not, following the etymology of the word, "what lies underneath," imagine that this supplement resides "beneath" the series of its manifestations. Sociology offers a much better definition of substance with its notion of *institution*, that which is above a series of entities and makes them act as a whole. Yes, at the end of the nineteenth century, "the airborne germs" has become a whole, an organized and systematic body of practice that cannot be shattered. But this solidity, this wholeness, is to be accounted for by the fact that it is now institutionalized. "Substance" can now be redefined as the supplement of solidity and unity given to a series of phenomena by their routinization and black-boxing, and wrongly attributed to something lying below everything and possessing another life. The advantage of the notion of institution is that it is not difficult to entertain the idea that it has a history, a beginning and an end. With the notion of institution to account for their solidity and the notion of technical project[12] to account for their local deployment, natural facts become firmly attached to their spatiotemporal envelopes and stop hovering over their own bodies like ghosts.

12. Project, by opposition to object, is an original ontological state that has been well documented by recent history and sociology of technology. See above and, for instance, Wiebe Bijker, *Of Bicycles, Bakelites, and Bulbs: Toward a Theory of Sociotechnical Change* (Cambridge: MIT Press, 1995).

This reworking of the notion of substance is crucial because it points to a phenomenon that is badly accounted for by history of science: how do phenomena *remain in existence* without a law of inertia? Why can't we say that Pasteur is right and Pouchet wrong? Well, we can say it, but on the condition of making very precise the institutional mechanisms that are still at work to maintain the asymmetry between the two positions. In whose world are we now living? That of Pasteur or that of Pouchet? I don't know about you, but for my part, I live inside the Pasteurian network, every time I eat pasteurized yogurt, drink pasteurized milk, or swallow antibiotics. In other words, even to account for a lasting victory, one does not have to grant extrahistoricity to a research program that would suddenly, at some breaking or turning point, need *no* further upkeep. One simply has to go on historicizing and localizing the network and finding who and what make up its descendants. In this sense I partake in the "final" victory of Pasteur over Pouchet, in the same way that I partake in the "final" victory of republican over autocratic modes of governments by voting in the last presidential election instead of abstaining or refusing to be registered. To claim that such a victory requires no more work, no more action, no more institution, would be foolish. I can simply say that I live in this continued history.[13] To claim that the everywhere and always of such events cover the whole spatiotemporal manifold would be at best an exaggeration. Step away from the networks, and completely different definitions of yogurt, milk, and forms of government will appear and this time, not spontaneously . . .

GRANTING HISTORICITY TO OBJECTS

This solution, which is obvious for human-made historical events such as republics and for technological artifacts, seems awkward at first when applied to natural events because we do not want to share historicity with the nonhumans mobilized by the natural sciences. Under the influence of their antiempiricist fights, social historians of science understand by the expression "plasticity of natural facts" only the debates that humans agents have *about* them. Pasteur and Pouchet disagree about the interpretation of facts because, so the historians say, those facts are *underdetermined* and cannot, contrary to the claims of empiricists, force rational minds into assent. So the first task of social historians and social constructivists, following Hume's

13. See Isabelle Stengers, *L'invention des sciences modernes* (Paris: La Découverte, 1993), for this Whiteheadian argument on descendance and heritage. This is a pragmatist argument except that pragmatism is extended to things, and no longer limited to human relations with things.

line of attack, was to show that we, the humans, faced with dramatically underdetermined matters of fact, have to enroll other resources to reach consensus—our theories, our prejudices, our professional or political loyalties,
our bodily skills, our standardizing conventions, etc. In their view, matters
of fact had to be banned forever from narrative about scientific success, because either they were too underdetermined to shut down a controversy, or,
worse, they could appear as the now bygone dispute closers of the realist
tradition.

This tack, which looked reasonable at first, turned out to be at best a gross
exaggeration of the abilities of social scientists to account for the closure of
disputes, and at worst a devastating move delivering the new field of social
historians straight into the teeth of Faffner and Fasolt. Why? Because social
historians had to accept that historicity, like the now-dismantled apartheid
in South African buses, was "for humans only," matters of fact playing no
role at all in the controversy human agents have about them. Just what the
dragons had roared all along . . . The acquiescence of the two archenemies,
social constructivists and realists, to the very same metaphysics for opposed
reasons has always been for me a source of some merriment.

A completely different source of plasticity and agitation can however be
easily discovered; it is the one that resides in the matters of fact themselves.
There is nothing in nature, in the series of causes and consequences, that
dictates forever what ferments are supposed to do, to be, and how they have
to behave once existence is defined as an event and that substances are redis-
tributed into associations and relations. The germs carried by the air in Pasteur's rue d'Ulm air pump experiment are certainly not *the same* as those
eggs that spontaneously appear at Rouen in Pouchet's flasks. They have to
be the same only if a *sub*stance having no time and space is supposed to endure *under* the passing attributes that humans detect through their passing
interpretations. But this is precisely the philosophy of existence that historians of science do not like to apply when offering their narratives of human, technological, and social-historical events. Applied to things, such a
reluctance makes as much sense. Asking where the germs of the air of Paris
were in 1864 at the rue d'Ulm, *before* 1864 and *away* from the rue d'Ulm,
for instance in Rouen, has about as much meaning as asking where Pasteur
was before he was born, and where the Second Empire was under Louis
Philippe's reign. Answer: they were *not* there. To be sure, they had ascen-
dants and predecessors, but those bear only family resemblances to them
and relied on different associations.

It is only the threat of relativism, in the version advocated by the two
dragons, and the threat of realism, in the version social constructivists have
fought for twenty years, that forced us to expect a *better* answer, an answer

that would either *not* use the humans—nature being made of ahistorical objects—*nor* use the nonhumans—consensus being reached by human and social factors only. The joint historicity of humans and nonhumans appears to be, to my eyes at least, the totally unexpected discovery collectively made over two decades by historians and sociologists of science. It forces philosophy, which had so heavily relied on a definition of truth-value *superior* to the collective production of history—either by defending it or by dismantling it—to become *realist again,* but through a completely different route, that is, by extending historicity and sociability to nonhumans.

That this discovery could not be made by "straight" historians is obvious, since "that Noble Dream of Objectivity" forced them to deal with a human history full of noise and furors, which took place *inside a natural background* of naturalized entities that they took for granted. Only our tiny subprofession, dealing at once with the "human element" and the former "natural context," had to push the philosophy of history a little bit further, until it reached the point where the very distribution of roles into what does and what does not have history was performed. This point, to be made philosophically consistent, requires, to be sure, an enormous effort in collaboration with ontology, metaphysics, and the cognitive sciences. But to ignore or deny its existence would seem a pity now that so much has been achieved. Constructivism and realism are two *synonyms,* every builder knows that, but the differences between what does and what does not have a history has managed to transform, through the years, a constructivist position about natural entities into a critical, skeptical, and even deconstructionist position. Strange paradox of our intellectual history.

CONCLUSION: FREEING SCIENCE FROM POLITICS

I do not claim, in this chapter, to have presented philosophical arguments but simply to have cleared the intermediary zone between the narratives of the best practice of historians of science and science studies, on the one hand, and the ontological problems that should now be tackled to make sense of the historicity of things, on the other. What has, I hope, been made clearer is the question of the spatiotemporal envelope of phenomena.

If the enormous work of retrofitting that requires history telling, textbook writing, instrument making, body training, creation of professional loyalties and genealogies, is ignored, then the question "Where were the microbes before Pasteur?" takes on a paralyzing aspect that stupefies the mind for a minute or two. After a few minutes, however, the question becomes empirically answerable: Pasteur also took care to *extend* his local

production into other times and spaces and to make the microbes the substrate of others' unwitting action; the French surgeons take great pains to
bring the mummy into direct contact with the hospital network so as to expand the existence of the Koch bacillus to span the three-thousand-year
stretch and to be made visible inside the brittle bones. Yes, there are *substances* that have been there all along, but on the condition that they are
made the substrate of activities, in the past as well as in space.[14] The always-
everywhere might be reached, but it is costly, and its localized and temporal
extension remains visible all the way. This can be made clearer through a
look at figure 10.3.

When we say that Ramses II died of tuberculosis, we now know, almost
automatically, that we should account for this extension of 1892 Koch bacillus onto the corpse of someone who has been dead for more than three millennia by taking into account the bringing of the mummy in 1976 to the
surgical table of a high-tech bacteriologist. Yes, the bacillus has been there
all along, but only *after* the sanitary flight to Paris that allowed "our scientists" to retrofit all of Egyptian history with a Pharaoh that, *from now on,*
coughs and spits Koch's bacilli, even when disputing with Moses about how
long the Ten Plagues will last . . . It might take a while before juggling effortlessly with those timings, but there is no logical inconsistency in talking
about the extension in time of scientific networks, no more than there are
discrepancies in following their extension in space. It can even be said that
the difficulties in handling those apparent paradoxes are small compared to
the smallest of those offered by quantum mechanics or cosmology.

A few elements should now be clear in this dialogue between history and
philosophy.

- If the historicity of humans is treated separately from the ahistoricity of nonhumans, then the principle of symmetry (Bloor's one,
  which fights whiggism) cannot be fully enforced.
- If a substance is added that would lie under the relations of any entity—human or nonhuman, individual or collective—then distor-
  tions will appear immediately in the rendering of their history, the
  substance being unable to have the same timing and the same spread
  as its properties, one floating at no cost in time while the others are
  stuck inside the precise envelope of their flesh-and-blood networks;
  this distortion will produce artifactual differences among "making

14. So there are two practical meanings now given to the word "substance"; one is the institution that holds together a vast array of practical setups, as we saw above, and the other one
is the retrofitting work that situates a more recent event as that which "lies beneath" an older
one.

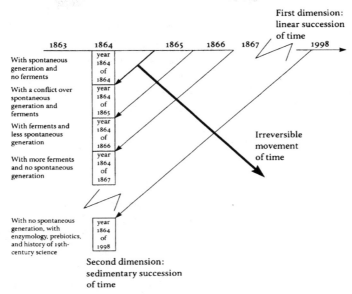

Figure 10.3.   Time's arrow is the result of two dimensions, not one: the first dimension, the linear succession of time, always moves forward (1865 is *after* 1864); the second one, sedimentary succession, moves backward (1865 occurs *before* 1864). When we ask the question "Where was the ferment before 1865?" we do not reach the top segment of the column that makes up the year 1864, but only the transverse line that marks the contribution of the year 1865 to the elaboration of the year 1864. This, however, implies no idealism or backward causation, since time's arrow always moves irreversibly forward. (From Bruno Latour, *Pandora's Hope* [Cambridge: Harvard University Press, 1999], 171; copyright © 1999 by the President and Fellows of Harvard College. Reprinted by permission of Harvard University Press)

up," "inventing," "discovering," "constructing," "socially constructing," "deconstructing," etc.

• If existence and reality are detached at some turning point from the institutional practice that enforces them, and relayed from there on by a mysterious law of inertia, then it becomes impossible to extend the empirical research of historians to the stabilization, routinization, and standardization of "definitively" existing entities, in space *as well as* in time. For any entity to gain definitive access to existence, a deep rearrangement in space and time has to be worked out practically.

• If a sharp demarcation between existing and nonexisting objects is requested, in the manner made popular by the philosophy of language, then the differentiation of the envelopes of various networks can no longer be made empirically clear, the battle for existence and nonexistence obfuscating the subtle explorations of *partial existences*. Demarcation, it should be underlined, is the moral, philosophical, and historical enemy of differentiation. The claim to morality made by demarcationists is entirely unwarranted since, on the contrary, relativism is the only way to pay the full cost of the extension in space and time of truth-values *and the maintenance* thereof.

• To avoid the dangers of relativism, especially those of having majority rule imposed in matters of knowledge, realists had to push matters of fact into nonhistorical nature *limiting* history to society and human passions; to avoid the dangers of realism, especially those of creating a suprasocial and suprahistorical scientific authority, social constructivists had to *abstain* from using matters of fact to account for the closure of historical controversies in science; the result was to imagine either that a nonhistorical and noncollective judge was necessary for differentiating knowledge claims, or that social history should never use things-in-themselves, except to debunk their claims to closure and expose their plasticity. However, as soon as historicity and socialization are extended to *all* members of collectives, the twin limits of relativism and realism are alleviated, as well as the strange metaphysics or political philosophy they thought necessary to endorse. As Whitehead shows in his cosmology, realism and relativism are synonymous expressions.

By this contribution, intermediary between philosophy and history of science—or better, ontology and the theory of history of science—I hope to have followed the intent of this volume and opened at least some conversations about the philosophy of history that would do justice to the more scholarly work presented in the other essays. A fascinating question to tackle now would be to understand why, if I am right in thinking that the thoroughgoing historicization here offered is neither inconsistent nor in danger of being morally bankrupt, it is nonetheless so difficult to entertain and so perilous to defend. What is especially puzzling to me is that many natural scientists have already rendered the world itself part of history, not only the living organisms of Darwinian theory but also cosmology.[15] Why

---

15. See the classic books of Stephen Jay Gould, esp. *Wonderful Life: The Burgess Shale and the Nature of History* (New York: W. W. Norton, 1989). It would probably be interesting to enter into a conversation with "evolutionary epistemology" at this point, for instance David L. Hull, *Science as a Process: An Evolutionary Account of the Social and Conceptual Development of Science* (Chicago: University of Chicago Press, 1988).

is time, if it is a good enough repository for animal bodies, for particles, for Big Bangs, not deemed stable enough for the knowledge claims made about those entities themselves? As if something else were needed, an Above and Beyond that could hold society and morality together? Something that, for purely contingent reasons, happens to be mixed up with the history of science, but is in no way related to the question of describing the sciences and accounting for their progress and demise. What progress could we make if we could disentangle the political question of maintaining social order from that of describing the history of the sciences? What step forward could be taken if we could depoliticize the sciences from the heavy burden that epistemology and Higher Superstitions have imposed on them for purely political reasons . . . ?

Cytoplasmic Particles

THE TRAJECTORY OF A
SCIENTIFIC OBJECT

INTRODUCTION

Let me start with an anecdote. In March 1956, some of the protagonists of cytoplasmic particle research, which I will describe in this essay, met at a CIBA Foundation Symposium on "The Influence of Ionizing Radiation on Cell Metabolism" in London. In the discussions, the obscure role of ribonucleic acids in the test tube synthesis of proteins came up again and again. During one of the sessions, Waldo Cohn from the Oak Ridge National Laboratory, a former colleague of Paul Zamecnik at the Massachusetts General Hospital, made a short intervention. Cohn had been working on nucleic acids for almost two decades. "This might be an appropriate time to make a few comments on nucleic acids," he said. Then he summarized his experience by quoting a phrase from a colleague, the nucleic acid expert Masson Gulland, who, had he survived his train accident in 1947,[1] might have spared the history of molecular biology two of its greatest heroes, James Watson and Francis Crick: "I think," Cohn went on, "it was Gulland who said that 'Nucleic acids are not compounds, they are methods of preparation.'"[2]

This essay is based on material presented in more detail in Hans-Jörg Rheinberger, *Toward a History of Epistemic Things: Synthesizing Proteins in the Test Tube* (Stanford: Stanford University Press, 1997). The paper is dedicated to the memory of Georges Canguilhem.

1. Keith L. Manchester, "Did a Tragic Accident Delay the Discovery of the Double Helical Structure of DNA?" *TIBS* 20 (March 1995): 126–28.
2. Discussion on Ernest F. Gale, "Nucleic Acids and Amino Acid Incorporation," in *CIBA Foundation Symposium on Ionizing Radiations and Cell Metabolism*, ed. G. E. W. Wolstenholme and Cecilia M. O'Connor (Boston: Little, Brown, 1956), 174–84, at 183.

This statement of a scientist not belonging to the category of notorious and naive realists, whose philosophical defenders have mobilized all their resources against the specter of constructivism, instrumentalism, and relativism, could make them little bit more cautious and less arrogant. The majority of scientists know perfectly well that they are working with transitory objects. At the end of his history of genetics, *La logique du vivant*, François Jacob asks himself, and this is more than a mere rhetorical formula to end a book, since it virtually opens a new chapter: "Today the world is messages, codes, information. Which dissection will displace our objects tomorrow and recompose them in a new space? What new Russian puppet will emerge from it?"[3] We historians then should ask ourselves on two levels, that of the peculiar historical character of scientific objects, or epistemic things, and that of our historical narratives, what it means to speak of scientific objects and what it means to organize our stories around them.[4]

To start with, I will briefly contextualize my study on the trajectory of cytoplasmic particles for those who are not familiar with this kind of object and this type of research, which came to be located at the junction of different disciplines of the twentieth-century life sciences, including cytomorphology, biochemistry, and molecular biology. Indeed, I argue that it is precisely at the intersection of these disciplinary structures and their respective techniques of representation and intervention that the object of my inquiry emerged and acquired contours. The story of this paper begins in cytochemistry and cytomorphology, that is, the study of the composition and form of cellular components. The story formed part of a cancer research program with all its medical alliances and institutional facilities and affiliations. But this did not remain so. Via differential reproduction, via the implementation of skills, tracing techniques, and instruments, such as laboratory rats, radioactive amino acids, biochemical model reactions, centrifuges, and technical expertise, these submicroscopic particles gained a momentum of their own. In a rapidly changing landscape of an emerging new biology, they became progressively disconnected from cancer research and its medical context, where they had been rooted. Instead, through several unprecedented shifts, they ended up becoming part of a biochemical analysis of protein synthesis in the test tube. Finally, in yet another surprising move, they provided one of the experimental tools for solving the central puzzle of molecular biology around 1960: the genetic code.

3. François Jacob, *La logique du vivant* (Paris: Gallimard, 1970), 345.
4. Rheinberger, *Toward a History of Epistemic Things*.

Most of the research area covered in this paper has so far received little attention from historians of biology or medicine. One of the reasons for this neglect is that the pathway of objects like microsomes cannot easily be reconstructed in terms of conceptual shifts that could be considered as being paradigmatic. This renders such fields resistant to a historiography occupied with theoretical breakthroughs. Of course, we can describe the rise of molecular biology as an encompassing reconfiguration of genetics in terms of molecular mechanisms of information transfer. But I claim that we cannot understand the dynamics of this reconfiguration if we neglect the moves that took place on the material level of object formation. We cannot understand these dynamics if we neglect the disseminating power of epistemic things that permanently manifest themselves in as yet unthought-of ways in the future. This power derives, not least of all, from the fragmented structure of a particular culture of experimental representation with its drive to make biological processes manipulable in the test tube.

Let me now come back to the epistemological issues I alluded to at the beginning. First, what is peculiar about the *historical* character of epistemic things such as the ones described in this paper? That ultrastructural cytoplasmic particles, microsomes, or ribosomes, exemplary objects of twentieth-century biology, owe their careers to methods of preparation needs no further emphasis in addition to the details that follow. Instead, let me anticipate a passage of Michael Polanyi that forms part of my conclusion: "To trust that a thing we know is real is to feel that it has the independence and power for manifesting itself in yet unthought of ways in the future."[5] Polanyi claims that the reality of a scientific object lies in its prospective history. The force and the reason of epistemic objects thus lies in the conjectures of what they might become, all while what they are going to be cannot be anticipated. Such entities, then, have a peculiar, paradoxical time structure characterized by "recurrence" in the sense Gaston Bachelard conveyed to this notion.[6] These research entities, for the very same reason, do not belong to the realm of objectivity in the sense of representing something independent from our manipulations. But they do not belong to the realm of deliberate construction either. The mode of scientific existence peculiar to such entities derives precisely from their resistance, resilience, and recalcitrance rather than from their malleability in the framework of our constructive and purposive ends. "[It] is difficult not to acknowledge that the

5. Michael Polanyi, Duke Lectures (1964), microfilm, University of California, Berkeley 1965, Library Photographic Service, 4th lecture, pp. 4–5; quoted in Marjorie Grene, *The Knower and the Known* (Washington: Center for Advanced Research in Phenomenology and University Press of America, 1984), 219.
6. Gaston Bachelard, *The New Scientific Spirit* (Boston: Beacon Press, 1984).

work of [biologists], their construction of models and theories is constrained not only by the techniques at their disposal and the conceptual tools they use, but most of all, by the surprising results they obtained."[7] Scientific objects come into existence as a result of unprecedented events that time and again subvert the finite capacities of imagination of a scientist who remains always embedded in a particular thinking frame and a local experimental culture. They remain objects of research as long as they have the power to manifest themselves in yet unthought-of ways in the future. And they pass away as scientific objects as soon as they lose their recalcitrance, either because they become black-boxed and are no longer questioned, or because they become marginalized as a result of other unprecedented events in related fields.

My second point concerns a metahistorical question. What does it mean for a historical narrative to focus on scientific objects? First, if we decide to follow the development of "epistemic objects," taken in the restricted sense of material research objects, rather than pursuing the development of concepts, disciplines, institutions, or individual researchers, we have to locate ourselves between boundaries: boundaries between representational techniques, experimental systems, established academic disciplines, institutionalized programs, individualized projects. Second, in following the path of epistemic things we have to abandon cherished classifications. Does the following study belong to the history of cancer research, of cytomorphology, of biochemistry, or of molecular biology? Is it a prehistory of protein synthesis? It belongs to all of these—and to none. It locates itself on a cross-disciplinary level, one that I think is worthy of being explored further. Third, talking about the trajectory of research objects gives voice to things as active participants in a conquest of transindividual dimensions in which the subjects concerned with these things are not the only players.

In the main part of the paper, I describe a sequence of representations that were generated in the search for biologically active, "purified" microsomes. These particles were one of *the* major objects of inquiry in the research field that developed around cell-free protein synthesis. Each of these representational interventions brought with it a particular mode of visualization, highlighting either physical aspects of the stipulated particle, such as shape and diameter under the electron microscope or weight as determined from velocity sedimentation; chemical aspects such as the amount of protein and nucleic acid present under different solubilization conditions;

7. Michel Morange, "The Developmental Gene Concept: History and Limits," in *The Concept of the Gene in Development and Evolution,* ed. Peter Beurton, Raphael Falk, and Hans-Jörg Rheinberger (Cambridge: Cambridge University Press, in press).

biochemical aspects such as amino acid incorporation as determined by radioactive tracing kinetics; and views from molecular biology such as the ribosome as a template. The list could be continued. What distinguished these modes of coming into being from mere biophysical and biochemical technicalities was that they always needed to remain bound to biological *function* in order to be taken as representative arguments in the realm of an inquiry into a biological object. The intricacy of the case, however, was such that these conjectured mechanisms of biological function themselves belonged to the realm of the emergent properties of the investigated particles, and therefore could not act as a stable referent: Their elucidation was just the central issue to which the research efforts were all directed.

It is not, then, the relation between representation and an imaginary object-referent as some thing-in-itself that makes this kind of empirical process of shaping scientific objects work. Rather, it is the match and mismatch among different representations, ideally *independent* of each other, all the way up from physics to biological function, that gives those who are working with these traces that sense and feeling of a conquest of reality, without which the fact that anybody should have been willing to get engaged in such messy laboratory work at all is incomprehensible. Reality, here, becomes a second-order concept that arises as an attribute at the intersection of alternative representations.[8] Scientific objects, not things per se, but objects insofar as they are targets of epistemic activity, are unstable concatenations of representations. At best, they become stabilized for some historically bounded period. It is not that there is no materiality there before such objects come into being, or that they would vanish altogether and shrink to nothing on their way into the future. But they can become, within a particular scientific context, altogether marginal, because nobody expects them to be generators of unprecedented events any more. They can also become transplanted and grafted onto other realms, of which the realm of technology is only one example. This, of course, can silence them as objects of research. In order to understand the strange and fragile reality of scientific objects in the long run, it is crucial to understand this double movement of becoming central and being rendered marginal, this concatenation and displacement within a particular epistemic field, and among other fields of human agency. Nowhere is the realm of the scientific real itself a closed space.

To sum up: recombination and reshuffling, bifurcation and hybridization within and among particular modes of representation is a prerequisite

---

8. Ian Hacking, *Representing and Intervening* (Cambridge: Cambridge University Press, 1983), 136.

for researchers to produce unprecedented epistemic events. Such events would not happen if the lines of descent of their objects were bred too pure. They must remain hybrids in order to remain generators of surprise.

So must, to come to my final introductory remark, the movement of narration. If *we*, historians, want to know what a particular epistemic thing was representing at a given time in the past, the material signifiers of the experimental game will already have turned it into something that, at that time, it could not (yet) have been. Let me refer in this context to Georges Canguilhem, the late doyen of French historical epistemology.[9] Canguilhem, and I think rightly so, warns the historian: "The past of a science of today should not be confounded with that science in its history."[10] That statement leaves us with a decision to make. Canguilhem was of the strong opinion that historical epistemology defines itself as looking at the past of a science of today. Many current historians of science might prefer to identify themselves with looking at a science in its history.

SETTING THE STAGE

The career of the little particles initially derived from the cytoplasm of higher organisms' cells, and later from bacteria as well, is remarkable in its convolutions. So is, in many respects, the role these objects played in the making of molecular biology. The changing fate of these tiny cytoplasmic granules is far too complex to be traced in its entirety, with all its capillary ramifications and anastomoses, in this survey. Instead, I will focus on what have come to be perceived as the major transitions in their trajectory qua scientific objects between 1935 and 1965. At this time, they had become stably connected, as ribosomes, to the cellular pathway of making proteins.

The quirks and breaks that mark the search around cytoplasmic particles, the thwarted plans, the chaotic moves at the front line of advanced techniques, the intrusions, displacements, and reappearances, in short everything that makes up the work at the experimental divide between the known and the unknown, tend to disappear in such a condensed historical account. There is only one excuse for this: To focus one's narrative on the trajectory of an object makes visible to the historian a recurrent continuum, "a formal sequence of linked solutions."[11] The biography and genealogy of things are told from the point of view of those selected insights that

9. Canguilhem died a few days before this paper was presented, in September 1995.

10. Georges Canguilhem, *Idéologie et rationalité dans l'histoire des sciences de la vie* (Paris: Vrin, 1981), 15.

11. George Kubler, *The Shape of Time: Remarks on the History of Things* (New Haven: Yale University Press, 1962), 33–39.

have marked their transformation. Once the decision has been made for this kind of narration, we cannot escape its implicit recurrence. But we can try to remain aware of the contingent character of the historian's objects of discourse. Exploiting a source of historical insight must not blind us to the choices on which it relies.

Let me add another brief remark before I start. It seems to me that it might be appropriate to make a distinction between scientific objects, or epistemic things as I would like to call them, and the experimental systems that allow the scientists to intervene with, to shape, and to represent them in one way or another. Experimental systems embed scientific objects into a broader field of material scientific culture and practice, including the realm of instrumentation and inscription devices as well as the model organisms to which these objects are generally connected, and the fluctuating concepts to which they are bound. I hope that the usefulness of this distinction will become clear as the story unfolds.

EMERGENC(I)E(S)

Two events mark the beginning of the dissection of the cell's cytoplasm in the test tube toward the end of the 1930s: the emergence of an invasive epistemic object, that is, a tumor-producing agent, from cancer research, and the introduction of a powerful new instrument: the ultracentrifuge.[12]

Disappointed by the results of his biochemical efforts to purify the filterable agent causing the sarcomas in chicken that Peyton Rous had first observed in 1910,[13] Albert Claude,[14] at James Murphy's pathology department of the Rockefeller Institute in New York, turned to the ultracentrifuge in 1936. News of the high-speed sedimentation of Rous's filterable tumor-causing agent had reached him from England.[15] Claude was on the

12. For the broader historical context see Hans-Jörg Rheinberger, "From Microsomes to Ribosomes: 'Strategies' of 'Representation,'" *Journal of the History of Biology* 28 (1995): 49–89. The following sections are based mainly on chapters 4 and 6 of Rheinberger, *Toward a History of Epistemic Things.*

13. Peyton Rous, "A Sarcoma of Fowl Transmissible by an Agent Separable from Tumor Cells," *Journal of Experimental Medicine* 13 (1911): 397–411. See also Ilana Löwy, "Variances in Meaning in Discovery Accounts: The Case of Contemporary Biology," *Historical Studies in the Physical and Biological Sciences* 21 (1990): 87–121; Ton van Helvoort, "Viren als Krebserreger: Peyton Rous, das 'infektiöse Prinzip' und die Krebsforschung," in *Strategien der Kausalität: Konzepte der Krankheitsverursachung im 19. und 20. Jahrhundert,* ed. Christoph Gradmann and Thomas Schlich (Pfaffenweiler: Centaurus, 1999), 187–228.

14. On Albert Claude see Jean Brachet, "Notice sur Albert Claude," Académie Royale de Belgique, Annuaire 1988, Palais des Académies, Brussels, 1988, 93–135.

15. John C. G. Ledingham and William E. Gye, "On the Nature of the Filterable Tumour-Exciting Agent in Avian Sarcomata," *Lancet* 228(I) (1935): 376–77; James McIntosh, "The

lookout for new technology, and he was eager to grapple with the new in-strument. Already the first results were extremely encouraging. The high-speed pellet derived from infected tissue displayed an enrichment of the agent by a factor of approximately three thousand. This was over two or-ders of magnitude more than he had achieved in the preceding years with conventional biochemical methods. In parallel experiments, however, Claude pelleted down a fraction from normal embryonic chicken tissue that, in its chemical and physical characteristics, could not be distinguished from the fraction containing the agent, with the sole but decisive biological difference that it was not infective.

Two interpretations were possible at this point. Either the main con-stituent of the tumor fraction, besides the agent itself, represented a "pre-cursor of the chicken tumor principle." The idea of an endogenous rather than a viral origin of the chicken sarcoma, in the form of a cellular precur-sor, had indeed been one of the motives for Murphy, toward the end of the 1920s, to resume the chicken agent research where Rous had left it around 1915. The alternative possibility was that the main constituent of the tumor fraction simply represented "inert elements existing also in nor-mal cells."[16]

The momentary impossibility of deciding between these two options first haunted Claude, then intrigued him, and finally, within a few years, led him away from Rous's tumor agent, which had kept him busy for almost a decade. This is thus a typical displacement of an epistemic object, induced by the incorporation of a new instrument into an existing experimental sys-tem, and by the inability to incorporate the resulting new finding into the existing framework. An alternative option had come into play. Claude had introduced the technique of differential ultracentrifugation into his system in order to isolate a submicroscopic "principle" responsible for cancer. Now the technique promised to open the door to fractionating the cytoplasm of normal cells—a new high-tech cytology emerged on the horizon. Accord-ing to many of those who knew him, Claude was a meticulous technical tin-kerer.[17]

By means of differential centrifugation, Claude started to unfold the cy-toplasm into a new space of representation: a space for the production, char-acterization, isolation, and purification of subcellular structures. For about

Sedimentation of the Virus of Rous Sarcoma and the Bacteriophage by a High-Speed Cen-trifuge," *Journal of Pathology and Bacteriology* 41 (1935): 215–17.

16. Albert Claude, "A Fraction from Normal Chick Embryo Similar to the Tumor-Produc-ing Fraction of Chicken Tumor I," *Proceedings of the Society for Experimental Biology and Medicine* 39 (1938): 398–403, at 402.

17. Hubert Chantrenne, personal communication, May 28, 1996.

a hundred years, cytomorphology had been the domain of observation by light microscopy and the corresponding preparative methods of fixation and of staining cells in situ, that is, within the context of the respective tissues. Besides the nucleus, the most characteristic feature of the eukaryotic cell, "mitochondria" had been visualized for many decades within a basophilic, more or less homogeneous cytoplasmic ground substance, the "ergastoplasm."[18] Destroying cells in order to reveal cellular structures appeared to many traditionally minded cytologists at the time nonsensical if not simply absurd.

Claude reported on his ongoing work at a Cold Spring Harbor Symposium on "Genes and Chromosomes" in 1941. If such a genetic context may appear strange to us today, it reminds us of the wider scientific environment in which the first generation of small cytoplasmic particles gained identity: the largely forgotten context of cytoplasmic inheritance, or of plasmagenesis, as opposed to chromosomal or nuclear inheritance.[19] At the beginning, Claude identified the small particles, which settled at the bottom of the test tube after one hour of centrifugation at $18000 \times g$, with the cytologically well characterized "mitochondria," or fragments thereof.[20] Although their size was below the power of resolution of light microscopy, they were still within the realm of a dark-field microscope, where the particles appeared as small clusters of reflecting points. Their overall chemical composition appeared remarkable in that besides lipids making up about half of their mass, they contained a major portion of proteins and, especially, significant amounts of ribonucleic acid (RNA).

THE RIBONUCLEIC ACID CONNECTION

The chemical composition of these cytoplasmic particles was intriguing, in particular their RNA, a compound that at that time still often came under the heading of zymonucleic acid, according to the organism in which it was thought to be most abundant: yeast. The report from the Rockefeller Institute quickly caught the attention of Jean Brachet at the Free University of Brussels.[21] In the preceding decade, Brachet had focused his interest on de-

18. Lars Ernster and Gottfried Schatz, "Mitochondria: A Historical Review," *Journal of Cell Biology* 91 (1981): 227s–255s.

19. For more on this topic, see Jan Sapp, *Beyond the Gene: Cytoplasmic Inheritance and the Struggle for Authority in Genetics* (Oxford: Oxford University Press, 1987).

20. Albert Claude, "Particulate Components of Cytoplasm," *Cold Spring Harbor Symposia on Quantitative Biology* 9 (1941): 263–71, at 265.

21. On Jean Brachet, see Hubert Chantrenne, "Notice sur Jean Brachet," Académie Royale de Belgique, Annuaire 1990, Palais des Académies, Brussels, 1990, 3–87.

veloping methods of differential histochemical staining, for DNA and RNA in particular, on quantifying these components in different tissues, cellular compartments, and animal species, and on approaching embryogenesis by cytochemical means.[22] His work during the 1930s on developing sea urchin eggs established the omnipresence of RNA, until then believed to be found only in plants, in fungi, and to some extent in the pancreas.[23] By combining enzymatic RNA digestion with a specific RNA-staining procedure based on methyl green-pyronine, Brachet had reached the conclusion that RNA was predominately located in the nucleolar structures of nuclei and in the ergastoplasm, and that cells actively engaged in protein synthesis were especially rich in RNA.[24] "The conclusion to which we are led," he said, "namely that the pentose nucleic acids intervene in protein synthesis according to a mechanism still obscure at present, is in perfect agreement with all the facts established so far."[25]

There was another coincidence. Brachet's colleague at the University of Liège, André Gratia, together with André Paillot from the Station de Zoologie Agricole du Sud-Est de Saint-Genis-Laval (Rhône), had essentially made the same fortuitous observation that had led Claude into cytoplasmic particle research—although Gratia and Paillot were working on a completely different experimental system that was connected to the silk industry. By means of an air-driven Henriot-Huguenard centrifuge, they were investigating the silkworm virus that was assumed to cause jaundice disease. While pelleting down the presumed virus particles from tissue homogenates, they found the same fine granules in healthy cells, with the sole difference, just as in the case of Claude, that the material derived from healthy tissue was not infective.[26]

With the help of the ultracentrifuge in Emile Henriot's laboratory basement in Brussels[27] and financial aid from the Fonds National de la

22. Richard Burian, "Exploratory Experimentation and the Role of Techniques in the Work of Jean Brachet, 1938–1952," *History and Philosophy of the Life Sciences* 19 (1997): 27–45.

23. For review, see Jean Brachet, "The Metabolism of Nucleic Acids during Embryonic Development," *Cold Spring Harbor Symposia on Quantitative Biology* 12 (1947): 18–27, at 18.

24. Jean Brachet, "La localisation des acides pentosenucléiques dans les tissus animaux et les oeufs d'Amphibiens en voie de développement," *Archives de Biologie* 53 (1942): 207–57.

25. Ibid., 239.

26. André Paillot and André Gratia, "Application de l'ultracentrifugation à l'isolement du virus de la grasserie des vers à soie," *Comptes Rendus Hebdomadaires de la Société de Biologie* 90 (1938): 1178–80.

27. Hubert Chantrenne, "Jean Brachet (1909–1988)," in *Selected Topics in the History of Biochemistry: Personal Recollections, III,* ed. G. Semenza and R. Jaenicke, Comprehensive Biochemistry, vol. 37 (Amsterdam: Elsevier, 1990), 201–13.

Recherche Scientifique, Brachet and his colleague Raymond Jeener, together with the young doctoral student Hubert Chantrenne, embarked on
a program of isolating what they called "cytoplasmic particles of macromolecular dimensions." In Brussels, the particles were investigated in the
context of protein synthesis and connected to cell differentiation.[28] Although Belgium was occupied by Nazi Germany in May 1940, the group
managed to analyze a considerable variety of tissues from different animals
by combining the new technique of ultracentrifugation with the previously elaborated methods of differential staining and ribonuclease digestion. But the working conditions became increasingly difficult. Scientific
journals from England and America were no longer available.[29] In November 1941, after the German administration required that Jewish professors
be expelled from the university, all teaching stopped. Soon thereafter, the
University of Brussels was closed, and the group was dismantled.[30]

In a preliminary but comprehensive set of papers published in 1943, Brachet, Chantrenne, and Jeener essentially reached the following conclusions: In adult cells, almost all of the cytoplasmic RNA was located in the
macromolecular granules. Moreover, these granules were associated with a
multitude of enzymes with either hydrolytic or respiratory functions. Theorizing along the lines of the then-current ideas about protein synthesis as
a reversal of proteolysis, Brachet assumed that the respiratory enzymes
would somehow funnel the necessary energy into the process, whereas the
hydrolytic enzymes, including several peptidases, would catalyze, in a reversal of their usual action, peptide bonds. In order to make the reaction
proceed in this opposite direction, the RNA would in a way trap the synthesized peptides and so remove them from equilibrium. Support of this conception came from the additional observation that in specialized cells, such
as insulin-producing pancreas cells or hemoglobin-producing blood cells,
appreciable amounts of these cell-specific proteins accompanied the cytoplasmic particles in the centrifuge.

In the course of this work, Brachet expressed his doubts about Claude's
hypothesis of an identity between the RNA-containing granules and the

28. Jean Brachet and Raymond Jeener, "Recherches sur des particules cytoplasmiques
de dimensions macromoléculaires riches en acide pentosenucléique. I. Propriétés générales,
relations avec les hydrolases, les hormones, les protéines de structure," *Enzymologia* 11
(1943–45): 196–212; Hubert Chantrenne, "Recherches sur des particules cytoplasmiques de
dimensions macromoléculaires riches en acide pentosenucléique. II. Relations avec les ferments respiratoires," *Enzymologia* 11 (1943–45): 213–21; Raymond Jeener and Jean Brachet,
"Recherches sur l'acide ribonucléique des levures," *Enzymologia* 11 (1943–45): 222–34.
29. Hubert Chantrenne, personal correspondence, 14 December 1995.
30. Chantrenne, "Notice sur Jean Brachet," 206–7.

mitochondria. He made, however, no attempts to sort his particles out into two or more different fractions, owing, as he put it later, "to war difficulties and lack of equipment."[31] Indeed, from 1942 until the end of the occupation, Brachet had no laboratory at his disposal and therefore was prevented from doing further experiments.[32]

CALIBRATIONS

Such was not the case with Claude. At the time Brachet's results were published, Claude had given up his initial identification of the particles with mitochondria. Under slightly different buffering and centrifugation conditions, the resuspended small-particle sediment no longer yielded particulate material of the estimated size of mitochondria. Claude renamed the small particles "microsomes."[33] Today, it appears unquestionable to us that the ultracentrifuge had been a competent tool for the structural dissection of the cytoplasm in the 1940s. In these early years, however, the new technology produced new entities and complicated rather than clarified the traditional cytological questions. It required a decade of standardization, and of connecting the new space of representation to classical cytological and biochemical knowledge, to redeem the role of the instrument.

This work was Claude's, and the new scientific object, the microsomes, bore the stamp of his name. In collaboration with several other biochemists, cytochemists, and enzymologists from Rockefeller, among them Rollin Hotchkiss, George Hogeboom, and Walter Schneider, Claude went on, comparatively unhindered by war impediments, to work out generally applicable conditions for a quantitative separation of the mitochondria and other cytoplasmic vesicles from the microsomes, and to subject the fractions to what the group called "biochemical mapping," or enzyme mapping.[34] As soon as a smooth procedure based on sucrose centrifugation had been established for recovering virtually intact mitochondria,[35] it turned

31. Jean Brachet, "The Localization and the Role of Ribonucleic Acid in the Cell," *Annals of the New York Academy of Sciences* 50 (1949): 861–69, at 863.

32. Instead, he wrote his first book. Jean Brachet, *Embryologie Chimique* (Liège: Desoer, 1944).

33. Albert Claude, "The Constitution of Protoplasm," *Science* 97 (1943): 451–56.

34. George E. Palade, "Intracellular Distribution of Acid Phosphatase in Rat Liver Cells," *Archives of Biochemistry* 30 (1951):, 144–58, at 144.

35. George H. Hogeboom, Walter C. Schneider, and George E. Palade, "Cytochemical Studies of Mammalian Tissues. I. Isolation of Intact Mitochondria From Rat Liver; Some Biochemical Properties of Mitochondria and Submicroscopic Particulate Material," *Journal of Biological Chemistry* 172 (1948): 619–35.

out that the majority of the respiratory enzymes previously found on the "small particles" went with the mitochondria. The enzyme pattern displayed by the microsomes, however, was poor, irregular, and not pointing in any specific direction. As far as their possible function was concerned, Claude, at the end of the 1940s, rather vaguely speculated that they might serve in the "anaerobic mechanism," or else that they were an "intermediate in the energy transfer for various synthetic reactions."[36]

CHALLENGES

It was only after World War II that Brachet's group rejoined in Brussels and resumed regular laboratory work . Hubert Chantrenne took up the problem of the size and uniformity of the material for which by then the term "microsomes" had gained wide currency. Working on mouse liver homogenates and refining the centrifugation conditions with the reestablished air-driven Henriot-Huguenard centrifuge, Chantrenne separated five different fractions, thus casting doubt on Claude's sharp dichotomy between mitochondria and microsomes. Chantrenne's fractions gradually differed in their chemical constitution with respect to RNA and enzymatic activity, but qualitatively, they exhibited quite comparable features. Chantrenne came to the conclusion that "it seems that one can partition the granules in as many groups as one wishes, and nothing in our experiments and observations indicates that there exist neat demarcation lines between the different groups of granules."[37] Coming back to his earlier observations of "free" ribonucleic acid in the cytoplasm of yeast, he now speculated that the particle continuum reflected a gradual growth process and that "initially 'free' ribonucleic acid becomes incorporated into sedimentable particles in the course of development. It could well be that the ribonucleic acid gets associated with little particles that grow in the course of development."[38] Were the microsomes, as far as their particulate identity was concerned, nothing else than arbitrary slices of a cytoplasmic continuum? Their boundaries appeared to depend on centrifugation conditions rather than on anything of an intrinsic biological significance. They were "methods of preparation."

In line with Chantrenne's musings, Brachet himself for a while pursued the idea that the microsomes might play a role in tissue differentiation dur-

---

36. Albert Claude, "Studies on Cells: Morphology, Chemical Constitution, and Distribution of Biochemical Function," *Harvey Lectures* 43 (1950): 121–64, at 163.

37. Hubert Chantrenne, "Hétérogénéité des granules cytoplasmiques du foie de souris," *Biochimica et Biophysica Acta* 1 (1947): 437–48, at 445.

38. Ibid., 447.

ing embryogenesis. The analogy to RNA viruses and the idea of cytoplasmic inheritance via plasmagenes not only loomed large in the background, but acted as a quite explicit perspective in his account of the RNA-containing macromolecules. Together with John Rodney Shaver from the University of Pennsylvania, Brachet initiated a program of testing the assumption of a morphogenetic activity of these granules in the induction of the nervous system. They injected isolated microsomes from various embryonic tissues into cleaving eggs of amphibians. Contrary to their expectation, however, after a tedious series of trials, they had to state that, with respect to induction, "our results have thus far been negative."[39]

Neither Claude nor Brachet stayed at the forefront of revealing the mechanistic mystery of the microsomes, that is, their biological function during the 1950s. Claude's strength was the calibration of instruments and the standardization of preparation procedures. In fact, during his last years at Rockefeller and before returning to Belgium in 1949, he turned the resolving power of the electron microscope back on his earlier chicken tumor agent. Brachet, on the other hand, appears to have been obsessed with the possible involvement of RNA in protein synthesis. His overwhelming interest in morphogenesis, however, made him concentrate on the role of microsomes in embryology.

A NEW TECHNIQUE: AMINO ACID TRACING

At the beginning of the 1950s, reports from Henry Borsook's laboratory at Caltech,[40] from Tore Hultin in Sweden,[41] from Norman Lee and Robert Williams at Harvard,[42] and from Elizabeth Keller of Paul Zamecnik's laboratory at the Massachusetts General Hospital in Boston,[43] all based on the introduction of radioactive or heavy isotope–derived amino acids as tracers

39. Jean Brachet and John Rodney Shaver, "The Injection of Embryonic Microsomes into Early Amphibian Embryos," *Experientia* 5 (1949): 204–5, at 205; John Rodney Shaver and Jean Brachet, "The Exposition of Chorioallantoic Membranes of the Chick Embryo to Granules from Embryonic Tissue," *Experientia* 5 (1949): 235.

40. Henry Borsook, Clara L. Deasy, Arie J. Haagen-Smit, Geoffrey Keighley, and Peter H. Lowy, "Incorporation of C^{14}-Labeled Amino Acids into Proteins of Fractions of Guinea Pig Liver Homogenate," *Federation Proceedings* 9 (1950): 154–55.

41. Tore Hultin, "Incorporation in Vivo of ^{15}N-Labeled Glycine into Liver Fractions of Newly Hatched Chicks," *Experimental Cell Research* 1 (1950): 376–81.

42. Norman D. Lee, Jean T. Anderson, Ruth Miller, and Robert H. Williams, "Incorporation of Labeled Cystine into Tissue Protein and Subcellular Structures," *Journal of Biological Chemistry* 192 (1951): 733–42.

43. Elizabeth B. Keller, "Turnover of Proteins of Cell Fractions of Adult Rat Liver in Vivo," *Federation Proceedings* 10 (1951): 206.

of the protein metabolism in animal tissues, began to strengthen the view of the microsomal particles as being the primary topological structures involved in the making of peptide bonds. Isotopically labeled amino acids had become available to a wider scientific community right after World War II. Exploring the tracing capacity of radioactive amino acids, similar to high-speed centrifugation, also had to go through half a decade of trouble-shooting, calibration, and experimental tinkering before it became a reliable tool for investigating the mechanisms of peptide bond formation. But with it, and in the context of experimental systems quite different from their original characterization, the small particles, or microsomes, hitherto operationally defined in terms of sedimentation, chemical composition, and enzyme mapping, became experimentally linked to protein synthesis systems in vivo and in vitro soon thereafter.[44]

FROM MICROSOMES TO RIBOSOMES

Meanwhile, Chantrenne's caveat about the heterogeneity of microsome preparations from animal cells had received ample support through what appeared to be a bewildering zoo of cytoplasmic particles of all sorts of sizes, enzymes attached to them, and composition. The search for biologically active, "purified" microsomes—if such things could be assumed to exist at all—advanced to one of *the* major issues in the development of protein synthesis systems based on cell extracts of animals, mostly rats, and plants.

Solubilization

For purification, Paul Zamecnik and his colleague John Littlefield at the Massachusetts General Hospital came to rely on the detergent sodium deoxycholate.[45] The detergent solubilized the protein-lipid aggregates of the microsomal fraction. Upon treatment with solubilizer, Littlefield was able to sediment small particles at high speed (105000 × g). Yet what he recovered from the detergent-insoluble sediment in terms of RNA-rich "ribonucleoprotein," in its RNA/protein composition, largely depended upon the concentration of the solubilizer. Since the solubilization procedure resulted

---

44. Philip Siekevitz, "Uptake of Radioactive Alanine in Vitro into the Proteins of Rat Liver Fractions," *Journal of Biological Chemistry* 195 (1952), 549–65; Elizabeth B. Keller and Paul C. Zamecnik, "Anaerobic Incorporation of C¹⁴-Amino Acids into Protein in Cell-Free Liver Preparations," *Federation Proceedings* 13 (1954): 239–40; see also Hans-Jörg Rheinberger, "Experiment and Orientation: Early Systems of in Vitro Protein Synthesis," *Journal of the History of Biology* 26 (1993): 443–71.

45. John W. Littlefield, Elizabeth B. Keller, Jerome Gross, and Paul C. Zamecnik, "Studies on Cytoplasmic Ribonucleoprotein Particles from the Liver of the Rat," *Journal of Biological Chemistry* 217 (1955): 111–23.

in abolishing all subsequent incorporation activity in the test tube, there was no functional correlate to the preparative definition. In this situation, alternative criteria were introduced in order to derive a "robust" particle by a new round of triangulation. It involved a steadily increasing circle of scientists comprising virologists, cytologists, biochemists, biophysicists, and cancer researchers.

Electron Microscopy

One of these representations operated on size and shape. With it, the search for microsomal function merged with another line of research: that of comparative in situ and in vitro studies of the cellular ultrastructure through electron microscopy. The seminal work of Albert Claude and Ernest Fullam at the Rockefeller Institute on mitochondria can be taken as its starting point.[46] Through a series of further investigations driven by new specimen-embedding techniques,[47] and of microtomes making it possible to cut sections as thin as 20 to 50 nanometers,[48] Claude's former coworker Keith Porter introduced, in 1953, a new cytoplasmic structure that he called the cell's "endoplasmic reticulum."[49] Two years later, Porter's colleague George Palade, by using an ensemble of advanced specimen preparation techniques, was able to visualize small, electron-dense particles on the surface of the endoplasmic reticulum in situ.[50] Philip Siekevitz, who had achieved the first fractionated, cell-free protein synthesis system in Paul Zamecnik's laboratory, joined Palade in 1954. He added his biochemical expertise to the work at the Rockefeller Institute that aimed at a correlation of the "cytochemical concepts" of microsomal particles with "morphological concepts" derived from electron microscopy.[51] In the course of these stud-

46. Albert Claude and Ernest F. Fullam, "An Electron Microscope Study of Isolated Mitochondria: Method and Preliminary Results," *Journal of Experimental Medicine* 81 (1945): 51–61. For a historical assessment, see Nicolas Rasmussen, "Mitochondrial Structure and the Practice of Cell Biology in the 1950s," *Journal of the History of Biology* 28 (1995): 381–429.

47. Keith R. Porter, "Observations on a Submicroscopic Basophilic Component of Cytoplasm," *Journal of Experimental Medicine* 97 (1953): 727–49; George E. Palade and Keith R. Porter, "Studies on the Endoplasmic Reticulum. I. Its Identification in Cells in Situ," *Journal of Experimental Medicine* 100 (1954): 641–56.

48. Keith R. Porter and Joseph Blum, "A Study in Microtomy for Electron Microscopy," *Anatomical Records* 117 (1953): 685–710.

49. Keith R. Porter, "Observations on a Submicroscopic Basophilic Component of Cytoplasm," *Journal of Experimental Medicine* 97 (1953): 727–49.

50. George E. Palade, "A Small Particulate Component of the Cytoplasm," *Journal of Biophysical and Biochemical Cytology* 1 (1955): 59–68.

51. George E. Palade and Philip Siekevitz, "Liver Microsomes: An Integrated Morphological and Biochemical Study," *Journal of Biophysical and Biochemical Cytology* 2 (1956): 171–200, at 171–72.

ies, the microsomal fraction derived from centrifugation studies came to be identified with fragments of the endoplasmic reticulum to which small, electron-dense particles were attached. The visualization of such particles within the cell produced a kind of representational resonance that can be seen as characteristic of what scientists, in shaping their epistemic objects, call "independent evidence" in a process of "artifact containment."

As a consequence of the distinction between an endoplasmic reticulum and small, dense particles attached to it, attempts were undertaken to separate these particles from the remainder of the crude fraction in terms of "pure particles." A number of laboratories obtained postmicrosomal fractions by various treatments, such as washing with sucrose solutions, "aging" at various temperatures, incubation in the presence of versene, and by the aforementioned deoxycholate treatment. In turn, the deoxycholate treatment led to a change of the particles' cytochemical representation: For being now composed roughly half of RNA and half of protein, they became referred to as "ribonucleoprotein (RNP) particles." These structures became the emblem and exemplar of cytoplasmic RNA, although the supernatant fluid left after pelleting down the microsomes invariably also contained RNA—approximately 10 percent of the cell's total RNA, about which nobody cared at that time.[52]

In contrast to the rough microsome fraction, which contained chunks of irregularly shaped granular material, the washed particles appeared relatively homogeneous when inspected without further treatment under the electron microscope.[53] Yet the use of electron microscopy brought with it serious operational problems connected to the vicissitudes of specimen preparation. Due to differences in preparation, Littlefield's particles measured between 19 and 33 nanometers—which in itself was a quite considerable variation—whereas Palade's osmium-treated particles were only 10 to 15 nanometers in diameter.[54] The problem of size thus could not be solved within the representational and, quite obviously, interventional space of the electron microscope alone.

Velocity Sedimentation

Besides electron microscopy, the characterization of these "macromolecules" involved electrophoretic mobility and velocity sedimentation. The latter technique of representation brought into play separation patterns of particles over time as well as sedimentation coefficients calculated from analytical ul-

---

52. Palade and Siekevitz, "Liver Microsomes."
53. Littlefield et al., "Studies on Cytoplasmic Ribonucleoprotein Particles."
54. Palade, "A Small Particulate Component."

tracentrifugation. Mary Petermann and her coworkers at the Sloan Ketter-
ing Institute in New York did pioneering work in velocity sedimentation of
"macromolecules" derived from malignant tissue between 1952 and 1954.[55]
Littlefield's particles appeared as a major peak with a velocity constant (S) of
47 in the optical record. This peak was similar to the main macromolecular
component already described by Petermann and her colleagues for rat liver. A
broader peak running in front of the 47S particle disappeared after treatment
of the material with deoxycholate. But there was also a smaller peak trailing
behind the 47S particle, which did not disappear under the same conditions.
Was the ribonucleoprotein portion of the microsomal fraction again then intrinsically heterogenous? Once more, the question could not be answered
within the framework of this representational technique alone.

For none of the representations was there a conceivable, ready-made referent at hand concerning the shape and composition of the scientific object
under preparation. It gradually took form from correlating and superpos-
ing representations derived from a wide variety of different biophysical,
biochemical, and biological techniques. But since the material was no
longer active in the test tube after the different isolation procedures, there
was no functional point of reference for comparison either. The experimental representations partly matched, partly they interfered with and thus
eliminated each other. The deoxycholate particle entered the field of test
tube protein synthesis around 1953. It occupied the stage for three years,
then it became obsolescent and disappeared from the experimental record
because no way could be ventured to render it functionally active.

Within this context of epistemic transformations, preparation procedures played an eminent role, and the corresponding terminology faith-
fully reflected the operational character of the resulting entities. The
different means and modes of representation came to mutually interact:
choice of material, instruments of inspection, physical separation, biochemical dissection. Following a lingering trajectory, these representations
eventually led to concepts that could be linked to either subcellular mor-

55. Mary L. Petermann and Mary G. Hamilton, "An Ultracentrifugal Analysis of the
Macromolecular Particle from Normal and Leukemic Mouse Spleen," *Cancer Research* 12
(1952): 373–78; Mary L. Petermann, Nancy A. Mizen, and Mary G. Hamilton, "The Macromolecular Particles of Normal and Regenerating Rat Liver," *Cancer Research* 13 (1953), 372–
75; Mary L. Petermann, Mary G. Hamilton, and Nancy A. Mizen, "Electrophoretic Analysis of
the Macromolecular Nucleoprotein Particles of Mammalian Cytoplasm," *Cancer Research* 14
(1954): 360–66; Mary L. Petermann and Mary G. Hamilton, "A Stabilizing Factor for Cytoplasmic Nucleoproteins," *Journal of Biophysical and Biochemical Cytology* 1 (1955): 469–72.
For bacterial extracts, see also Howard K. Schachman, Arthur B. Pardee, and Roger Y. Stanier,
"Studies on the Macromolecular Organization of Microbial Cells," *Archives of Biochemistry
and Biophysics* 38 (1952): 245–60.

phology or biological function, but the representations did not necessarily merge. For instance, in the intact cell, an in situ distinction could be made between membrane-bound and free small particles. None of the research groups involved, however, was able to prepare a cell homogenate that would retain this distinction. All centrifugation procedures yielded an inseparable mixture of free and vesicle-bound particles. This was especially disappointing for all in vitro workers, since the distinction had led to far-reaching speculations about the differential function of these two sorts of granules: the membrane-bound particles were supposed to be related to tissue-specific protein production, whereas the free particles were thought to maintain the general, housekeeping protein metabolism.[56]

Ribosomes as Templates

Between 1940 and 1955, the small cytoplasmic particles had changed from a sedimentable mitochondrial entity no longer visible in the conventional microscope, to a microsome with its presumed function as a plasmagene, to a morphogenetic unit, to an operationally defined granular cytoplasmic constituent that was deoxycholate-insoluble or NaCl-insoluble, and finally to a ribonucleoprotein particle consisting half of protein and half of RNA, readily visible under the electron microscope, and topologically connected to peptide bond formation. Gradually, following Brachet's early assumptions,[57] the RNA moiety of the particles had attracted more and more attention, and around 1955, it became generally seen as providing the "template" upon which the amino acids assembled to protein threads.

In 1958, Howard Dintzis coined the term "ribosome" for purified microsomes virtually devoid of membrane fragments.[58] During the following years, this neologism, disseminated by Richard Roberts,[59] made its way into the laboratories and into the literature. Although the biological reason for changing the name was not quite obvious, the new designation clearly reflected, in addition to preparation routines, a physiological function. The "ribosome" began to subvert the biochemically characterized protein syn-

56. Keller and Zamecnik, "Anaerobic Incorporation of C^{14}-Amino Acids"; Littlefield et al., "Studies on Cytoplasmic Ribonucleoprotein Particles."

57. At the same time and independently from Brachet, Torbjörn Caspersson from Stockholm had come to similar conclusions. Torbjörn Caspersson, "Studien über den Eiweißumsatz der Zelle," *Naturwissenschaften* 29 (1941): 33–43.

58. Howard Dintzis to Wim Möller, 22 August 1989; Wim Möller, personal communication, 22 May 1992.

59. Richard B. Roberts, introduction to *Microsomal Particles and Protein Synthesis*, ed. Richard B. Roberts (New York: Pergamon Press, 1958), vii–viii; Richard B. Roberts, "Ribosomes. A. General Properties of Ribosomes," in *Studies of Macromolecular Biosynthesis*, ed. Richard B. Roberts (Washington: Carnegie Institution, 1964), 147–68, at 148.

thesis systems as a part of what Francis Crick had baptized the "central *dogma*" of molecular biology[60]—the notion that the genetic information makes its way from DNA to RNA to protein and that, once in the protein, it cannot get back to DNA. The ribosome became the synonym for an RNA intermediate in the overarching process of gene expression. With that, it entered the realm of molecular biology.

THE "TWO-SUBUNIT" PICTURE OF THE RIBOSOME

With respect to their physical parameters, the protein-synthesizing particles continued to change their appearance. After having been successfully homogenized and freed from the endoplasmic reticulum, they split up again. Around 1956 and after many trials, Fu-Chuan Chao and Howard Schachman from Wendell Stanley's virus laboratory at Berkeley found that yeast microsomes sedimented with a velocity constant of 80 and dissociated into two unequal components of 60S and 40S.[61] A year or so later, Petermann and their coworkers took apart 78S liver ribosomes into 62S and 46S particles.[62] At the same time, Alfred Tissières and James Watson, at Harvard, had started to work with *Escherichia coli* ribosomes and had their particles sediment with 70S. They were able to dissociate them reversibly into a 50S and a 30S portion.[63] Gradually, in the course of years of painstaking isolation attempts, in which such comparatively simple procedures as sucrose gradient centrifugation came to occupy center stage, the longstanding confusion about the size of the RNP particles cleared up, and people realized that the secret of stabilization chiefly resided in the concentration of magnesium ions. Results with a variety of particles from other sources began to point to two distinguishing features: bacterial particles (roughly 70S) were consistently smaller than their eukaryotic counterparts (roughly 80S), but both could be separated into something that began to be recognized as a small and a large subunit.

60. Francis H. C. Crick, "On Protein Synthesis," *Symposia of the Society for Experimental Biology London* 12 (1958): 138–63, at 153.

61. Fu-Chuan Chao and Howard K. Schachman, "The Isolation and Characterization of a Macromolecular Ribonucleoprotein from Yeast," *Archives of Biochemistry and Biophysics* 61 (1956): 220–30.

62. Mary L. Petermann, Mary G. Hamilton, M. Earl Balis, Kumud Samarth, and Pauline Pecora, "Physicochemical and Metabolic Studies on Rat Liver Nucleoprotein," in *Microsomal Particles and Protein Synthesis*, ed. R. B. Roberts (London: Pergamon Press, 1958), 70–75.

63. Alfred Tissières and James D. Watson, "Ribonucleoprotein Particles from Escherichia coli," *Nature* 182 (1958): 778–80; Alfred Tissières, James D. Watson, David Schlessinger, and B. R. Hollingworth, "Ribonucleoprotein Particles from Escherichia coli," *Journal of Molecular Biology* 1 (1959): 221–33.

FROM EUKARYOTES TO BACTERIA, FROM BIOCHEMISTRY TO MOLECULAR BIOLOGY

The concept of microsomes containing a stable template that had become widely accepted toward the end of the 1950s had emerged from systems based on animal cells: a heritage of cancer research in which the vast majority of the protein synthesis work between 1945 and 1955 was embedded. This concept was clearly incompatible with the observations on the association of an unstable RNA with bacterial protein synthesis[64] and the synthesis of phage.[65] Basically, however, none of those working on cells from higher organisms knew what to do with these findings. For them, the ribosome represented "a stable factory already containing an RNA transcript of DNA."[66] Implicit in this vision was a kind of "one microsome—one enzyme" hypothesis, meaning that a particular ribosome was engaged in the fabrication of a particular, specific protein. Moreover, bacterial in vitro systems were considered unreliable in the leading circles of protein synthesis workers. They saw them as metabolically uncontrollable systems in which virtually everything was possible.[67]

This situation was bound to change quite abruptly with the emergence in different, but all-bacterial, experimental contexts, of a short-lived RNA, distinct from ribosomal RNA and crucial in the synthesis of proteins. It came to be known as "messenger RNA" (mRNA). Such molecules were characterized by Jacques Monod and François Jacob at the Pasteur Institute in Paris in the course of their genetic studies of enzyme induction in *E. coli*,[68] by François Gros and other coworkers of Watson at Harvard studying the RNA turnover of *E. coli* cells,[69] and by Heinrich Matthaei and Marshall Nirenberg at the National Institutes of Health, who used *E. coli* extracts and a synthetic ribonucleic acid to pin down the first word of the ge-

64. Ernest F. Gale and Joan Folkes, "The Assimilation of Amino Acids by Bacteria. 21. The Effect of Nucleic Acids on the Development of Certain Enzymic Activities in Disrupted Staphylococcal Cells," *Biochemical Journal* 59 (1955): 675–84.

65. Lazarus Astrachan and Elliot Volkin, "Properties of Ribonucleic Acid Turnover in T2-Infected Escherichia coli," *Biochimica et Biophysica Acta* 29 (1958): 536–44.

66. Mahlon Hoagland, *Toward the Habit of Truth* (New York: W. W. Norton, 1990), 107.

67. Robert B. Loftfield, "The Biosynthesis of Protein," *Progress in Biophysics and Biophysical Chemistry* 8 (1957): 348–86, at 375–77.

68. François Jacob and Jacques Monod, "Genetic Regulatory Mechanisms in the Synthesis of Proteins," *Journal of Molecular Biology* 3 (1961): 316–56; Sidney Brenner, François Jacob, and Matthew Meselson, "An Unstable Intermediate Carrying Information from Genes to Ribosomes for Protein Synthesis," *Nature* 190 (1961): 576–81.

69. François Gros, H. Hiatt, Walter Gilbert, Chuck G. Kurland, R. W. Risebrough, and James D. Watson, "Unstable Ribonucleic Acid Revealed by Pulse Labeling of E. coli," *Nature* 190 (1961): 581–85.

netic code.[70] This work went along with the establishment of a reliable test tube protein synthesis system based on bacterial cell homogenates.[71]

In the course of this work, the ribosome again changed its identity. It mutated from a template for protein synthesis to a machinery for reading the codewords of gene-derived messenger RNAs and for translating them into peptides. Its former template RNA now took on the role of a structural scaffold holding together a giant multiprotein enzyme involved in the fabric of specific peptide bonds. This picture remained unchallenged for the next twenty years. (It became fluid again with the realization, at the beginning of the 1980s, that ribonucleic acids can act as enzymes.[72] The ensuing suspicion that ribosomal RNA might be involved enzymatically in peptide bond formation changed, once again, the picture of the ribosomal particle.)

With the beginning of a deliberate manipulation of viral and especially synthetic messenger RNAs, the stage was set for a molecular dissection of ribosomal function.[73] Most of the features of the initiation,[74] the repeated cycle of peptide bond formation (elongation),[75] and the termination of protein synthesis[76] were outlined in more and more sophisticated and reduced

70. Marshall W. Nirenberg and J. Heinrich Matthaei, "The Dependence of Cell-Free Protein Synthesis in E. coli upon Naturally Occurring or Synthetic Polyribonucleotides," *Proceedings of the National Academy of Sciences of the United States of America* 47 (1961): 1588–1602.

71. Marvin R. Lamborg and Paul C. Zamecnik, "Amino Acid Incorporation into Protein by Extracts of E. coli," *Biochimica et Biophysica Acta* 42 (1960): 206–11; J. Heinrich Matthaei and Marshall W. Nirenberg, "The Dependence of Cell-Free Protein Synthesis in E. coli upon RNA Prepared from Ribosomes," *Biochemical and Biophysical Research Communications* 4 (1961): 404–8; Alfred Tissières, David Schlessinger, and François Gros, "Amino Acid Incorporation into Proteins by E. coli Ribosomes," *Proceedings of the National Academy of Sciences of the United States of America* 46 (1960): 1450–63.

72. Kelly Kruger, Paula J. Grabowski, Arthur J. Zaug, Julie Sands, Daniel E. Gottschling, and Thomas R. Cech, "Self-splicing RNA: Autoexcision and Autocyclization of the Ribosomal RNA Intervening Sequence of Tetrahymena," *Cell* 31 (1982): 147–57.

73. James D. Watson, "Involvement of RNA in the Synthesis of Proteins," *Science* 140 (1963): 17–26; Fritz Lipmann, "Messenger Ribonucleic Acid," *Progress in Nucleic Acid Research* 1 (1963): 135–61.

74. George Brawerman, "Role of Initiation Factors in the Translation of Messenger RNA," *Cold Spring Harbor Symposia on Quantitative Biology* 34 (1969): 307–12.

75. Sidney Pestka, "Translocation, Aminoacyl-oligonucleotides, and Antibiotic Action," *Cold Spring Harbor Symposia on Quantitative Biology* 34 (1969): 395–410; Philip Leder, Alberto Bernardi, David Livingston, Barbara Loyd, Donald Roufa, and Lawrence Skogerson, "Protein Biosynthesis: Studies Using Synthetic and Viral mRNAs," *Cold Spring Harbor Symposia on Quantitative Biology* 34 (1969): 411–17; Herbert Weissbach, Nathan Brot, D. Miller, M. Rosman, and R. Ertel, "Interaction of Guanosine Triphosphate with E. coli Soluble Transfer Factors," *Cold Spring Harbor Symposia on Quantitative Biology* 34 (1969): 419–31.

76. Thomas Caskey, Edward Scolnick, Richard Tompkins, Joseph Goldstein, and Gregory Milman, "Peptide Chain Termination, Codon, Protein Factor, and Ribosomal Requirements," *Cold Spring Harbor Symposia on Quantitative Biology* 34 (1969): 479–91.

partial in vitro systems based on extracts of E. coli and operating on the synthetic messenger RNA polyuridylic acid or variants thereof.[77] This work went along with the deciphering of the genetic lexicon. In this process, the ribosomal particles assumed both the role as an *object* of research and as a *tool* for the elucidation of another epistemic object: the code.

In the context of pursuing ribosomal function, and after the mRNA concept had been established, gentle isolation of messenger-ribosome complexes had become a matter of priority in the early 1960s. Larger particles appeared on sucrose gradient patterns as well as on electron-microscopic images. They were variously termed "ribosomal clusters,"[78] "active complexes,"[79] "ergosomes,"[80] or "aggregated ribosomes,"[81] before the term "polysomes" was coined[82] and came into general use. Polysomes appeared to consist of strings of ribosomes translating a particular messenger RNA. Special isolation procedures were required to prevent them from breaking down during fractionation. Once again, the stabilized picture of the two-subunit single ribosome turned out to be an abstraction that corresponded to a state of function in the test tube rather than to a state of function in the cell.

Once the bacterial ribosome had been firmly inserted into the functional network of protein synthesis, which in turn had become inserted into the overarching process of gene expression and thus into the context of molecular biology, it gained additional epistemic significance as a model object for the study of the molecular interactions between proteins and ribonucleic acids, and as a first target for the ultimate dream of reduction in molecular biology—the atomic resolution of a cellular organelle: the mapping of its genetic makeup, the primary, secondary, and tertiary structure of its components, its quaternary shape in space, and its reconstitution from the components in the test tube. To tell the story of this endeavor would take

77. For a recent survey see Alexander Spirin, "Ribosome Preparation and Cell-Free Protein Synthesis," in *The Ribosome. Structure, Function, and Evolution*, ed. Walter E. Hill, Peter B. Moore, Albert Dahlberg, David Schlessinger, Roger A. Garrett, and Jonathan R. Warner (Washington: American Society for Microbiology, 1990), 56–70.

78. Jonathan R. Warner, Alexander Rich, and Cecil E. Hall, "Electron Microscope Studies of Ribosomal Clusters Synthesizing Hemoglobin," *Science* 138 (1962): 1399–1403.

79. Walter Gilbert, "Polypeptide Synthesis in E. coli. I. Ribosomes and the Active Complex," *Journal of Molecular Biology* 6 (1963): 374–88.

80. F. O. Wettstein, Theophil Staehelin, and Hans Noll, "Ribosomal Aggregate Engaged in Protein Synthesis: Characterization of the Ergosome," *Nature* 197 (1963): 430–35.

81. Alfred Gierer, "Function of Aggregated Reticulocyte Ribosomes in Protein Synthesis," *Journal of Molecular Biology* 6 (1963): 148–57.

82. Jonathan R. Warner, Paul M. Knopf, and Alexander Rich, "A Multiple Ribosomal Structure in Protein Synthesis," *Proceedings of the National Academy of Sciences of the United States of America* 49 (1963): 122–29.

another paper. It would have to tell the tale, not of a few laboratories, but of a thirty-year effort of a whole scientific community around the world, the "ribosomologists," who turned the power of neutron beams, X rays, advanced electron microscopes, and the whole arsenal of sequencing and genetic engineering techniques on their particles, and who still argue, in terms of a molecular, mechanistic picture, about how precisely their object of desire performs its central task—the making of a peptide bond.

CONCLUSION: A HISTORY FROM EPISTEMIC THINGS

I come back to my remarks on scientific objects, experimental systems, and model organisms. Scientific objects, such as the cytoplasmic particles described in this paper, are embedded, both synchronically and diachronically, in different experimental systems bounded by different instruments and tracing devices, and these systems in turn can both be derived from and rely on different model organisms. In these contexts, they get shaped and reshaped, and take on different meanings. It is these contexts that channel the emergence, the persistence, and the obsolescence of scientific objects. By looking at the experimental networks in and through which scientific objects move and that they constitute at the same time through their movement, we arrive at a view of material, experimental cultures and at a feeling for the nature of the respective epistemic practices quite different from the traditional view through the lens of disciplines. The cytoplasmic particles I have been describing *grosso modo* belong neither to virology, nor to cytology, biochemistry, microbiology, or molecular biology alone—their trajectory traverses all of them. As boundary objects in a sense similar to that in which Ilana Löwy uses the notion,[83] they constitute trajectories that define a space quite different from and extending beneath the disciplinary coordinates that make academic institutions work: the space of laboratory cultures, their equipment,[84] their experimental systems,[85] their organisms.[86] We have only started to explore this space.[87] There appears still a long way to go from telling the history of ideas, of scientists, of disciplines, of institu-

83. Ilana Löwy, "The Strength of Loose Concepts: Boundary Concepts, Federative Experimental Strategies and Disciplinary Growth: The Case of Immunology," *History of Science* 30 (1992): 371–95.

84. Nicolas Rasmussen, *Picture Control: The Electron Microscope and the Transformation of American Biology, 1940–1959* (Stanford: Stanford University Press, 1997).

85. Rheinberger, *Toward a History of Epistemic Things.*

86. Robert E. Kohler, *Lords of the Fly: Drosophila Genetics and the Experimental Life* (Chicago: University of Chicago Press, 1994).

87. Peter Galison, *Image and Logic: A Material Culture of Microphysics* (Chicago: University of Chicago Press, 1997).

tions, to telling a history from epistemic things. If the process of gaining experimental knowledge is to be understood as a discourse that has shaped the modern sciences, and whose special relation to the real remains an issue, then it is worth trying to understand its "objectivity" in terms of the peculiar "objecticity" it confers on its objects. I would like to come back to Michael Polanyi, whose contributions to science studies still await their proper due: "[This] capacity of a thing to reveal itself in unexpected ways in the future, I attribute to the fact that the thing observed is an aspect of a reality, possessing a significance that is not exhausted by our conception of any single aspect of it. To trust that a thing we know is real is, in this sense, to feel that it has the independence and power for manifesting itself in yet unthought of ways in the future."[88]

The new is the result of spatiotemporal singularities. Experimental systems are precisely the arrangements that allow cognitive spatiotemporal singularities to emerge. The generic reality of epistemic things is their capacity to give rise to unprecedented events. The study of scientific objects within their experimental systems should convince us that these systems are "machines for making the future."[89] Which means, to conclude with Jacques Derrida, that they are elements of a "differential typology of forms of iteration"[90] that still awaits elaboration.

88. Polanyi, Duke Lectures; quoted in Grene, *The Knower and the Known*, 219.
89. François Jacob, *La statue intérieure* (Paris: Editions Odile Jacob, 1987), 13.
90. Jacques Derrida, "Signature événement contexte," in *Marges de la philosophie* (Paris: Editions de Minuit, 1972), 365–93, at 389.

CONTRIBUTORS

JED Z. BUCHWALD
Dibner Institute for the History of Science and Technology
Massachusetts Institute of Technology
Room E56-100
38 Memorial Drive
Cambridge, Massachusetts 02139

LORRAINE DASTON
Max Planck Institute for the History of Science
Wilhelmstrasse 44
10117 Berlin
Germany

RIVKA FELDHAY
The Cohn Institute for the History and Philosophy of Science and Ideas
Tel Aviv University
IL-69978 Tel Aviv
Israel

JAN GOLDSTEIN
Department of History
University of Chicago
1128 East 59th Street
Chicago, Illinois 60637

GÉRARD JORLAND
Centre de Recherches Historiques
E.H.E.S.S.
54, Boulevard Raspail
F-75006 Paris
France

DORIS KAUFMANN
Max Planck Institute for the History of Science
Wilhelmstrasse 44
10117 Berlin
Germany

BRUNO LATOUR
Centre de Sociologie de l'Innovation
Ecole Nationale Superieure des Mines
60, Boulevard Saint-Michel
F-75006 Paris, Cedex 06
France

THEODORE M. PORTER
Department of History
University of California, Los Angeles, Box 951473
6265 Bunche Hall
Los Angeles, California 90095

HANS-JÖRG RHEINBERGER
Max Planck Institute for the History of Science
Wilhelmstrasse 44
10117 Berlin
Germany

MARSHALL SAHLINS
Department of Anthropology
University of Chicago
1126 East 59th Street
Chicago, Illinois 60637

PETER WAGNER
Department of Political and Social Sciences
European University Institute
Badia Fiesolana
Via Dei Roccettini 9
I-50016 San Domenico di Fiesole
Italy

INDEX

Abegg, Johann, 72

adverse selection, 227–30, 233, 237

aesthetics: Jouffroy's theory, 111–12, 114; utility and, 123n. 11

African culture, 186–88, 191, 199–200

Agrippa, Cornelius, 24n. 34

Albert of Saxony, 46, 47

alchemy, 37

algebra, 42, 43, 54, 56

Allais, Maurice, 129

âme: vs. *moi,* 86, 91, 91n. 11, 97–98, 99, 109, 113

amino acid tracing, 274, 283–84

Anderson, Benedict, 149

Angebert, Caroline, 114n. 68

anomalies: in economic theory, 131; in microsome research, 12. *See also* preternatural philosophy

anomalous dispersion, 212n. 6, 213, 219, 221

Ansell, Charles, 228

anthropology. *See* culture

applied metaphysics, 1–2, 3

Archimedean mechanics, 47, 49, 50, 52–53, 64, 65

Aristotle: applied metaphysics and, 1, 3, 5, 13; on chance, 18; distinction between philosophy and history, 17n. 4; economics of, 118–20, 122, 123; on falling bodies, 45–47, 49, 50, 51, 52, 64–65; mathematical entities and, 55; moral-political order and, 135, 148; on regularities, 15–16, 17, 20, 27, 28–29; on wonder, 28

arithmos, 42, 55

Arrow, Kenneth, 131

the artificial: vs. the natural, 20, 24

astrology, 37

atoms: existence of, 205; Kelvin's vortices, 218; STM images of, 203–5, 206–7

Avicenna, 18, 25

Bachelard, Gaston, 2, 117, 272

Bacon, Francis: on anomalies, 16, 18, 20, 21, 29, 33; on causes, 26, 27, 28, 29; on natural magic, 23, 24n. 34; on wonder, 31

Baker, Keith Michael, 135n. 5

Banez, Dominic, 63

Baradère, Germain, 110n. 61

Béguin, Albert, 67

Benjamin, Walter, 167

Bernoulli, Daniel, 127

Binswanger, Ludwig, 67–68

black-boxing, 257, 262, 273. *See also* grey boxes

Blancanus, Josephus, 60–61, 64

blood pressure: life insurance and, 8, 236, 241–42, 245–46

Boas, Franz, 165, 166, 167

Bonnett, Earl C., 244

Borsook, Henry, 283

bourgeoisie: *bürgerliche* identity, 7, 68–69, 84; German Counter-Enlightenment and, 163, 164, 165; the *moi* and, 92, 107, 110–11, 111n. 64, 113, 114–16; postcolonial, 199. *See also* class

Boyle, Robert, 34, 37–38, 39, 40

Brachet, Jean, 278–81, 282–83, 288

Brettel, Caroline, 188n. 80

Browne, Thomas, 23, 30, 37

Bruner, Edward, 186, 186n. 73

Buchwald, Jed Z., 8, 11

bürgerliche identity, 7, 68–69, 84
Buridan, 47
Butler, Joseph, 96

Cabanis, Pierre-Jean-Georges, 80–81
Cabral, Amilcar, 199
Canguilhem, Georges, 275
capitalism: cultural self-consciousness and, 197, 199, 202; culture concept and, 159, 160, 160n. 4, 161, 162, 164, 168; local cultures' responses to, 170, 172, 193, 197
Cardano, Girolamo: vs. natural philosophers, 34; on the preternatural, 18, 20, 26, 27, 28, 29; on wonder, 29–30, 31, 32
Carus, Carl Gustav, 82
Carus, Friedrich August, 82–83
Casaubon, Meric, 30
Castle, Terry, 108n. 57
Cauchy, Augustin-Louis, 212n. 6, 216–17, 218, 219
causal explanations: in natural philosophy, 34–35, 37, 41; in preternatural philosophy, 17, 18, 20, 24–26, 28, 29, 34–35; social-deterministic, 138, 144–45, 146, 156
center of gravity, 9, 45, 47–49, 50–53, 56, 64
chance: rare phenomena and, 18, 20, 21, 23, 24
Chantrenne, Hubert, 280, 282, 284
Chao, Fu-Chuan, 289
Charleton, Walter, 33
chemistry: nineteenth-century schemata in, 210n. 5, 210–11; Pasteur's ferment and, 258–59
Christiansen, Christian, 219
"civilization": vs. "culture," 158, 163, 166–67
class: Cousinian *moi* and, 7, 114–16; culture and, 161, 162, 201; labor theory of value and, 128; "society" and, 137. *See also* bourgeoisie
Claude, Albert, 276–78, 279, 280–82, 283, 285
Clavius, Christopher, 45, 60, 62, 63, 65
Clifford, James, 167
Cohn, Waldo, 270
colonialism: cultural self-consciousness and, 197–98, 199, 201, 202; culture concept and, 160, 160n. 3, 161, 162–63, 168; translocal societies and, 187, 190, 190n. 82. *See also* imperialism

Commandino, Federico, 48, 50, 50n. 12
community, 143, 152. *See also* culture; society
computing technology: life insurance and, 244
Comte, Auguste, 36, 132
Condillac, Étienne Bonnot de, 93, 95–96; Cousin on, 99, 101, 103, 108, 109; curriculum associated with, 110, 110n. 61; on economics, 121, 122; on memory, 95, 95n. 17; on the *moi*, 95, 95n. 18, 98–99, 99n. 29, 103, 109, 110, 110n. 61
Condorcet's paradox, 131
consciousness. *See* self
constructionism. *See* social constructionism
contract theories: vs. Herder's culture concept, 165
conventionalism, 14
Copernicans: Guldin vs., 9, 49, 52, 64, 66
Coste, Pierre, 99–100, 100n. 31
Counter-Enlightenment, Germanic, 163n. 12, 163–65, 167
Cournot, Antoine-Augustin, 122, 127, 128, 131
Cousin, Victor: on Descartes, 91n. 11; gender bias of, 114, 114n. 68; institutionalization of the *moi* by, 92, 109–11, 111n. 63, 112–14, 113n. 67; as lecturer, 102n. 36, 102–3, 112–13; *moi*-centered philosophy of, 7, 92–93, 101–7; moral philosophy of, 101, 113, 115; on Pascal, 100n. 32; politics of, 101–2, 106, 115–16; on private property, 7, 106–7, 116; psychological method of, 104–5, 110–11, 115; vs. sensationalism, 93, 99, 101, 102, 103–4, 108–9; "sentiment of personality" and, 111, 112, 114; tripartite division of consciousness, 104–6, 105n. 46, 108. *See also* eclectic philosophy; *moi*
Cowley, Abraham, 36
Crick, Francis, 270, 289
cultural enhancement, 169–70
culturalism, 192–202; agency of indigenous peoples in, 193–94; capitalism and, 197, 199, 202; colonialism and, 197–98, 199, 201, 202; defined, 192; diasporas and, 192–93; global culture and, 201–2; of Kayapo, 194–96; of Maya, 201; Mixtec, 201n. 111; of Wánai, 196n. 99; of westernized persons, 198–200
cultural salience, 6–7

culture, 158–202; in American anthropology, 165–66, 166n. 17; in British anthropology, 166, 186–87; broad usage of, 159n. 2; cannot disappear, 158, 171; vs. "civilization," 158, 163, 166–67; defined, 158, 193; difference and, 160, 160n. 4, 161n. 5, 161n. 7, 161–63, 163n. 12; disappearance of indigenous cultures and, 159, 167–69; in French anthropology, 166–67; in German anthropology, 167; Herder's concept of, 162, 162n. 10, 163n. 12, 164, 165; imperialism and, 159, 162, 163, 168, 184, 197; vs. nature, 3; new global organization and, 201–2; racism and, 159, 160, 160n. 4, 162; self-consciousness about (*see* culturalism); "society" and, 141, 143, 150, 150n. 30, 151, 155; transformations of, 10, 169–71, 174n. 38; urbanization and, 185–89, 186n. 73, 187n. 78, 191n. 84, 191n. 86, 191–92. *See also* capitalism; colonialism; translocal society
Cuvier, Georges, 4–5
cytoplasmic inheritance, 278, 283, 288
cytoplasmic particles, 270–94; amino acid tracing and, 274, 283–84; cancer research and, 271, 276, 290; cross-disciplinary character of, 12, 271, 273, 293; emergence as epistemic object, 276–78; historical character of, 270–73, 275–76, 293–94; methods of preparation and, 270, 272, 282, 287–88; microsomes, 12, 272, 273, 281–86, 288, 289, 290; mitochondria, 278, 281–82, 285, 288; polysomes, 292; representations of, 271, 273–75, 276, 277–78, 287; ribosomes, 272, 274, 275, 288–93; in test tube protein synthesis, 291; ultracentrifuge experiments, 276–81, 286–88, 289

Daston, Lorraine, 8–9
Davidson, Wolf, 79
Davis, Audrey, 241
death: as scientific object, 7–8. *See also* life insurance
Della Porta, 32
demand curves, 122, 127, 128–29. *See also* supply and demand
democracy: Cousin's rejection of, 102, 114–16; Tocqueville on, 137

demographic research, 142, 143
demons, 20–21, 23–24, 25
Derrida, Jacques, 294
Descartes, René: algebra of, 43–44; dualism of, 35; Foucault on, 90n. 10; mechanical philosophy of, 17, 34, 36; the *moi* and, 86, 91, 91n. 11, 93; preternatural philosophy and, 24, 27; on wonder, 31
Destutt de Tracy, Antoine-Louis-Claude, 107–8
determinism: social, 138, 144–45, 146, 156. *See also* causal explanations; free will
developman, 171–76, 174n. 36, 190, 197
Dictionnaire des sciences philosophiques, 86, 87–89, 92, 93, 102
Dintzis, Howard, 288
Dirks, Nicholas, 160n. 3
discovery: vs. invention, 2–5, 13, 267
dispersion of light, 216, 217; anomalous, 212n. 6, 213, 219, 221
DNA (deoxyribonucleic acid), 279, 289. *See also* genetic code
dreams, German Enlightenment discourse on, 67–85; *bürgerliche* identity and, 7, 68–69, 84; dissociation of the self and, 73–75, 77, 81, 83; Freud and, 67, 83–84; moral boundaries and, 7, 71–73, 74n. 27, 79–80, 81; neglected by historiography, 67–68; physical causes and, 75, 80–81, 83–84; public accounts in, 69–70, 71–77, 78, 85; self-knowledge and, 68, 70–71, 73, 77, 80; therapeutic significance and, 81–83; three main approaches in, 78–83
dreams, prophetic, 20
Drude, Paul, 221, 222, 224
Dufay, Charles, 40, 41
Dumont, Louis, 161n. 7
Dupleix, Scipion, 19n. 11, 20
Durkheim, Emile: anthropology and, 166, 175; "society" and, 141, 143, 144n. 18, 145, 147, 155
Dwight, Edwin W., 238

Eakins, O. M., 233, 238
eclectic philosophy, Cousinian, 102, 103, 106n. 50, 109; class bias of, 114–16; in educational system, 110, 111, 112n. 65, 114; gender bias of, 114, 114n. 68; private property and, 107. *See also* Cousin, Victor; *moi*

economics: perfect competition as concept in, 152n. 31; "society" and, 137, 139, 143, 143n. 17, 144, 154, 155. *See also* value theories

electricity: early theories of, 26, 34–35, 40, 41

electrocardiography: life insurance and, 242–44, 245, 246

electromagnetic field, 221, 222

electron microscopy: of cytoplasmic particles, 273, 283, 285–86, 288, 292

electron theory, 222

elementary particles, 209–10, 214–15, 219n. 12

Elkan, Walter, 191

embeddedness of scientific objects, 12–13, 273, 276

Encyclopédie, 86, 87

endoplasmic reticulum, 285, 286, 289

Enlightenment. *See* Counter-Enlightenment, Germanic; dreams, German Enlightenment discourse on; *moi*; natural philosophy

epistemic objects: displacement of, 277; experimental systems and, 276, 293, 294; historical character of, 271, 272; independent evidence and, 286; trajectories of, 273, 287–88, 293–94; unexpected manifestations of, 11–12, 272, 294. *See also* scientific objects

epistemology: historical, 275; neo-Kantian, 13–14; political significance of, 269; in preternatural philosophy, 27–29, 35–36

Erfahrungsseelenkunde, 7, 67, 68, 77; journal of, 69, 71–72, 73–77

Eschenmayer, Carl August, 77

ESP (extrasensory perception): life insurance and, 244

ether: continuous, 218–19, 220; Kelvin on, 214; Lorentz theory, 222n. 16; as particulate lattice, 214–15, 216–17, 218, 220, 220n. 14, 221; wave optics and, 215–17

existence of scientific objects: of atoms, 205; defined as an event, 264; institutions and, 255, 257, 262–63, 267; laboratory practice and, 253–54, 258; nature and, 3, 255, 257, 260, 264, 265, 268; networks of production and, 12–13, 250–51, 258, 261, 262, 263, 266; of numbers, 59–60; paradigms and, 258, 259, 262; partial, 268; performances and, 260, 262; prag-

matism and, 263n. 13; recalcitrance and, 272–73; relative, 13, 251–53, 256–57, 259–60; of "society," 10, 134n. 4, 148, 148n. 25, 152n. 31, 153; spatiotemporal envelopes and, 250–51, 253, 256, 259–63, 265; technical projects and, 251, 262, 262n. 12. *See also* historicity of scientific objects

extrasensory perception (ESP): life insurance and, 244

"fact": semantic transformations of, 4

facts: historicity of, 255, 257, 259, 263; networks of production of, 250, 251, 262; social constructivism and, 263–64, 268

Feldhay, Rivka, 9

Fellows, Haynes Harold, 243

Ferguson, James, 192–93

Ficino, Marsilio, 18, 25

Fisher, John W., 241, 242

FitzGerald, George, 218

Flaubert, Gustave, 88

Fonseca, Pedro da, 59–60

Fontenelle, Bernard de, 38, 39, 40

Foucault, Michel, 90n. 10, 116

Franck, Adolphe, 102

Frank, Robert, 243

Franklin, Benjamin, 35

free will: dreams and, 71, 75; Kant on, 70–71; of natural magicians and demons, 24; salvation and, 61, 63. *See also* causal explanations; social determinism; will

French Revolution: the *moi* and, 101–2; "society" and, 10, 135n. 5, 135–36, 140, 143

Fresnel, Augustin-Jean, 216, 217

Freud, Sigmund, 67, 83–84

Fullam, Ernest, 285

functionalism: in anthropology, 161, 186n. 71; in social theory, 142

Galen, 18

Galileo, 31

gender distinctions: in Cousinian philosophy, 110, 111n. 63, 111–12, 114, 114n. 68, 116; in Mendi exchange system, 174n. 36. *See also* women

gene expression, 289, 292

general magnitudes, 42, 43–44, 52, 53, 54, 59. *See also* mathematical entities

genetic code, 12, 271, 290–91, 292

Gesner, Conrad, 37

Gifford, Edward Winslow, 184
Gluckman, Max, 186–87, 187n. 78
God: miracles and, 20, 32; possible beings and, 57–58
Goldstein, Jan, 6–7
Gratia, André, 279
gravity: center of, 45, 47–49, 50–53, 56, 64; Hooke on, 36
Green, George, 217
Greenblatt, Stephen, 168
Gregory, Chris, 170, 172
grey boxes, 254n. 6. See also black-boxing
Gros, François, 290
Grumet, Robert, 170
Guizot, François, 92n. 12, 101, 111, 111n. 64
Guldin, Paulus, 9, 45–54, 56; Aristotle's physics and, 45–47, 49, 50, 51, 52; center of gravity and, 45, 47–49, 50–53, 56, 64; Jesuit culture and, 56, 64–66; mathematical entities and, 51–54, 66; motion of earth and, 48–50, 51, 52, 53, 56, 64–66
Gulland, Masson, 270
Gupta, Akhil, 192–93

Hart, Keith, 188–89
Hau'ofa, Epeli, 171, 176–81, 184, 189
Hayer, Hubert, 96–97, 97n. 23, 99
Hegel, G. W. F.: Cousin influenced by, 101; history of science and, 117; "society" and, 137, 138, 138n. 10, 143
Heilbron, Johan, 134n. 3
Helmholtz, Hermann, 219, 220, 221, 222–24, 223n. 18
Henriot, Emile, 279
Herder, Johann Gottfried von: "culture" and, 162, 162n. 10, 163n. 12, 164, 165; empirical psychology and, 69
Hertz, Heinrich, 221, 222, 225
high blood pressure: life insurance and, 8, 236, 241–42, 245–46
high-energy physics, 209–10, 214, 215, 219n. 12
historical epistemology, 275
historical realism, 251
historicity of scientific objects, 2–3, 5, 6, 13, 14, 251, 264–65, 266–69; Koch's bacillus as example, 248–51, 266; spontaneous generation as example, 253, 255–58, 260, 263–64. See also existence of scientific objects

history: vs. philosophy, 17
history of science: biased picture in, 117–18; ontology and, 13–14
Hobbes, Thomas, 164
Hogeboom, George, 281
Holden, Edgar, 231–32
Hooke, Robert, 36
Hotchkiss, Rollin, 281
Hountondji, Paulin, 200
Hultin, Tore, 283
human nature: German Enlightenment and, 68, 70–71, 76, 78, 81, 84, 85
Humboldt, Wilhelm von, 161n. 7, 163n. 12
Hume, David: on causation, 36; on miracles, 21n. 18; on personal identity, 96, 96n. 20, 98, 98n. 28; Smith's theory of value and, 123; social constructivism and, 263–64
Hunter, Arthur, 234, 236–37, 242

identity. See self
Idéologues, sensationalist, 107, 108, 110
imagination: dreams and, 7, 72, 73, 74, 78, 79; eighteenth century ambivalence toward, 4–5; in Enlightenment natural philosophy, 35, 41; in Herderian anthropology, 165; in Hume's theory of personal identity, 96; in preternatural philosophy, 18, 19, 26, 29, 30
imperialism, 159, 162, 163, 168, 184, 197. See also colonialism
incommensurability: about cause of Ramses' death, 249; about spontaneous generation, 261
information: in genetic code, 271, 272; for insurance industry, 227
instrumentalism, 14, 271
insurance. See life insurance
invention: vs. discovery, 2–5, 13, 267

Jacob, François, 271, 290
Jakob, Ludwig Heinrich, 81
James, William, 15
Janes, Craig, 183
Jeener, Raymond, 280
Jesuit culture: mathematical discourse in, 44–45, 54, 56, 59–64, 65–66; moral theology in, 61, 63
Johnson, Harry M., 152n. 31
Johnson, Samuel, 39

Jolly, Margaret, 174n. 38
Jorland, Gérard, 10–11
Jouffroy, Théodore, 111–12, 112n. 65, 114

Kant, Immanuel: on dreams, 72; history of
    science and, 13–14; on personal free-
    dom, 70–71; on sociability, 133
Kaufman, Doris, 6–7
Kayapo culture, 194–96
Keller, Elizabeth, 283
Kelvin, Lord (William Thomson), 212–14,
    218, 219–20
Keynes, J. M., 120, 125, 131
Klein, Jacob, 42–44, 54–56, 62
Koch's bacillus, 248–51, 266
Krüger, Johann Gottlob, 78, 79
Kultur theories, 163n. 12, 163–65, 167

laboratory practice: spontaneous generation
    and, 253–54, 258
labor theory of value, 123–27, 130
Lakau, Andrew, 199
language: social science and, 141, 155
Latour, Bruno, 8, 12–13, 134n. 4, 148n. 25
Lavoisier, Antoine, 35
Lederman, Rena, 171–76
Lee, Norman, 283
Leibniz, Gottfried Wilhelm, 36
Lenz, Carl Gotthold, 72, 74n. 27
Leontief, Wassily, 130
Lévi-Strauss, Claude, 167, 170
liberalism, classical, 135, 138, 143, 145,
    157
Liceti, Fortunio, 20
Liebig, Justus, 258, 259
life insurance, 226–46; adverse selection
    and, 227–30, 233, 237; agent's role in,
    231, 232–34, 238, 240, 242; creation of
    scientific objects for, 7–8, 226–27, 245–
    46; forms of information for, 227–30;
    medical tests and, 7–8, 230, 237, 239–
    44, 245–46; numerical method for, 234–
    39; objectivity and, 226–27, 230, 237,
    239, 245–46; statistical methods in, 227,
    231, 234, 235, 236, 238–39, 245; trust
    and, 8, 228, 230, 231, 245; for women,
    233
life tables, 227, 228
Lignac, abbé de, 97–98, 99
Lindstrom, Lamont, 198
Littlefield, John, 284, 286, 287

Locke, John: French critiques of, 96–97,
    97n. 23, 98, 101; French translations of,
    99–101, 100n. 31; Herderian anthropol-
    ogy vs., 164, 165; on the microscope, 36;
    on private property, 107; on the self, 93–
    94, 94n. 16, 106
Lorentz, Hendrik Antoon, 221–22, 222n. 16,
    223, 224, 225
Lounsberry, R. L., 234
Löwy, Ilana, 293
Lycosthenes, Conrad, 25

MacCullagh, James, 218
Mach, Ernst, 14, 36
MacIver, Robert, 148–49
Macpherson, Cluny, 183
madness. See mental disorders
magic, natural, 23–24, 24n. 34, 32, 37
magnetism: early theories of, 26, 34, 40
magneto-optics, 221, 224
Maimon, Salomon, 73, 74, 74n. 27
Malebranche, Nicholas, 35
Malinowski, Bronislaw, 166, 167–68
Malthus, Thomas, 125, 126
Marcus, George, 184
marginal utility theory, 120, 127–30
Martin, Charles F., 239
marvels, 20–21, 29, 30, 32, 38, 39, 41
Marx, Karl: on community vs. money, 174;
    culturalism and, 201; social class and, 137,
    138; value theories and, 126, 127, 128
mathematical entities: existence of num-
    bers, 59–60; in Guldin's discourse, 9, 51–
    54, 66; in Jesuit discourse, 56–57, 59–64,
    65–66; symbolic number as, 42–45, 52,
    54–57; Vieta on, 54
Matthaei, Heinrich, 290
Mauss, Marcel, 90, 90n. 10
Maya culturalism, 201
Mayhew, Leon, 150–52
M'Bokola, Elikia, 200
medical tests: life insurance and, 7–8, 230,
    237, 239–44, 245–46
memory: Condillac on, 95, 95n. 17, 99, 103;
    Cousin on, 103; Locke on, 93–94
Mendelssohn, Moses, 76, 78
Mendi people, 171–76
mental disorders: Freud on, 83–84; German
    Enlightenment discourse on, 69–70, 74–
    75, 77, 78–79, 83, 85; Locke on, 94n. 16
Merrill, Mike, 172–73, 175

mesmerism, 35, 41
messenger RNA, 290–92
metaphysics: applied, 1–2, 3
microphysics, 203–25; continuous ether in, 218–19, 220; ether lattice in, 214–15, 216–17, 218, 220, 220n. 14, 221; existence of atoms in, 203–7; Kelvin's molecular dynamics and, 212–14; nineteenth-century practice in, 210–12; nineteenth-century schemes in, 215; origin of modern practice in, 219n. 12, 219–25; productivity of tools in, 11; twentieth-century practice in, 207–10; twentieth-century schemes in, 215; wave optics and, 215–17
microscope: electron, 273, 283, 285–86, 288, 292; light, 36, 278; scanning tunneling, 203–5, 206–7, 208, 209–11
microsomes, 12, 272, 273, 281–86, 288, 289, 290
miracles, 17, 20, 21, 28, 29, 32, 35
Mitchell, Clyde, 187
mitochondria, 278, 281–82, 285, 288
Mixtec culturalism, 201n. 111
model organisms, 293
Mohl, Robert von, 10, 136–37, 138–40, 141n. 14, 142–43; on social causality, 144–45; twentieth-century social thought and, 146, 148, 149; "weak" concept of society and, 154
moi: vs. *âme,* 86, 91, 91n. 11, 97–98, 99, 109, 113; bourgeoisie and, 92, 107, 110–11, 111n. 64, 113, 114–16; Condillac on, 95, 95n. 18, 98–99, 99n. 29, 103, 109, 110, 110n. 61; of Cousinian philosophy, 7, 92–93, 101–7; Destutt de Tracy on, 107–8; in *Encyclopédie* and *Dictionnaire,* 86, 87, 91, 92–93, 102; Guizot on, 111, 111n. 64; institutionalization of, 92, 109–11, 111n. 63, 112–14, 113n. 67; Locke's French translators and, 99–101, 100n. 31; Locke's theory of personal identity and, 96–97, 97n. 23, 98, 101; Pascal on, 99n. 29, 100, 100n. 32; post-Revolutionary society and, 101–2, 107, 115–16; self as scientific object and, 6, 89–91; of women, 7, 111n. 63, 111–12, 114, 114n. 68, 116. *See also* Cousin, Victor; eclectic philosophy, Cousinian

molecular biology: central dogma of, 289; cytoplasmic particles and, 271, 272, 274, 275, 289, 292–93
Molina, Louis. *See* Molinist theology
Molinet, Claude, 37
Molinist theology, 57, 61, 63, 65
Molyneux, William, 93
monetarist paradigm, 131
Monod, Jacques, 290
moral-political order, 134, 134n. 4, 137–38
moral responsibility: Cousinian view of, 101, 113, 115; dreams and, 71–73, 74n. 27, 79–80, 81; Locke on, 94, 94n. 16
Morgan, Lewis Henry, 167
Morgenstern, Oskar, 129
Morigi, Paolo, 31
Moritz, Karl Philipp, 69–70, 71, 73, 79
motion of the earth, 9, 44, 48–50, 51, 52, 53, 56, 64–66
mRNA, 290–92
multilocal culture. *See* translocal society (multilocal culture)
Murphy, James, 276, 277

the natural, 20, 21
natural magic, 23–24, 24n. 34, 32, 37
natural philosophy: causal explanations in, 34–35, 37, 41; vs. natural history, 17n. 4; new metaphysics of, 37, 38; new sensibility of, 37–41; preternatural philosophy and, 8–9, 17–18, 28, 33–36; replicability of findings and, 40–41; symbolic number and, 44. *See also* preternatural philosophy
natural theology, 39
nature: vs. culture, 3; existence of scientific objects and, 3, 255, 257, 260, 264, 265, 268
networks of production, 12–13, 250–51, 258, 261, 262, 263, 266
Neumann, Franz, 219
Neumann, John von, 129
New Guinea: cultural transformation in, 171–76, 174n. 36, 190, 197
Newton, Isaac: alchemy and, 37; causal hypotheses of, 34, 36; metaphysics of, 37
Nirenberg, Marshall, 290
nucleic acids: as transitory objects, 270–71. *See also* DNA; RNA

numbers: existence of, 59–60; symbolic, 42–45, 52, 54–57. *See also* mathematical entities
Nye, Mary Jo, 210, 210n. 5

"objective": semantic transformation of, 4
objectivity: of experimental knowledge, 294; incentives to, 226; independent of manipulations, 272; in insurance underwriting, 226–27, 230, 237, 239, 245–46
occult properties, 18, 21n. 20, 22, 23, 32; natural philosophers and, 17, 34, 36, 36n. 94
occult sciences, 37
ontology: historicity of scientific objects and, 13–14; of preternatural philosophy, 20–26, 33–34
Oresme, Nicole, 17–18

Paillot, André, 279
Palade, George, 285, 286
Palissy, Bernard, 18
Panckoucke, Charles-Joseph, 87–88
Pappus, 50, 50n. 12
Paracelsus, 24n. 34
paradigms: in cytoplasmic particle research, 272; in economics, 120, 122, 131; existence of scientific objects and, 258, 259, 262
Paré, Ambroise, 22
Pareto optimality principle, 129
Parsons, Talcott, 147–48, 149, 149n. 27, 150, 155
partial existences, 268. *See also* relative existence
Pascal, Blaise, 99n. 29, 100, 100n. 32
Pasteur, Louis, 252–54, 255–62, 258n. 9, 261n. 11, 263–64, 265–66
Patton, J. Allen, 238
Perera, Benedictus, 60
performances: scientific objects and, 260, 262
personal identity. *See* self
Petermann, Mary, 287, 289
phantasmagorias, 108–9
philosophes: Herder's culture concept and, 164, 165; post-Revolutionary philosophy and, 89, 92
physics: Jesuit mathematicians and, 53, 56, 60, 65, 66; symbolic number and, 42, 43, 44, 45, 56. *See also* microphysics
physiocrats, 120–22, 123, 124

plasmagenes, 278, 283, 288
Pliny, 21n. 20, 25, 37
Pockels, Friedrich, 73
Polanyi, Michael, 272, 294
polysomes, 292
Pomponazzi, Pietro, 18, 20, 24, 24n. 34, 29, 34, 35
Porter, Keith, 285
Porter, Theodore M., 7–8
positivism: in anthropology, 158; history of science and, 14; in nineteenth-century philosophy, 36, 89, 89n. 8, 111
possible beings: in Jesuit discourse, 57–59, 61, 62
potential theory, 218, 220
Pouchet, Félix-Archimède, 252–54, 255–62, 258n. 9, 261n. 11, 263–64
practice: microphysics as, 207–10, 219, 225
pragmatism: existence of scientific objects and, 263n. 13
preternatural philosophy, 16–19; causal explanations in, 17, 18, 20, 24–26, 28, 29, 34–35; courtly magnificence and, 31–32; demise of, 33–41; divine portents and, 8, 23, 25; epistemology of, 27–29, 35–36; examples of objects in, 16–17, 19–20; natural hierarchy in, 33; natural philosophy and, 8–9, 17–18, 28, 33–36; ontology of, 20–26, 33–34; sensibility of, 8, 29–33. *See also* natural philosophy
private property: in classical liberalism, 135; Cousin on, 7, 106–7, 116; Destutt de Tracy on, 107–8; Locke on, 107
probability: life insurance and, 227; truth and, 13. *See also* statistical methods
protein synthesis. *See* cytoplasmic particles

Quesnay, François, 121–22
quotidian objects, 2, 13

racism: culture concept and, 159, 160, 160n. 4, 162
rationalism: of Cousin, 102
Rayleigh, Baron (John William Strutt), 213
realism: about economic concepts, 128; historical, 251; about scientific objects, 2–3, 5, 14, 251, 264–65, 268; of scientists, 271; about universals, 4, 56. *See also* social constructionism
reality: of cultures, 10; embeddedness and, 12–13, 273; historicity and, 251, 255,

257, 261, 267; in Jesuit philosophy, 57–61, 62; manifests itself in the future, 11–12, 272, 294; neo-Kantian epistemology of, 13–14; networks of production and, 12–13, 259, 261; as relative property, 1, 13; representations and, 274. *See also* existence of scientific objects
reason: Cousin on, 7, 104–6, 106n. 50, 108, 113, 114; German Enlightenment and, 68, 71, 75, 78, 79; Herder on, 165
Redfield, Robert, 186, 186n. 73
reductive explanation, 210, 212
Reid, Thomas, 96
Reil, Johann Christian, 77, 80
relative existence, 13, 251–53, 256–57, 259–60. *See also* partial existences
relativism, 252, 255, 264–65, 268, 271
reswitching of techniques paradox, 130
reversion of capital paradox, 130
Rheinberger, Hans-Jörg, 8, 11–12
Rhine, J. B., 244
ribonucleic acid. *See* RNA
ribonucleoprotein. *See* RNP
ribosomes, 272, 274, 275, 288–93
Ricardo's theory of value, 125–27, 128
RNA (ribonucleic acid): in embryogenesis experiments, 283; as enzyme, 291; messenger, 290–92; in ribonucleoprotein, 284, 286, 287, 288, 289; as template, 288–89, 290, 291; in ultracentrifuge experiments, 278–81, 282, 284, 286
RNP (ribonucleoprotein), 284, 286, 287, 288, 289
Roberts, Richard, 288
Robertson, Etienne-Gaspard, 108, 109n. 58
Robinson, Joan, 130
Rockwell, T. H., 238
Rogers, Oscar, 234, 235–37, 239, 240, 241, 242
romanticism: "society" and, 141, 143
Rothacker, Erich, 145n. 20
Rous, Peyton, 276, 277
Ryan, Dawn, 192

Sahlins, Marshall, 9–10
Salisbury, Richard, 169–70, 172
Samoan diaspora, 181–84
Samuelson, Paul A., 129
Scaliger, Julius Caesar, 28
scanning tunneling microscope (STM), 203–5, 206–7, 208, 209–11

Schachman, Howard, 289
Schneider, Walter, 281
Schott, Gaspar, 18
science studies, 250, 251, 259, 265, 294
scientific objects: criteria for, 15–16; cross-disciplinary trajectories of, 12, 273, 293; embeddedness of, 12–13, 273, 276; emergence and passing away of, 1, 5, 9–10, 13, 117–18, 131, 273; productivity of, 10–12; vs. quotidian objects, 2, 13; realism about, 2–3, 5, 14, 251, 264–65, 268; as representations, 274–75, 276; salience of, 6–9. *See also* epistemic objects; existence of scientific objects; historicity of scientific objects; mathematical entities
secrecy, 31–33, 37. *See also* trust, life insurance and
self: German Enlightenment, 69, 70, 73–76; Hume on, 96, 96n. 20, 98, 98n. 28; Locke on, 93–94, 94n. 16, 106; nineteenth-century French (see *moi*); in sociology, 147; twentieth-century theories of, 90, 90n. 10
self-consciousness: cultural, 163, 177–78, 192–202; of the *moi*, 86, 91; physiological approach to, 80
self-knowledge: German Enlightenment and, 68, 70–71, 73, 77, 80
Sellmeier, Wolfgang von, 219–20, 220n. 14
sensationalist philosophy: vs. bourgeois social stability, 113; Catholic Cartesians vs., 96–98; Cousin's view of, 93, 99, 101, 102, 103–4, 108–9; curriculum based on, 109–10, 110n. 61; of Destutt de Tracy, 107–8; Herder vs., 165; of Hume, 96; of Locke, 93–94, 96, 102; phantasmagorias and, 108–9. *See also* Condillac, Étienne Bonnot de
"sentimental pessimism," 168, 169, 170, 193
sexual acts: "against nature," 20
Shaver, John Rodney, 283
Siekevitz, Philip, 285
Simmel, Georg, 146–47, 149, 149n. 27
Smith, Adam: "society" and, 137; value theories and, 121, 122–25, 123n. 11, 126
social class. *See* class
social constructionism: about scientific objects, 2–3, 248, 263–65, 267, 268, 271; in sociology, 147. *See also* realism
social determinism, 138, 144–45, 146, 156

social order: scientific realism and, 269; social sciences and, 149, 149n. 27
society, 132–57; causal efficacy of, 138, 144–
45, 146, 156; cultural analyses of, 141,
143, 150, 150n. 30, 151, 155; doubts of
Weber and Simmel, 146–47; economic
relations and, 137, 139, 143, 143n. 17,
144, 154, 155; emergence of, 132–38,
153–54, 154n. 34; empirical vs. normative tensions about, 138, 140–41, 145–
46, 152, 153, 157; existence of, 10, 134n.
4, 148, 148n. 25, 152n. 31, 153; future of
social analysis and, 156–57; Herder on,
164; individual and, 138, 143–44, 145,
150n. 29, 155–56; language about, 132,
133, 148, 153; organicist theories of, 142,
145n. 20, 155; passing away of, 152–53,
154–57; vs. state, nineteenth-century,
133, 134, 138–41, 139n. 11, 143, 145,
146; vs. state, twentieth-century, 149,
155; in twentieth-century sociology,
147n. 23, 147–53, 150n. 29, 150n. 30
socionature, 258, 259, 261
soul: âme, 86, 91, 91n. 11, 97–98, 99, 109,
113; German Enlightenment models of,
71, 72, 74, 75, 77, 78, 80
Spalding, Johann Joachim, 74–76, 78
Spate, O. H. K., 174
spatiotemporal envelopes, 250, 253, 256,
259–63, 265
spatiotemporal singularities, 294
species of magnitude, 48, 52, 53, 54, 55, 56
sphygmomanometer, 8, 241–42, 246
spontaneous generation, 252–54, 255–64,
258n. 9, 261n. 11, 267
Stanley, Wendell, 289
state: in Herderian anthropology, 165; vs.
"society," nineteenth-century, 133, 134,
138–41, 139n. 11, 143, 145, 146; vs. "society," twentieth-century, 149, 155
"state sciences," 136, 138, 139, 140, 142
statistical methods: in life insurance, 227,
231, 234, 235, 236, 238–39, 245; in social
research, 142, 143
Stein, Lorenz von, 137
Stendhal, 108–9
Stengers, Isabelle, 263n. 13
Stevin, Simon, 43–44
STM (scanning tunneling microscope),
203–5, 206–7, 208, 209–11
Stokes, George, 213, 218

structuralism: in anthropology, 166n. 17,
186n. 71
Strüpell, Adolf, 84
Suarez, Franciscus, 57–59, 60, 61–62, 63, 64
the subject: Foucault on, 90n. 10
substance: historicity of objects and, 257,
260, 262–63, 264, 266n. 14, 266–67
the supernatural, 17, 20, 21, 23, 25
supply and demand, 121. *See also* demand
curves
Sutter, F. K., 181
symbolic interactionism, 147
symbolic number, 9, 42–45, 52, 54–57. *See
also* mathematical entities
systems theory: in sociology, 142, 147, 152,
155

tableau économique, 120–22, 125, 128
Taylor, Charles, 90n. 10
Taylor, H. F., 243
technical projects, 251, 262, 262n. 12
technological artifacts, 250, 263
Thomism. *See* Jesuit culture
Thornton, W. E., 243
Tiedemann, Dieterich, 77–78
Tissières, Alfred, 289
Tocqueville, Alexis de, 137
Toennies, Ferdinand, 143
Tongan cultural transformation, 176–81,
179n. 52, 179n. 53, 183, 184, 188
tradition: cultural change and, 174, 176; cultural self-consciousness and, 198, 200; in
Herderian anthropology, 165
translocal society (multilocal culture), 170,
171, 171n. 26; in antiquity, 185n. 68; circular migration and, 185n. 69, 188n. 80,
191; generalizations about, 189–93;
Samoan, 181–84; Tongan, 176–81,
179n. 52, 179n. 53, 183, 184, 188; urbanization and, 185–89, 186n. 73, 187n. 78,
191n. 84, 191n. 86, 191–92
Treitschke, Heinrich von, 136–37, 139n. 11,
140, 141, 145n. 20
trust: life insurance and, 8, 228, 230, 231, 245
truth-value: probability and, 13; relativism
and, 265, 268
Turner, Terence, 171, 193n. 90, 193–96
Tylor, E. B., 158, 166

ultracentrifuge, 276–81, 286–88, 289
Ungerleider, Harry, 243, 244

universals: realism about, 4, 56
the unnatural, 20, 20n. 16
utility theory of value: Aristotelian, 118–
    20, 122, 123; marginal, 120, 127–30;
    paradox of, 123; Smith's rejection of,
    123, 123n. 11

Valerio, Luca, 48
value theories, 117–31; Aristotelian utility,
    118–20, 122, 123; marginal utility, 120,
    127–30; paradoxes of, 126–27, 128,
    130–31; passing away of, 11, 129–31;
    physiocratic paradigm, 120–22, 123,
    124; replaced by monetarist paradigm,
    131; Ricardo's embodied labor theory,
    125–27, 128; Smith's labor theory, 122–
    25, 126, 130; unbiased history of science
    and, 117–18
van der Waals, Johannes Diderik, 223, 224
Varnhagen von Ense, Rahel, 72
Veit, Joseph, 73
velocity sedimentation, 273, 286–88, 289
Velsen, Van, 188
Vieta, Franciscus, 43–44, 52, 54
Villalpando, Juan Baptista, 53
vitalism: Pasteur's ferment and, 258–59
Voltaire, 35

Wagner, Peter, 9–10
Wallis, John, 44
Wánai culture, 196n. 99
Warren, Kay, 201
Watanabe, John, 201
Watson, James, 270, 289, 290

wave optics, 215–17
Weber, Max, 141, 146, 147, 149, 150n. 29,
    152n. 31
Weber, Wilhelm, 223, 223n. 18, 224
Weeks, Rufus W., 235
White, Leslie, 158, 166
Whitehead, Alfred North, 263n. 13, 268
Wicksell effect, 130
will: in Cousinian philosophy, 7, 104, 105,
    106, 108, 111–13, 114; Destutt de Tracy
    on, 107–8; dreams and, 74, 78, 80; in so-
    ciology, 143. *See also* free will
Williams, Robert, 283
Wilson, Frank, 244
Wise, Norton, 121
witchcraft, 24
women: Cousinian *moi* and, 7, 111n. 63,
    111–12, 114, 114n. 68, 116; life insur-
    ance for, 233; as property, 135; as suscep-
    tible to imagination, 26, 35. *See also*
    gender distinctions
wonder, 28, 29–33, 36, 37, 38–39
Wood, Arthur B., 233
worldviews: in Herderian anthropology,
    165; vs. reality, 13
Wunderkammern, 19, 22, 27, 33

X rays: for life insurance testing, 243–44; in
    microphysics, 222, 224

Zamecnik, Paul, 270, 283, 284, 285
Zeeman, Pieter, 221, 224
Zippenfeld, Peter, 203–4, 205, 209
Zöllner, K. F., 223n. 18